Learning Resource Centre

Park Road, Uxbridge, Middlesex UB8 1NQ
Renewals: 01895 853344

Please return this item to the LRC on or before the last Date stamped below:

620.78

Electronic Instrumentation and Measurements

Second Edition

David A. Bell
Lambton College of Applied Arts and Technology
Sarnia, Ontario, Canada

Published by:
David A. Bell 1693 Trinity Cr. Sarnia Ontario Canada N7S 5P8

OXFORD
UNIVERSITY PRESS

8 Sampson Mews, Suite 204, Don Mills, Ontario M3C 0H5
www.oupcanada.com

Oxford University Press is a department of the University of Oxford.
It furthers the University's objective of excellence in research, scholarship,
and education by publishing worldwide in

Oxford   New York

Auckland   Cape Town   Dar es Salaam   Hong Kong   Karachi
Kuala Lumpur   Madrid   Melbourne   Mexico City   Nairobi
New Delhi   Shanghai   Taipei   Toronto

With offices in
Argentina   Austria   Brazil   Chile   Czech Republic   France   Greece
Guatemala   Hungary   Italy   Japan   Poland   Portugal   Singapore
South Korea   Switzerland   Thailand   Turkey   Ukraine   Vietnam

Oxford is a trade mark of Oxford University Press
in the UK and in certain other countries

Published in Canada
by Oxford University Press

Library and Archives Canada Cataloguing in Publication

Bell, David A., 1930–
Electronic instrumentation and
measurements / David A. Bell. — 2nd ed.

Includes index.
ISBN 978–0–96–837052–0

1. Electronic Measurements. 2. Electronic Instruments. I. Title.

TK7878.B48 1994   621.3815'48–dc20   93-26157

Cover image: NASA

This book is printed on permanent (acid-free) paper ∞.

Contents

3   ELECTROMECHANICAL INSTRUMENTS   29

4   ÀNALOG ELECTRONIC VOLT-OHM-MILLIAMMETERS   86

Preface

The objectives of this book are to explain the operation, performance, and applications of the most important measuring instruments normally encountered in an electronics laboratory, and to discuss electronics measuring techniques. An understanding of electrical fundamentals and transistor circuit operation is assumed.

Because digital instruments are (generally) much more accurate, more versatile, tougher, and less expensive than analog instruments, they are rapidly replacing analog instruments. Therefore, analog instrument coverage is reduced, and treatment of digital instruments is greatly expanded in the second edition of this book.

Starting with SI units and measurement errors, the text progresses through electromechanical instruments, analog electronic instruments, digital voltmeters and frequency meters; to resistance, inductance, and capacitance measurement techniques. The specialized instruments investigated in the latter half of the book include analog oscilloscopes, digital storage oscilloscopes, signal generators, waveform analyzers, and graphic recording instruments. Instrument calibration is also explained.

The content of this book has been heavily influenced by those who reviewed the first edition and/or the manuscript for the second edition. I would very much like to receive comments on the second edition from users of the book.

David Bell

Units, Dimensions, and Standards

1

Objectives

You will be able to:

1. Discuss the three fundamental mechanical units in the SI system, define the basic SI mechanical derived units, and identify the various metric prefixes.
2. Define the SI units for the following electrical and magnetic quantities: current, charge, emf, resistance, conductance, magnetic flux, flux density, inductance, capacitance.
3. Explain the two SI temperature scales.
4. Convert from non-SI to SI units, and determine the dimensions of various quantities.
5. Define the various measurement standards and their applications.

Introduction

Before standard systems of measurement were invented, many approximate units were used. A long distance was often measured by the number of *days* it would take to ride a horse over the distance; a horse's height was measured in *hands;* liquid was measured by the *bucket* or *barrel.*

With the development of science and engineering, more accurate units had to be devised. The English-speaking peoples adopted the *foot* and the *mile* for measuring distances, the *pound* for mass, and the *gallon* for liquid. Other nations followed the lead of the French in adopting a *metric system,* in which large and small units are very conveniently related by a factor of 10.

With the increase of world trade and the exchange of scientific information between nations, it became necessary to establish a single system of units of measurement that would be acceptable internationally. After several world conferences on the matter, a metric system which uses the *meter, kilogram,* and *second* as fundamental units has now been generally adopted around the world. This is known, from the French term "système international," as the *SI* or *international system.*

1-1 SI MECHANICAL UNITS

Fundamental Units

The three basic units in the SI system are:

Unit of *length:*   the **meter** (m)*

Unit of *mass:*   the **kilogram** (kg)

Unit of *time:*   the **second** (s)

These are known as *fundamental units.* Other units derived from the fundamental units are termed *derived units.* For example, the unit of area is *meters squared (m^2),* which is derived from meters.

The *meter* was originally defined as 1 ten-millionth of a meridian passing through Paris from the north pole to the equator. The *kilogram* was defined as 1000 times the mass of 1 cubic centimeter of distilled water. The *liter*[†] is 1000 times the volume of 1 cubic centimeter of liquid. Consequently, 1 liter of water has a mass of 1 kilogram. Because of the possibility of error in the original definitions, the meter was redefined in terms of atomic radiation. Also, the kilogram is now defined as the mass of a certain platinum-iridium standard bar kept at the International Bureau of Weights and Measures in France. The second is, of course, 1/(86 400) of a mean solar day, but it is more accurately defined by atomic radiation.

Unit of Force

The SI unit of force is the newton[‡] (N), defined as that force which will give a mass of 1 kilogram an acceleration of 1 meter per second per second.

When a body is to be accelerated or decelerated, a force must be applied proportional to the desired rate of change of velocity, that is, proportional to the acceleration (or deceleration).

$$\boxed{\begin{array}{c} \textbf{Force} = \textbf{mass} \times \textbf{acceleration} \\ F = ma \end{array}} \qquad (1\text{-}1)$$

When the mass is in kilograms and the acceleration is in m/s^2, the foregoing equation gives the force in newtons.

*Canadian spelling is metre.

[†]Canadian spelling is litre.

[‡]Named for the great English philosopher and mathematician Sir Isaac Newton (1642–1727).

If the body is to be accelerated vertically from the earth's surface, the *acceleration due to gravity* (*g*) must be overcome before any vertical motion is possible. In SI units:

$$g = 9.81 \text{ m/s}^2 \qquad (1\text{-}2)$$

Thus, a mass of 1 kg has a gravitational force of 9.81 N.

Work

When a body is moved, a force is exerted to overcome the body's resistance to motion.

*The **work** done in moving a body is the product of the force and the distance through which the body is moved in the direction of the force.*

$$\text{Work} = \text{force} \times \text{distance}$$
$$W = Fd \qquad (1\text{-}3)$$

*The SI unit of work is the **joule** * (**J**), defined as the amount of work done when a force of 1 newton acts through a distance of 1 meter.*

Thus, the *joule* may also be termed a *newton-meter*. For the equation $W = Fd$, work is expressed in joules when F is in newtons and d is in meters.

Energy

Energy is defined as the capacity for doing work.

Energy is measured in the same units as work.

Power

Power is the time rate of doing work.

If a certain amount of work W is to be done in a time t, the power required is

$$power = \frac{work}{time}$$

$$P = \frac{W}{t} \qquad (1\text{-}4)$$

*The SI unit of power is the **watt** † (**W**), defined as the power developed when 1 joule of work is done in 1 second.*

For $P = W/t$, P is in watts when W is in joules and t is in seconds.

*Named after the English physicist James P. Joule (1818–1899).
†Named after the Scottish engineer and inventor James Watt (1736–1819).

1-2 SCIENTIFIC NOTATION AND METRIC PREFIXES

Scientific Notation

Very large or very small numbers are conveniently written as a number multiplied by 10 raised to a power:

$$100 = 1 \times 10 \times 10$$
$$= 1 \times 10^2$$
$$10\ 000 = 1 \times 10 \times 10 \times 10 \times 10$$
$$= 1 \times 10^4$$
$$0.001 = \frac{1}{10 \times 10 \times 10}$$
$$= \frac{1}{10^3}$$
$$= 1 \times 10^{-3}$$
$$1500 = 1.5 \times 10^3$$
$$0.015 = 1.5 \times 10^{-2}$$

Note that in the SI system of units, spaces are used instead of commas when writing large numbers. Four-numeral numbers are an exception. One thousand is written as *1000,* while ten thousand is *10 000.*

Metric Prefixes

Metric prefixes and the letter symbols for the various multiples and submultiples of 10 are listed in Table 1-1, with those most commonly used with electrical units shown in bold type. The prefixes are employed to simplify the writing of very large and very small quantities. Thus, *1000 Ω* can be expressed as *1 kilohm,* or *1 kΩ.* Here *kilo* is the prefix that represents *1000,* and *k* is the symbol for kilo. Similarly, *1 × 10⁻³ A* can be written as *1 milliampere,* or *1 mA.*

Engineering Notation

As already discussed, *1 kΩ* is *1 × 10³ Ω*, and *1 mA* is *1 × 10⁻³ A*. Note also from Table 1-1 that *1 × 10⁶ Ω* is expressed as *1 MΩ*, and *1 × 10⁻⁶ A* can be written as *1 μA.* These quantities, and most of the metric prefixes in Table 1-1, involve multiples of *10³* or *10⁻³*. Quantities that use *10³* or *10⁻³* are said to be written in *engineering notation.* A quantity such as *1 × 10⁴ Ω* is more conveniently expressed as *10 × 10³ Ω*, or *10 kΩ.* Also, *47 × 10⁻⁴ A* is best written as *4.7 × 10⁻³ A,* or *4.7 mA.* For electrical calculations, engineering notation is more convenient than ordinary scientific notation.

TABLE 1-1 SCIENTIFIC NOTATION AND METRIC PREFIXES

Value	Scientific notation	Prefix	Symbol
1 000 000 000 000	10^{12}	tera	T
1 000 000 000	10^{9}	**giga**	**G**
1 000 000	10^{6}	**mega**	**M**
1 000	10^{3}	**kilo**	**k**
100	10^{2}	hecto	h
10	10	deka	da
0.1	10^{-1}	deci	d
0.01	10^{-2}	centi	c
0.001	10^{-3}	**milli**	**m**
0.000 001	10^{-6}	**micro**	**μ**
0.000 000 001	10^{-9}	**nano**	**n**
0.000 000 000 001	10^{-12}	**pico**	**p**

1-3 SI ELECTRICAL UNITS

Units of Current and Charge

Electric current (I) is a flow of charge carriers. Therefore, current could be defined in terms of the quantity of electricity (Q) that passes a given point in a conductor during a time of 1 s.

*The **coulomb*** *(C) is the unit of electrical charge or quantity of electricity.*

The coulomb was originally selected as the fundamental electrical unit from which all other units were derived. However, since it is much easier to measure current accurately than it is to measure charge, the unit of *current* is now the *fundamental electrical unit* in the SI system. Thus, the coulomb is a *derived unit,* defined in terms of the unit of electric current.

*The **ampere**[†] (A) is the unit of electric current.*

*The **ampere** is defined as that constant current which, when flowing in each of two infinitely long parallel conductors 1 meter apart, exerts a force of 2×10^{-7} newton per meter of length on each conductor.*

*The **coulomb** is defined as that charge which passes a given point in a conductor each second, when a current of 1 ampere flows.*

These definitions show that the coulomb could be termed an *ampere-second.* Conversely, the ampere can be described as a *coulomb per second:*

*Named after the French physicist Charles Augustin de Coulomb (1736–1806).
[†]Named after the French physicist and mathematician André Marie Ampère (1775–1836).

$$\boxed{\text{amperes} = \frac{\text{coulombs}}{\text{seconds}}} \qquad \text{(1-5)}$$

It has been established experimentally that *1 coulomb is equal to the total charge carried by 6.24×10^{18} electrons*. Therefore, the charge carried by one electron is

$$Q = \frac{1}{6.24 \times 10^{18}}$$

$$= 1.602 \times 10^{-19} \text{ C}$$

Emf, Potential Difference, and Voltage

The volt* (V) is the unit of electromotive force (emf) and potential difference.

> *The volt (V) is defined as the potential difference between two points on a conductor carrying a constant current of 1 ampere when the power dissipated between these points is 1 watt.*

As already noted, the coulomb is the charge carried by 6.24×10^{18} electrons. One joule of work is done when 6.24×10^{18} electrons are moved through a potential difference of 1 V. One electron carries a charge of $1/(6.24 \times 10^{18})$ coulomb. If only one electron is moved through 1 V, the energy involved is an *electron volt* (eV).

$$1 \text{ eV} = \frac{1}{6.24 \times 10^{18}} \text{ J}$$

The electron-volt is frequently used in the case of the very small energy levels associated with electrons in orbit around the nucleus of an atom.

Resistance and Conductance

The ohm[†] is the unit of resistance, and the symbol used for ohms is Ω; the Greek capital letter omega.

> *The ohm is defined as that resistance which permits a current flow of 1 ampere when a potential difference of 1 volt is applied to the resistance.*

The term *conductance (G)* is applied to the reciprocal of resistance. *The siemens[‡] (S) is the unit of conductance.*

---

*Named in honor of the Italian physicist Count Alessandro Volta (1745–1827), inventor of the voltaic pile.

[†]Named after the German physicist Georg Simon Ohm (1787–1854), whose investigations led to his statement of "Ohm's law of resistance."

[‡]Named after Sir William Siemens (1823–1883), a British engineer who was born Karl William von Siemens in Germany. The unit of conductance was previously the mho ("ohm" spelled backwards).

$$\boxed{\text{Conductance} = \frac{1}{\text{resistance}}}$$ (1-6)

Magnetic Flux and Flux Density

The weber* (Wb) is the SI unit of magnetic flux.

> *The weber is defined as the magnetic flux which, linking a single-turn coil, produces an emf of 1 V when the flux is reduced to zero at a constant rate in 1 s.*

The tesla† (T) is the SI unit of magnetic flux density.

> *The tesla is the flux density in a magnetic field when 1 weber of flux occurs in a plane of 1 square meter; that is, the tesla can be described as 1 Wb/m².*

Inductance

The SI unit of inductance is the henry‡ (H).

> *The inductance of a circuit is 1 henry, when an emf of 1 volt is induced by the current changing at the rate of 1 A/s.*

Capacitance

The farad§ (F) is the SI unit of capacitance.

> *The farad is the capacitance of a capacitor that contains a charge of 1 coulomb when the potential difference between its terminals is 1 volt.*

1-4 SI TEMPERATURE SCALES

There are two SI temperature scales, the *Celsius scale¶* and the *Kelvin scale.‖* The Celsius scale has 100 equal divisions (or *degrees*) between the freezing temperature and the boiling temperature of water. At normal atmospheric pressure, water freezes at 0°C (*zero degrees Celsius*) and boils at 100°C.

The Kelvin temperature scale, also known as the *absolute scale*, commences at absolute zero of temperature, which corresponds to −273.15°C. Therefore, 0°C is equal to 273.15 K, and 100°C is the same temperature as 373.15 K. A temperature difference of 1 K is the same as a temperature difference of 1°C.

*Named after the German physicist Wilhelm Weber (1804–1890).
†Named for the Croatian-American researcher and inventor Nikola Tesla (1856–1943).
‡Named for the American physicist Joseph Henry (1797–1878).
§Named for the English chemist and physicist Michael Faraday (1791–1867).
¶Invented by the Swedish astronomer and scientist Anders Celsius (1701–1744).
‖Named for the Irish-born scientist and mathematician William Thomson, who became Lord Kelvin (1824–1907).

1-5 OTHER UNIT SYSTEMS

In the traditional English-language (*American* and *Imperial*) systems of measurements, the fundamental mechanical units are the *foot* for length, the *pound* for mass, and the *second* for time. Other mechanical units derived from these are similar in both systems, with the exception of the units for liquid measure. The Imperial gallon equals approximately 1.2 U.S. gallons.

Before the SI system was adopted, *CGS systems* using the *centimeter, gram,* and *second* as fundamental mechanical units were employed for scientific purposes. There were two CGS systems: an electrostatic system and a magnetic system. Many CGS units were too small or too large for practical engineering applications, so *practical units* were also used.

When solving problems, it is sometimes necessary to convert from the traditional unit systems to SI units. Appendix 1 provides a list of conversion factors for this purpose.

Example 1-1

A bar magnet with a 1 inch square cross section is said to have a total magnetic flux of 500 maxwell. Determine the flux density in tesla.

Solution

From Appendix 1,

total flux,
$$\Phi = (500 \text{ maxwell}) \times 10^{-8} \text{ Wb}$$
$$= 5 \ \mu\text{Wb}$$

area,
$$A = (1 \text{ in.} \times 1 \text{ in.}) \times (2.54 \times 10^{-2})^2 \text{ m}^2$$
$$= 2.54^2 \times 10^{-4} \text{ m}^2$$

flux density,
$$B = \frac{\Phi}{A} = \frac{5 \ \mu\text{Wb}}{2.54^2 \times 10^{-4} \text{ m}^2}$$
$$= 7.75 \text{ mT}$$

Example 1-2

The normal human body temperature is given as 98.6°F. Determine the equivalent Celsius and Kelvin scale temperatures.

Solution

From Appendix 1,

$$\text{Celsius temperature} = \frac{°F - 32}{1.8} = \frac{98.7 - 32}{1.8}$$

$$= 37°C$$

$$\text{Kelvin temperature} = \frac{°F - 32}{1.8} + 273.15$$

$$= 310.15 \text{ K}$$

1-6 DIMENSIONS

Table 1-2 gives a list of quantities, quantity symbols, units, unit symbols, and quantity dimensions. The symbols and units are those approved for use with the SI system. To understand the dimensions column, consider the fact that the area of a rectangle is determined by multiplying the lengths of the two sides.

$$\text{area} = \text{length} \times \text{length}$$

The *dimensions* of area are $(\text{length})^2$

TABLE 1-2 SI UNITS, SYMBOLS, AND DIMENSIONS

Quantity	Symbol	Unit	Unit symbol	Dimensions
Length	l	meter	m	$[L]$
Mass	m	kilogram	kg	$[M]$
Time	t	second	s	$[T]$
Area	A	square meter	m²	$[L^2]$
Volume	V	cubic meter	m³	$[L^3]$
Velocity	v	meter per second	m/s	$[LT^{-1}]$
Acceleration	a	meter per second per second	m/s²	$[LT^{-2}]$
Force	F	newton	N	$[MLT^{-2}]$
Pressure	p	newton per square meter	N/m²	$[ML^{-1}T^{-2}]$
Work	W	joule	J	$[ML^2T^{-2}]$
Power	P	watt	W	$[ML^2T^{-3}]$
Electric current	I	ampere	A	$[I]$
Electric charge	Q	coulomb	C	$[IT]$
Emf	V	volt	V	$[ML^2T^{-3}I^{-1}]$
Electric field strength	ξ	volt per meter	V/m	$[MLT^{-3}I^{-1}]$
Resistance	R	ohm	Ω	$[ML^2T^{-3}I^{-2}]$
Capacitance	C	farad	F	$[M^{-1}L^{-2}T^4I^2]$
Inductance	L	henry	H	$[ML^2T^{-2}I^{-2}]$
Magnetic field strength	H	ampere per meter	A/m	$[IL^{-1}]$
Magnetic flux	Φ	weber	Wb	$[ML^2T^{-2}I^{-1}]$
Magnetic flux density	B	tesla	T	$[MT^{-2}I^{-1}]$

or
$$[area] = [L][L]$$
$$= [L]^2$$

Similarly,

$$[velocity] = \frac{[length]}{[time]} = \frac{[L]}{[T]}$$
$$= [LT^{-1}]$$

$$[acceleration] = \frac{[velocity]}{[time]} = \frac{[LT^{-1}]}{[T]}$$
$$= [LT^{-2}]$$

$$[force] = [mass] \times [acceleration] = [M][LT^{-2}]$$
$$= [MLT^{-2}]$$

$$[work] = [force] \times [distance] = [MLT^{-2}][L]$$
$$= [ML^2T^{-2}]$$

$$[power] = \frac{[work]}{[time]} = \frac{[ML^2T^{-2}]}{[T]}$$
$$= [ML^2T^{-3}]$$

For the electrical quantities, current is another fundamental unit. So electrical quantities can be analyzed to determine dimensions in the fundamental units of L, M, T, and I.

$$Charge = current \times time$$
$$[charge] = [I][T]$$
$$= [IT]$$

Example 1-3

Determine the dimensions of voltage and resistance.

Solution

From

$$P = EI$$

voltage,

$$E = \frac{P}{I} = \frac{[ML^2T^{-3}]}{[I]}$$
$$= [ML^2T^{-3}I^{-1}]$$

resistance,

$$R = \frac{E}{I} = \frac{[ML^2T^{-3}I^{-1}]}{[I]}$$
$$= [ML^2T^{-3}I^{-2}]$$

Working Standards

Electrical measurement standards are precise resistors, capacitors, inductors, voltage sources, and current sources, which can be used for comparison purposes when measuring electrical quantities. For example, resistance can be accurately measured by means of a Wheatstone bridge which uses a standard resistor (see Section 7-3). Similarly, standard capacitors and inductors can be employed in bridge (or other) methods to accurately measure capacitance and inductance.

The standard resistors, capacitors, and inductors usually found in an electronics laboratory are classified as *working standards*. Working standard resistors are normally constructed of manganin or a similar material, which has a very low temperature coefficient. They are available in resistance values ranging from 0.01 Ω to 1 MΩ, with typical accuracies of ±0.01% to ±0.1%. A working standard capacitor might be air dielectric type, or might be constructed of silvered mica. Available capacitance values are 0.001 μF to 1 μF with a typical accuracy of ±0.02%. Standard inductors are available in values ranging from 100 μH to 10 H with typical accuracies of ±0.1%. *Calibrators* provide standard voltages and currents for calibrating voltmeters and ammeters (see Section 12-3).

Standard Classifications

Measurement standards are classified in four levels: *international standards, primary standards, secondary standards,* and *working standards.* Thus, the working standards already discussed are the lowest level of standards.

International standards are defined by international agreements, and are maintained at the International Bureau of Weights and Measures in France. These are as accurate as it is scientifically possible to achieve. They may be used for comparison with primary standards, but are otherwise unavailable for any application.

Primary standards are maintained at institutions in various countries around the world, such as the National Bureau of Standards in Washington. They are also constructed for the greatest possible accuracy, and their main function is checking the accuracy of secondary standards.

Secondary standards are employed in industry as references for calibrating high-accuracy equipment and components, and for verifying the accuracy of working standards. Secondary standards are periodically checked at the institutions that maintain primary standards.

In summary, working standards are used as measurement references on a day-to-day basis in virtually all electronics laboratories. Secondary standards are more accurate than working standards, and are used throughout industry for checking working standards, and for calibrating high-accuracy equipment. Primary standards are more accurate than secondary standards. They are maintained to the highest possible accuracy by national institutions as references for calibrating secondary standards. International standards are maintained by international agreement, and may be used for checking primary standards.

REVIEW QUESTIONS

1-1 List the three fundamental SI mechanical units and unit symbols, and discuss their origin.

1-2 State the SI units and unit symbols for *force* and *work,* and define each unit.

1-3 State the SI units and unit symbols for *energy* and *power,* and define each unit.

1-4 List the names of the various metric prefixes and the corresponding symbols. Also, list the value represented by each prefix in scientific notation.

1-5 State the SI units and unit symbols for *electric current* and *charge,* and define each unit.

1-6 State the SI units and unit symbols for *electrical resistance* and *conductance,* and define each unit.

1-7 State the SI units and unit symbols for *magnetic flux* and *flux density,* and define each unit.

1-8 State the SI units and unit symbols for *inductance* and *capacitance,* and define each unit.

1-9 Name the two SI temperature scales, and identify the freezing and boiling temperatures of water for each scale.

1-10 List the various levels of measurement standards, and discuss the application of each classification.

PROBLEMS

1-1 Referring to the unit conversion factors in Appendix 1, perform the following conversions: **(a)** 6215 miles to kilometers, **(b)** 50 miles per hour to kilometers per hour, and **(c)** 12 square feet to square centimeters.

1-2 Determine how long it takes light to travel to earth from a star 1 million miles away if the speed of light is 3×10^8 m/s.

1-3 The speed of sound in air is 345 m/s. Calculate the distance in miles from a thunderstorm when the thunder is heard 5 s after the lightning flash.

1-4 A 140 lb person has a height of 5 ft 7 in. Convert these measurements into kilograms and centimeters.

1-5 A bar magnet has a cross section of 0.75 in. × 0.75 in. and a flux density of 1290 lines per square inch. Calculate the total flux in webers.

1-6 Calculate the Celsius and Kelvin scale equivalents of 80°F.

1-7 A ¼ horsepower electric motor is operated 8 hours per day for 5 days every week. Assuming 100% efficiency, calculate the kilowatthours of energy consumed in 1 year.

1-8 Determine the dimensions of *area, volume, velocity,* and *acceleration.*

1-9 Determine the dimensions of *force, work, energy,* and *power.*

1-10 Determine the dimensions of *charge, voltage,* and *resistance.*

1-11 Determine the dimensions of *capacitance* and *inductance.*

Measurement Errors

2

Objectives

You will be able to:

1. Define and explain the following types of errors that occur in measurements: gross, systematic, absolute, relative, random.
2. Explain and apply the following measurement terms: tolerance, accuracy, precision, resolution.
3. Determine the resultant error for various calculations involving instrument and component error combinations.
4. Use basic statistical methods for analyzing measurement errors.

Introduction

No electronic component or instrument is perfectly accurate; all have some error or inaccuracy. It is important to understand how these errors are specified and how they combine to create even greater errors in measurement systems. Although it is possible that in some cases errors might almost completely cancel each other out, the worst-case combination of errors must always be assumed.

Apart from equipment errors, some operator or observer error is inevitable. Also, even when equipment errors are very small, the system of using the instruments can introduce a *systematic error*. Errors of unexplainable origin are classified as *random errors*. Where accuracy is extremely important, some errors can be minimized by taking many readings of each instrument and determining mean values.

2-1 GROSS ERRORS AND SYSTEMATIC ERRORS

Gross errors are essentially human errors that are the result of carelessness. One of the most common errors is the simple misreading of an instrument. In Figure 2-1(a) the digital display of 32.5 mA (the range is 300 mA) might inadvertently be read as 32.5 A. Of course, many digital instruments (not all) display the measurement units alongside the indicated quantity. Thus, making this kind of error is less likely. Figure 2-1(b) shows an analog instrument with three scales: 0 to 25, 0 to 10, and 0 to 50. The 25 scale is used when the range selection switch is set to a multiple of 25, the 10 scale is for ranges that are a multiple of 10, and the 50 scale is for multiple of 50 ranges. Obviously, it is possible to read the wrong scale or, even when using the correct scale, to assume the wrong range multiple.

Sometimes a meter is read correctly but the reading is recorded incorrectly, or perhaps it is recorded in the wrong column in a table of measurements. Everyone makes these kinds of mistakes occasionally. They can be avoided only by taking care in using and reading all instruments, and by thinking about whether or not each measurement makes sense. Substituting instrument readings into an appropriate equation, or plotting a few points on a graph, also helps to check the validity of recorded quantities while measurements are still in progress.

Measurement errors will occur if the accuracy of an instrument has not been checked for some time, that is, if the instrument has not been calibrated (see Chapter 12). Errors will also occur with analog instruments if the pointer has not been mechanically zeroed before use. Analog ohmmeters must also be electrically zeroed for correct use (as well as mechanically zeroed). These kinds of errors can be termed gross errors, because

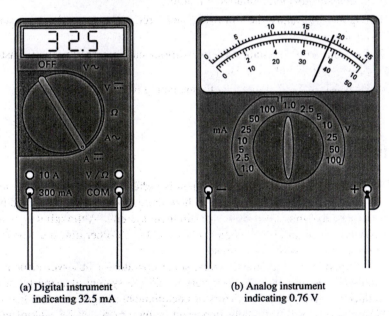

(a) Digital instrument
indicating 32.5 mA

(b) Analog instrument
indicating 0.76 V

Figure 2-1 Serious measurement errors can occur if an instrument is not read correctly. The digital instrument is on a 300 mA range, so its reading is in milliamperes. For the analog meter, the range selection must be noted, and the pointer position must be read from the correct scale.

they can be avoided with care. However, they might also be classified as *systematic errors,* because they are the result of the measurement system.

Other systematic errors occur because the measurement system affects the measured quantity. For example, when a voltmeter is employed to measure the potential difference between two points in a circuit, the voltmeter resistance may alter the circuit voltage (see Section 3-4). Similarly, an ammeter resistance might change the level of a current. Errors that are the result of instrument inaccuracy are also systematic errors. Where more than one instrument is involved, the errors due to instrument inaccuracy tend to accumulate. The overall measurement error is then usually larger than the error in any one instrument. This is explored in Section 2-4.

2-2 ABSOLUTE ERRORS AND RELATIVE ERRORS

If a resistor is known to have a resistance of 500 Ω with a possible error of ±50 Ω, the ±50 Ω is an *absolute error.* This is because 50 Ω is stated as an absolute quantity, *not* as a percentage of the 500 Ω resistance. When the error is expressed as a percentage or as a fraction of the total resistance, it becomes a *relative error.* Thus, the ±50 Ω is ±10%, relative to 500 Ω, or ±$^1/_{10}$ of 500 Ω. So the resistance can be specified as

$$R = 500 \ \Omega \pm 10\%$$

Percentages are usually employed to express errors in resistances and other electrical quantities. The terms *accuracy* and *tolerance* are also used. A resistor with a possible error of ±10% is said to be accurate to ±10%, or to have a tolerance of ±10% [see Figure 2-2(a)]. *Tolerance* is the term normally used by component manufacturers. Suppose that a voltage is measured as 20.00 V using an instrument which is known to have a ±0.02 V

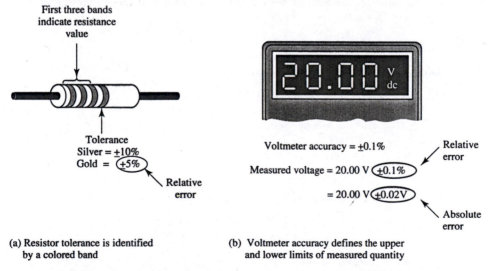

(a) Resistor tolerance is identified by a colored band

(b) Voltmeter accuracy defines the upper and lower limits of measured quantity

Figure 2-2 Percentage accuracy gives the relative error in a measured, or specified quantity. The absolute error can be determined by converting the percentage error into an absolute quantity.

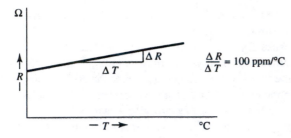

$$\frac{\Delta R}{\Delta T} = 100 \text{ ppm/°C}$$

Figure 2-3 Instead of percentages, errors can be expressed in parts per million (ppm) relative to the total quantity. Resistance change with temperature increase is usually stated in ppm/°C.

error. The measured voltage can be stated as 20.00 V ± 0.02 V. The 0.02 V is an absolute quantity, so it is an absolute error. But 0.02 V is also 0.1% relative to 20 V, so the measured quantity could be expressed as 20 V ± 0.1% [see Figure 2-2(b)], and now the error is stated as a relative error.

Another method of expressing an error is to refer to it in *parts per million (ppm)* relative to the total quantity. For example, the temperature coefficient of a 1 MΩ resistor might be stated as 100 ppm/°C, which means 100 parts per million per degree Celsius. One millionth of 1 MΩ is 1 Ω; consequently, 100 ppm of 1 MΩ is 100 Ω. Therefore, a 1°C change in temperature may cause the 1 MΩ resistance to increase or decrease by 100 Ω (see Figure 2-3).

Example 2-1

A component manufacturer constructs certain resistances to be anywhere between 1.14 kΩ and 1.26 kΩ and classifies them to be 1.2 kΩ resistors. What tolerance should be stated? If the resistance values are specified at 25°C and the resistors have a temperature coefficient of +500 ppm/°C, calculate the maximum resistance that one of these components might have at 75°C.

Solution

Absolute error = 1.26 kΩ − 1.2 kΩ = +0.06 kΩ

or ⠀⠀⠀⠀⠀⠀= 1.2 kΩ − 1.14 kΩ = −0.06 kΩ

⠀⠀⠀⠀⠀⠀⠀⠀= ±0.06 kΩ

$$Tolerance = \frac{\pm 0.06 \text{ k}\Omega}{1.2 \text{ k}\Omega} \times 100\%$$

⠀⠀⠀⠀⠀⠀= ±5%

Largest possible resistance at 25°C:

$$R = 1.2 \text{ k}\Omega + 0.06 \text{ k}\Omega$$

$$= 1.26 \text{ k}\Omega$$

Resistance change per °C:

$$500 \text{ ppm of } R = \frac{1.26 \text{ k}\Omega}{1\,000\,000} \times 500$$

$$= 0.63 \text{ }\Omega/\text{°C}$$

Temperature increase:

$$\Delta T = 75°C - 25°C$$
$$= 50°C$$

Total resistance increase:

$$\Delta R = 0.63 \ \Omega/°C \times 50°C$$
$$= 31.5 \ \Omega$$

Maximum resistance at 75°C:

$$R + \Delta R = 1.26 \ k\Omega + 31.5 \ \Omega$$
$$= 1.2915 \ k\Omega$$

2-3 ACCURACY, PRECISION, RESOLUTION, AND SIGNIFICANT FIGURES

Accuracy and Precision

When a voltmeter with an error of ±1% indicates exactly 100 V, the true level of the measured voltage is somewhere between 99 V and 101 V. Thus, the measurement *accuracy* of ±1% defines how close the measurement is to the actual measured quantity. The *precision* with which the measurement is made is not the same as the accuracy of measurement, although accuracy and precision are related.

Consider the digital voltmeter indication shown in Figure 2-4(a). For the 8.135 V quantity indicated, the last (right-side) numeral refers to millivolts. If the measured quantity increases or decreases by 1 mV, the reading becomes 8.136 V or 8.134 V, respectively. Therefore, the voltage is measured with a precision of 1 mV. For the analog voltmeter in Figure 2-4(b), the pointer position can be read to within (perhaps) one-fourth of the smallest scale division. Since the smallest scale division represents 0.2 V (on the 10 V

Thousanths of volts
(millivolts)

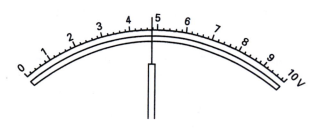

(a) Digital voltmeter display with a precision of 1 mV

(b) Analog instrument display with a precision of 50 mV

Figure 2-4 Measurement precision depends on the smallest change that can be observed in the measured quantity. A 1 mV change will be indicated on the digital voltmeter display above. For the analog instrument, 50 mV is the smallest change that can be noted.

range), one-fourth of the scale division is 50 mV. So 50 mV is the measurement precision of the analog instrument. Neither of these two measurements (digital or analog) takes account of the measurement accuracy.

Suppose that the digital voltmeter referred to above has an accuracy of ±0.2%. The measured voltage is 8.135 V ± 0.2%, or 8.135 V ± 16 mV, meaning that the actual voltage is somewhere between 8.119 V and 8.151 V. So, although the quantity is measured with a precision of 1 mV, the measurement accuracy is ±16 mV. The analog voltmeter in Figure 2-4(b) might have a typical accuracy of ±2% of full scale, or ±2% of 10 V. Thus, the measured quantity is 4.85 V ± 200 mV, that is, 4.65 V to 5.05 V. In this case, the measurement is made to a precision of 50 mV, but the measurement accuracy is ±200 mV.

The measurement precision for the digital and analog instruments discussed above might seem unimportant given the possible error due to the instrument accuracy. However, the instrument accuracy normally depends on the accuracy of internal components, and any error due to the measurement precision must be much smaller than that due to the specified accuracy of the instrument.

Resolution

The measurement precision of an instrument defines the smallest change in measured quantity that can be observed. This (smallest observable change) is the *resolution* of the instrument. In the case of the 10 V analog instrument scale that can be read to a precision of 50 mV, 50 mV is the smallest voltage change that can observed. Thus, the measurement resolution is 50 mV. Similarly, with the digital instrument, the measurement resolution is 1 mV.

Consider the *potentiometer* illustrated in Figure 2-5. The circuit symbol in Figure 2-5(a) illustrates a resistor with two terminals and a contact that can be moved anywhere between the two. The potentiometer construction shown in Figure 2-5(b) reveals that the movable contact slides over a track on one side of a number of turns of resistance wire. The contact does not slide along the whole length of the wire but *jumps* from one point on one turn of the wire to a point on the next turn. Assume that the total potentiometer resistance is 100 Ω and that there are 1000 turns of wire. Each turn has a resistance of

$$\frac{100 \ \Omega}{1000} = 0.1 \ \Omega$$

When the contact moves from one turn to the next, the resistance from any end to the moving contact changes by 0.1 Ω. It can now be stated that the resistance from one end to the moving contact can be adjusted from 0 to 100 Ω with a *resolution* of 0.1 Ω, or a resolution of 1 in 1000. In the case of the potentiometer, the resolution defines how precisely the resistance may be set. It also defines how precisely the variable voltage from the potentiometer moving contact may be adjusted when a potential difference is applied across the potentiometer.

Significant Figures

The number of significant figures used in a measured quantity indicate the precision of measurement. For the 8.135 V measurement in Figure 2-4(a), the four significant figures

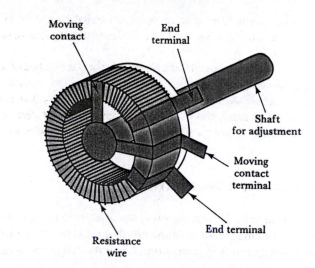

Moving
contact

End
terminal

Shaft
for adjustment

Moving
contact
terminal

End terminal

(a) Potentiometer
circuit symbol

Resistance
wire

(b) Potentiometer construction

Figure 2-5 A potentiometer consists of a resistance wire wound around an insulating former. The movable contact slides from one turn to the next, changing the resistance (from one end to the moving contact) in steps. The potentiometer resolution depends on the number of steps.

show that the measurement precision is 0.001 V, or 1 mV. If the measurement was made to a precision of 10 mV, the display would be 8.13 V or 8.14 V; that is, there would be only three significant figures.

In the case of a resistance value stated as 47.3 Ω, the actual value may not be exactly 47.3 Ω, but it is assumed to be closer to 47.3 Ω than it is to either 47.2 Ω or 47.4 Ω. The three significant figures show that measurement precision is 0.1 Ω. If the quantity was 47.3 kΩ, the measurement precision would be 0.1 kΩ, or 100 Ω. If 47.3 Ω is rewritten with two significant figures, it becomes 47 Ω, because 47.3 Ω is clearly closer to 47 Ω than it is to 48 Ω. If the quantity were written as 47.0 Ω, it would imply that the resistance is closer to 47 Ω than it is to 47.1 Ω, and in this case the zero (in 47.0 Ω) would be a significant figure.

Now consider the result of using an electronic calculator to determine a resistance value from digital measurement of voltage and current.

$$R = \frac{V}{I} = \frac{8.14\ V}{2.33\ mA} = 3.493\ 562\ 232\ k\Omega$$

Clearly, it does not make sense to have an answer containing 10 significant figures when each of the original quantities had only three significant figures. The only reasonable approach is to use the same number of significant figures in the answer as in the original quantities. So the calculation becomes

$$R = \frac{V}{I} = \frac{8.14\ V}{2.33\ mA} = 3.49\ k\Omega$$

The resistance is now stated to the same precision as the measured voltage and current. This calculation has not taken the accuracy of the voltmeter and ammeter into account (see Section 2-4).

As illustrated by the discussion above, the number of significant figures in a quantity defines the precision of the measuring instruments involved. No greater number of significant figures should be used in a calculation result than those in the original quantities. Where the quantities in a calculation have different precisions, the precision of the answer should not be greater than the least precise of the original quantities.

2-4 MEASUREMENT ERROR COMBINATIONS

When a quantity is calculated from measurements made on two (or more) instruments, it must be assumed that the errors due to instrument inaccuracy combine in the worst possible way. The resulting error is then larger than the error in any one instrument.

Sum of Quantities

Where a quantity is determined as the sum of two measurements, the total error is the sum of the absolute errors in each measurement. As illustrated in Figure 2-6(a),

$$E = (V_1 \pm \Delta V_1) + (V_2 \pm \Delta V_2)$$

giving
$$\boxed{E = (V_1 + V_2) \pm (\Delta V_1 + \Delta V_2)} \qquad (2\text{-}1)$$

Example 2-2

Calculate the maximum percentage error in the sum of two voltage measurements when $V_1 = 100 \text{ V} \pm 1\%$ and $V_2 = 80 \text{ V} \pm 5\%$.

Solution

$$V_1 = 100 \text{ V} \pm 1\%$$
$$= 100 \text{ V} \pm 1 \text{ V}$$

$$V_2 = 80 \text{ V} \pm 5\%$$
$$= 80 \text{ V} \pm 4 \text{ V}$$

$$E = V_1 + V_2$$
$$= (100 \text{ V} \pm 1 \text{ V}) + (80 \text{ V} \pm 4 \text{ V})$$
$$= 180 \text{ V} \pm (1 \text{ V} + 4 \text{ V})$$
$$= 180 \text{ V} \pm 5 \text{ V}$$
$$= 180 \text{ V} \pm 2.8\%$$

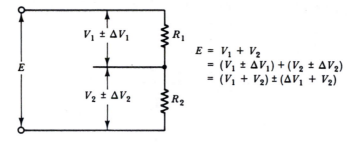

$$E = V_1 + V_2$$
$$= (V_1 \pm \Delta V_1) + (V_2 \pm \Delta V_2)$$
$$= (V_1 + V_2) \pm (\Delta V_1 + V_2)$$

(a) Error in sum of quantities equals sum of errors

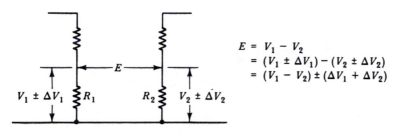

$$E = V_1 - V_2$$
$$= (V_1 \pm \Delta V_1) - (V_2 \pm \Delta V_2)$$
$$= (V_1 - V_2) \pm (\Delta V_1 + \Delta V_2)$$

(b) Error in difference of quantities equals sum of error

$$P = EI$$
$$P = (E \pm \Delta E) \times (I \pm \Delta I)$$
$$R = \frac{E \pm \Delta E}{I \pm \Delta I}$$

(c) Percentage error in product or quotient of
quantities equals sum of percentage errors

Figure 2-6 When measured quantities are combined to determine another quantity, the measurement errors must be assumed to combine in the way that gives the largest possible resultant error.

In Example 2-2, note that *the percentage error in the final quantity cannot be calculated directly from the percentage errors in the two measured quantities.*

Where two or more measured quantities are summed to determine a final quantity, the absolute values of the errors must be summed to find the total possible error.

Difference of Quantities

Figure 2-6(b) illustrates a situation in which a potential difference is determined as the *difference between two measured voltages.* Here again, *the errors are additive:*

$$E = V_1 - V_2$$
$$= (V_1 \pm \Delta V_1) - (V_2 \pm \Delta V_2)$$

$$\boxed{E = (V_1 - V_2) \pm (\Delta V_1 + \Delta V_2)} \qquad (2\text{-}2)$$

Example 2-3

Calculate the maximum percentage error in the difference of two measured voltages when $V_1 = 100 \text{ V} \pm 1\%$ and $V_2 = 80 \text{ V} \pm 5\%$.

Solution

$$\left. \begin{array}{l} V_1 = 100 \text{ V} \pm 1 \text{ V} \\ \\ V_2 = 80 \text{ V} \pm 4 \text{ V} \end{array} \right\} \quad \text{(as in Example 2-2)}$$

and

$$E = (100 \text{ V} \pm 1 \text{ V}) - (80 \text{ V} \pm 4 \text{ V})$$
$$= 20 \text{ V} \pm 5 \text{ V}$$
$$= 20 \text{ V} \pm 25\%$$

Example 2-3 demonstrates that *the percentage error in the difference of two quantities can be very large*. If the difference was smaller, the percentage error would be even larger. Obviously, measurement systems involving the difference of two quantities should be avoided.

Product of Quantities

When a calculated quantity is the product of two or more quantities, *the percentage error is the sum of the percentage errors in each quantity* [consider Figure 2-6(c)]:

$$P = EI$$
$$= (E \pm \Delta E)(I \pm \Delta I)$$
$$= EI \pm E \, \Delta I \pm I \, \Delta E \pm \Delta E \, \Delta I$$

Since $\Delta E \, \Delta I$ is very small,

$$P \simeq EI \pm (E \, \Delta I + I \, \Delta E)$$

$$\text{percentage error} = \frac{E \, \Delta I + I \, \Delta E}{EI} \times 100\%$$

$$= \left(\frac{E \, \Delta I}{EI} + \frac{I \, \Delta E}{EI} \right) \times 100\%$$

Measurement Errors   Chap. 2

$$= \left(\frac{\Delta I}{I} + \frac{\Delta E}{E} \right) \times 100\%$$

$$\boxed{\% \text{ error in } P = (\% \text{ error in } I) + (\% \text{ error in } E)} \qquad (2\text{-}3)$$

Thus, when a voltage is measured with an accuracy of ± 1%, and a current is measured with an accuracy of ±2%, the calculated power has an accuracy of ±3%.

Quotient of Quantities

Here again it can be shown that the percentage error is the sum of the percentage errors in each quantity. In Figure 2-6(c),

$$\boxed{\% \text{ error in } E/I = (\% \text{ error in } E) + (\% \text{ error in } I)} \qquad (2\text{-}4)$$

Quantity Raised to a Power

When a quantity A is raised to a power B, the percentage error in A^B can be shown to be

$$\boxed{\% \text{ error in } A^B = B(\% \text{ error in } A)} \qquad (2\text{-}5)$$

For a current I with an accuracy of ±3%, the error in I^2 is $2(\pm 3\%) = \pm 6\%$.

Example 2-4

An 820 Ω resistance with an accuracy of ± 10% carries a current of 10 mA. The current was measured by an analog ammeter on a 25 mA range with an accuracy of ±2% of full scale. Calculate the power dissipated in the resistor, and determine the accuracy of the result.

Solution

$$P = I^2 R$$
$$P = (10 \text{ mA})^2 \times 820 \ \Omega$$
$$= 82 \text{ mW}$$

$$\textit{error in } R = \pm 10\%$$

$$\textit{error in } I = \pm 2\% \text{ of } 25 \text{ mA}$$
$$= \pm 0.5 \text{ mA}$$
$$= \frac{\pm 0.5 \text{ mA}}{10 \text{ mA}} \times 100\%$$
$$= \pm 5\%$$

$$\text{\% error in } I^2 = 2(\pm5\%)$$
$$= \pm10\%$$

$$\text{\% error in } P = (\text{\% error in } I^2) + (\text{\% error in } R)$$
$$= \pm(10\% + 10\%)$$
$$= \pm20\%$$

Summary

For $X = A \pm B$,   error in $X = \pm [(\text{error in } A) + (\text{error in } B)]$

For $X = AB$,   % error in $X = \pm [(\text{\% error in } A) + (\text{\% error in } B)]$

For $X = A/B$,   % error in $X = \pm [(\text{\% error in } A) + (\text{\% error in } B)]$

For $X = A^B$,   % error in $X = \pm B(\text{\% error in } A)$

2-5 BASICS OF STATISTICAL ANALYSIS

Arithmetic Mean Value

When a number of measurements of a quantity are made and the measurements are not all exactly equal, the best approximation to the actual value is found by calculating the average value, or *arithmetic mean*, of the results. For *n* measured values of $x_1, x_2, x_3, \ldots, x_n$, the arithmetic mean is

$$\bar{x} = \frac{x_1 + x_2 + x_3 + \cdots + x_n}{n} \tag{2-6}$$

Determining the arithmetic mean of several measurements is one method of minimizing the effects of *random errors*. Random errors are the result of chance or accidental occurrences. They may be human errors produced by fatigue, or they may be the result of such events as a surge in ac supply voltage, a brief draft upon equipment, or a variation in frequency.

When determining the mean value of a number of readings, it is sometimes found that one or two measurements differ from the mean by a much larger amount than any of the others. In this case, it is justifiable to reject these few readings as mistakes and to calculate the average value from the other measurements. This action should not be taken when more than a small number of readings differ greatly from the mean. Instead, the whole series of measurements should be repeated.

Deviation

The difference between any one measured value and the arithmetic mean of a series of measurements is termed the *deviation*. The deviations $(d_1, d_2, d_3, \ldots, d_n)$ may be positive or negative, and the algebraic sum of the deviations is always zero. The *average devi-*

ation may be calculated as the average of the *absolute* values of the deviations (neglecting plus and minus signs). If the measured quantity is assumed to be constant, the average deviation might be regarded as an indicator of the measurement precision (see Example 2-5).

$$D = \frac{|d_1| + |d_2| + |d_3| + \cdots + |d_n|}{n}$$

(2-7)

Example 2-5

The accuracy of five digital voltmeters are checked by using each of them to measure a standard 1.0000 V from a calibration instrument (see Section 12-3). The voltmeter readings are as follows: $V_1 = 1.001$ V, $V_2 = 1.002$, $V_3 = 0.999$, $V_4 = 0.998$, and $V_5 = 1.000$. Calculate the average measured voltage and the average deviation.

Solution

From Equation 2-6,

$$V_{av} = \frac{V_1 + V_2 + V_3 + V_4 + V_5}{5}$$

$$= \frac{1.001 \text{ V} + 1.002 \text{ V} + 0.999 \text{ V} + 0.998 \text{ V} + 1.000 \text{ V}}{5}$$

$$= 1.000 \text{ V}$$

$$d_1 = V_1 - V_{av} = 1.001 \text{ V} - 1.000 \text{ V}$$
$$= 0.001 \text{ V}$$

$$d_2 = V_2 - V_{av} = 1.002 \text{ V} - 1.000 \text{ V}$$
$$= 0.002 \text{ V}$$

$$d_3 = V_3 - V_{av} = 0.999 \text{ V} - 1.000 \text{ V}$$
$$= -0.001 \text{ V}$$

$$d_4 = V_4 - V_{av} = 0.998 \text{ V} - 1.000 \text{ V}$$
$$= -0.002 \text{ V}$$

$$d_5 = V_5 - V_{av} = 1.000 \text{ V} - 1.000 \text{ V}$$
$$= 0 \text{ V}$$

From Equation 2-7,

$$D = \frac{|d_1| + |d_2| + |d_3| + |d_4| + |d_5|}{5}$$

$$= \frac{0.001 \text{ V} + 0.002 \text{ V} + 0.001 \text{ V} + 0.002 \text{ V} + 0}{5}$$

$$= 0.0012 \text{ V}$$

From Example 2-5, the average measured voltage is 1.000 V, and the average deviation from this is 1.2 mV. These figures could be used to determine the accuracy of measurement made on any of the five instruments.

Standard Deviation and Probable Error

As already discussed, measurement results can be analyzed by determining the arithmetic mean value of a number of measurements of the same quantity and by further determining the deviations and the average deviation. The *mean-squared value* of the deviations can also be calculated by first squaring each deviation value before determining the average. This gives a quantity known as the *variance*. Taking the square root of the variance produces the *root mean squared (rms)* value, also termed the *standard deviation* (σ).

$$\sigma = \sqrt{\frac{d_1^2 + d_2^2 + d_3^2 + \ldots + d_n^2}{n}} \tag{2-8}$$

For the case of a large number of measurements in which only random errors are present, it can be shown that the probable error in any one measurement is 0.6745 times the standard deviation:

$$\text{probable error} = 0.6745\,\sigma \tag{2-9}$$

Example 2-6

Determine the standard deviation and the probable measurement error for the group of instruments referred to in Example 2-5.

Solution

Equation 2-8,

$$\sigma = \sqrt{\frac{d_1^2 + d_2^2 + d_3^2 + \ldots + d_n^2}{n}}$$

$$= \sqrt{\frac{0.001^2 + 0.002^2 + 0.001^2 + 0.002^2 + 0^2}{5}}$$

$$= 0.0014 \text{V}$$

Equation 2-9, $\text{probable error} = 0.6745\,\sigma = 0.6745 \times 1.4 \text{ mV}$

$$= 0.94 \text{ mV}$$

REVIEW QUESTIONS

2-1 Explain gross errors and systematic errors. Give examples of each.

2-2 Define absolute errors and relative errors.

2-3 Discuss accuracy, precision, and resolution, and explain how they are related.

2-4 Explain the significance of the number of significant figures in a stated quantity.

2-5 Discuss the resultant error in calculations involving quantities with stated accuracies when the quantities are (a) added, (b) subtracted, (c) multiplied, (d) divided, and (e) one quantity is raised to the power of the other.

PROBLEMS

2-1 For the analog instrument in Figure 2-1(b), determine the meter reading when the selector switch is set to (a) 2.5 mA, (b) 5 V, and (c) 100 mA.

2-2 A batch of resistors that each have a nominal resistance of 330 Ω are to be tested and classified as ±5% and ±10% components. Calculate the maximum and minimum absolute resistance for each case.

2-3 The resistors in Problem 2-2 are specified at 25°C, and their temperature coefficient is −300 ppm/°C. Calculate the maximum and minimum resistance for these components at 100°C.

2-4 Estimate the measurement precision of the digital and analog instruments in Figure 2-1.

2-5 Estimate the measurement precision of the digital instrument in Figure 2-2(b).

2-6 A 1 kΩ potentiometer that has a resolution 0.5 Ω is used as a potential divider with a 10 V supply. Determine the precision of the output voltage.

2-7 Three of the resistors referred to in Problem 2-2 are connected in series. One has a ±5% tolerance, and the other two are ±10%. Calculate the maximum and minimum values of the total resistance.

2-8 A dc power supply provides currents to four electronic circuits. The currents are, 37 mA, 42 mA, 13 mA, and 6.7 mA. The first two are measured with an accuracy of ±3%, and the other two are measured with ±1% accuracy. Determine the maximum and minimum levels of the total supply current.

2-9 Two currents from different sources flow in opposite directions through a resistor. I_1 is measured as 79 mA on a 100 mA analog instrument with an accuracy of ±3% of full scale. I_2, determined as 31 mA, is measured on a digital instrument with a ±100 μA accuracy. Calculate the maximum and minimum levels of the current in R_1.

2-10 The voltages at opposite ends of a 470 Ω, ±5% resistor are measured as V_1 = 12 V and V_2 = 5 V. The measuring accuracies are ±0.5 V for V_1 and ±2% for V_2. Calculate the level of current in the resistor, and specify its accuracy.

2-11 A resistor R_1 has a potential difference of 25 V across its terminals, and a current of 63 mA. The voltage is measured on a 30 V analog instrument with an accuracy of ±5% of full scale. The current is measured on a digital instrument with a ±1 mA accuracy. Calculate the resistance of R_1 and specify its tolerance.

2-12 Calculate the maximum and minimum power dissipation in the resistor in Problem 2-10.

2-13 Determine the maximum and minimum power dissipation in the resistor in Problem 2-11.

2-14 A 470 Ω, ±10% resistor has a potential difference of 12 V across its terminals. If the voltage is measured with an accuracy of ±6%, determine the power dissipation in the resistor, and specify the accuracy of the result.

2-15 The output voltage from a precision 12 V power supply, monitored at intervals over a period of time, produced the following readings: $V_1 = 12.001$ V, $V_2 = 11.999$ V, $V_3 = 11.998$ V, $V_4 = 12.003$ V, $V_5 = 12.002$ V, $V_6 = 11.997$ V, $V_7 = 12.002$ V, $V_8 = 12.003$ V, $V_9 = 11.998$ V, and $V_{10} = 11.997$ V. Calculate the average voltage level, the mean deviation, the standard deviation, and the probable error in the measured voltage at any time.

2-16 Successive measurements of the temperature of a liquid over a period of time produced the following data: $T_1 = 25.05°C$, $T_2 = 25.02°C$, $T_3 = 25.03°C$, $T_4 = 25.07°C$, $T_5 = 25.55°C$, $T_6 = 25.06°C$, $T_7 = 25.04°C$, $T_8 = 25.05°C$, $T_9 = 25.07°C$, $T_{10} = 25.03°C$, $T_{11} = 25.02°C$, $T_{12} = 25.04°C$, $T_{13} = 25.02°C$, $T_{14} = 25.03°C$, and $T_{15} = 25.05°C$. Determine the average temperature, the mean deviation from average, the standard deviation, and the probable measurement error.

Electromechanical Instruments

3

Objectives

You will be able to:

1. Sketch the construction of a permanent-magnet moving-coil (PPMC) instrument, and explain its operation.
2. Show how PPMC instruments are used as galvanometers, dc ammeters, dc voltmeters, ac ammeters, and ac voltmeters.
3. Calculate appropriate shunt and series resistance values for given ammeter and voltmeter ranges, and determine instrument accuracy.
4. Sketch and explain the operation of series and shunt ohmmeter circuits. Explain ohmmeter scale shapes, and determine ohmmeter accuracy.
5. Sketch the front panel and scales for a typical volt-ohm-milliameter (VOM). Explain function and range selections, and discuss the use of the VOM.
6. Sketch and describe the construction of an electrodynamic instrument, and explain its dc and ac operation. Show how an electrodynamic instrument may be used as a voltmeter, an ammeter, and a wattmeter.

Introduction

The permanent-magnet moving-coil (PMMC) instrument consists basically of a lightweight coil of copper wire suspended in the field of a permanent magnet. Current in the wire causes the coil to produce a magnetic field that interacts with the field from the magnet, resulting in partial rotation of the coil. A pointer connected to the coil deflects over a calibrated scale, indicating the level of current flowing in the wire.

The PMMC instrument is essentially a low-level dc ammeter; however, with the use of parallel-connected resisors, it can be employed to measure a wide range of direct current levels. The instrument may also be made to function as a dc voltmeter by connecting appropriate-value resistors in series with the coil. Ac ammeters and voltmeters can be constructed by using rectifier circuits with PMMC instruments. Ohmmeters can be made from precision resistors, PMMC instruments, and batteries. Multirange meters are available that combine ammeter, voltmeter, and ohmmeter functions in one instrument.

The electrodynamic instrument is similar to the PMMC instrument except that it uses stationary coils instead of a permanent mangnet. The most important application of the electrodynamic instrument is as a wattmeter.

3-1 PERMANENT-MAGNET MOVING-COIL INSTRUMENT

Deflection Instrument Fundamentals

A deflection instrument uses a pointer that moves over a calibrated scale to indicate a measured quantity. For this to occur, three forces are operating in the electromechanical mechanism (or *movement*) inside the instrument: a *deflecting force*, a *controlling force*, and a *damping force*.

The *deflecting force* causes the pointer to move from its zero position when a current flows. In the *permanent-magnet moving-coil* (PMMC) instrument the deflecting force is magnetic. When a current flows in a lightweight moving coil pivoted between the poles of a permanent magnet [Figure 3-1(a)], the current sets up a magnetic field that interacts with the field of the permanent magnet. A force is exerted on a current-carrying conductor situated in a magnetic field. Consequently, a force is exerted on the coil turns, as illustrated, causing the coil to rotate on its pivots. The pointer is fixed to the coil, so it moves over the scale as the coil rotates.

The controlling force in the PMMC instrument is provided by spiral springs [Figure 3-1(b)]. The springs retain the coil and poiner at their zero position when no current is flowing. When current flows, the springs "wind up" as the coil rotates, and the force they exert on the coil increases. The coil and pointer stop rotating when the controlling force becomes equal to the deflecting force. The spring material must be nonmagnetic to avoid any magnetic field influence on the controlling force. Since the springs are also used to make electrical connection to the coil, they must have a low resistance. Phosphor bronze is the material usually employed.

As illustrated in Figure 3-2(a), the pointer and coil tend to oscillate for some time before settling down at their final position. A damping force is required to minimize (or damp out) the oscillations. The damping force must be present only when the coil is in motion; thus it must be generated by the rotation of the coil. In PMMC instruments, the damping force is normally provided by *eddy currents*. The coil former (or frame) is constructed of aluminum, a nonmagnetic conductor. Eddy currents induced in the coil former set up a magnetic flux that opposses the coil motion, thus damping the oscillations of the coil [see Figure 3-2(b)].

Two methods of supporting the moving system of a deflection instrument are illustrated in Figure 3-3. In the *jeweled-bearing* suspension shown in Figure 3-3(a), the pointed ends of shafts or *pivots* fastened to the coil are inserted into cone-shaped cuts in jewel (sapphire or glass) bearings. This allows the coil to rotate freely with the least possible

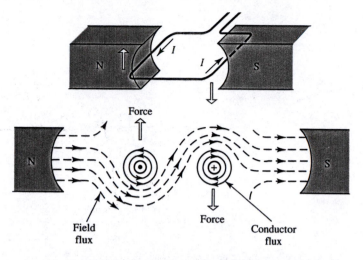

(a) The deflecting force in a PMMC instrument is provided by a current-carrying coil pivoted in a magnetic field.

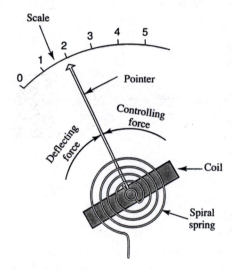

(b) The controlling force from the springs balances the deflecting force.

Figure 3-1 The deflecting force in a PMMC instrument is produced by the current in the moving coil. The controlling force is provided by spiral springs. The two forces are equal when the pointer is stationary.

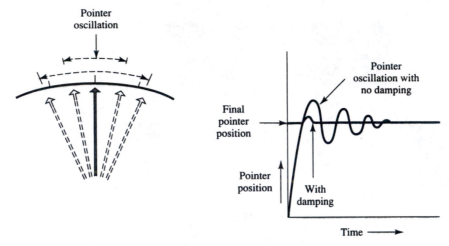

(a) Lack of damping causes the pointer to oscillate.

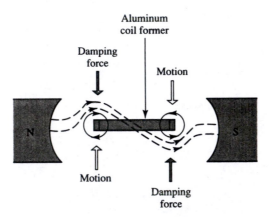

(b) The damping force in a PMMC instrument is provided
by eddy currents induced in the aluminum coil former
as it moves through the magnetic field.

Figure 3-2 A deflection instrument requires a damping force to stop the pointer oscillating about the indicated reading. The damping force is usually produced by eddy currents in a nonmagnetic coil former. These exist only when the coil is in motion.

friction. Although the coil is normally very lightweight, the pointed ends of the pivots have extremely small areas, so the surface load per unit area can be considerable. In some cases the bearings may be broken by the shock of an instrument being slammed down heavily upon a bench. Some jewel bearings are spring supported (as illustrated) to absorb such shocks more easily.

The *taut-band* method shown in Figure 3-3(b) is much tougher than jeweled-bearing suspension. As illustrated, two flat metal ribbons (phosphor bronze or platinum alloy) are

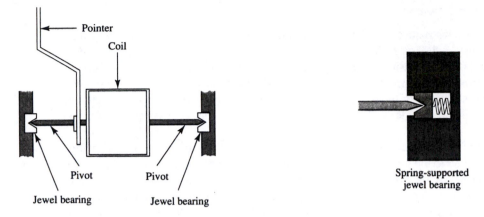

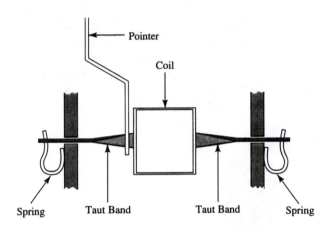

Spring-supported
jewel bearing

(a) Pivot and jewel-bearing suspension

(b) Taut-band suspension

Figure 3-3 The moving coil in a PMMC instrument may be supported by pivots in jeweled bearings, or by two flat metal ribbons held taut by springs. Taut-band suspension is the toughest and the most sensitive of the two.

held under tension by springs to support the coil. Because of the springs, the metal ribbons behave like rubber under tension. The ribbons also exert a controlling force as they twist, and they can be used as electrical connections to the moving coil. Because there is less friction, taut-band instruments can be much more sensitive than the jeweled-bearing type. The most sensitive jeweled-bearing instruments give full-scale deflection (FSD) with a coil current of 25 μA. With taut-band suspension FSD may be achieved with as little as 2 μA of coil current. The fact that the spring-mounted ribbon behaves as a rubber band makes the instrument extremely rugged compared to a jeweled-bearing instrument. If a jeweled-bearing instrument is dropped to a concrete floor from bench height, the bearings will almost certainly be shattered. A taut-band instrument is unlikely to be affected by a similar fall.

PMMC Construction

Details of the construction of a PMMC instrument or *D'Arsonval instrument* are illustrated in Figure 3-4. The main feature is a permanent magnet with two soft-iron *pole shoes*. A cylindrical soft-iron *core* is positioned between the shoes so that only very narrow air gaps exist between the core and the faces of the pole shoes. The lightweight moving coil is pivoted to move within these narrow air gaps. The air gaps are made as narrow as possible in order to have the strongest possible level of magnetic flux crossing the gaps.

Figure 3-4 also shows one of the two controlling spiral springs. One end of this spring is fastened to the pivoted coil, and the other end is connected to an adjustable *zero-position control*. By means of a screw on the instrument cover, the zero-position control can be adjusted to move the end of the spring. This allows the coil and pointer position to be adjusted (when no coil current is flowing) so that the pointer indicates exactly zero on the instrument scale.

Another detail shown in Figure 3-4 is one of (usually) two or three *counterweights* attached to the pointer. This is simply a machine screw along which a small screw-threaded weight can be adjusted. The counterweights provide correct mechanical balance of the moving system so that there is no gravitational effect on the accuracy of the instrument.

The PMMC instrument in Figure 3-5 illustrates a different type of construction. Instead of using a horseshoe-shaped permanent magnet, the permanent magnet is placed inside the coil (i.e., it replaces the soft-iron core shown in Figure 3-4). A thick cylindrical piece of soft iron surrounds the coil and the magnet. The magnetic flux flows across the air gaps and through the soft iron, and the coil sides move within the narrow air gaps. A

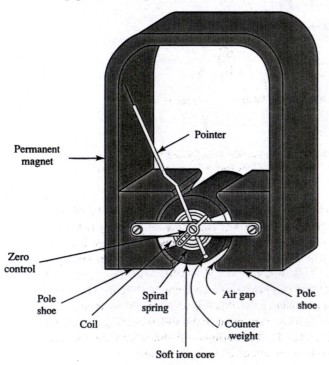

Permanent magnet

Pointer

Zero control

Pole shoe

Coil

Spiral spring

Air gap

Counter weight

Pole shoe

Soft iron core

Figure 3-4 A typical PMMC instrument is constructed of a *horseshoe* magnet, soft-iron pole shoes, a soft-iron core, and a suspended coil that moves in the air gap between the core and the pole shoes.

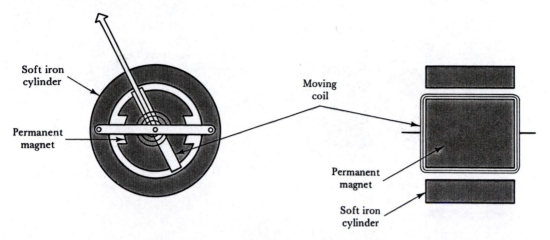

Figure 3-5 In a core-magnet PMMC instrument, the permanent magnet is located inside the moving coil, and the coil and magnet are positioned inside a soft-iron cylinder.

major advantage of this *core-magnet* type of construction is that the moving coil is shielded from external magnetic fields due to the presence of the soft-iron cylinder.

The current in the coil of a PMMC instrument must flow in one particular direction to cause the pointer to move (positively) from the zero position over the scale. When the current is reversed, the interaction of the magnetic flux from the coil with that of the permanent magnet causes the coil to rotate in the opposite direction, and the pointer is deflected to the left of zero (i.e., off-scale). The terminals of a PMMC instrument are identified as + and − to indicate the correct polarity for connection, and the instrument is said to be *polarized*. Because it is polarized, the PMMC instrument cannot be used directly to measure alternating current. Without rectifiers, it is purely a dc instrument.

Torque Equation and Scale

When a current I flows through a one-turn coil situated in a magnetic field, a force F is exerted on each side of the coil [Figure 3-6(a)]:

$$F = BIl \quad \text{newtons}$$

where B is the magnetic flux density in tesla, I is the current in amperes, and l is the length of the coil in meters.

Since the force acts on each side of the coil, the total force for a coil of N turns is

$$F = 2BIlN \quad \text{newtons}$$

The force on each side acts at a radius r, producing a deflecting torque:

$$T_D = 2BIlNr \quad \text{newton meters (N · m)}$$

$$= BIlN(2r)$$

$$\boxed{T_D = BIlND} \tag{3-1}$$

where D is the coil diameter [Figure 3-6(b)].

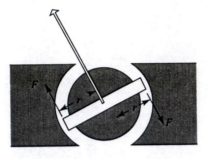

(a) Force F acts on each side of the coil

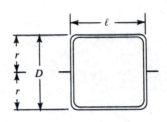

(b) Area enclosed by coil is $D \times \ell$

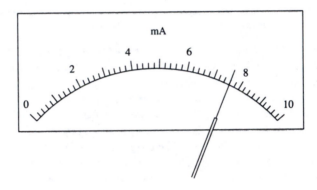

(c) Linear scale on a PMMC instrument

Figure 3-6 The deflecting torque on the coil of a PMMC instrument is directly proportional to the magnetic flux density, the coil dimensions, and the coil current. This gives the instrument a linear scale.

The controlling torque exerted by the spiral springs is directly proportional to the deformation or "windup" of the springs. Thus, the controlling torque is proportional to the actual angle of deflection of the pointer:

$$T_C = K\theta$$

where K is a constant. For a given deflection, the controlling and deflecting torques are equal:

$$K\theta = BlIND$$

Since all quantities except θ and I are constant for any given instrument, the deflection angle is

$$\boxed{\theta = CI}$$

(3-2)

where C is a constant.

Equation 3-2 shows that the pointer deflection is always proportional to the coil current. Consequently, the scale of the instrument is *linear,* or uniformly divided; that is, if 1 mA produces a 1 cm movement of the pointer from zero, 2 mA produces a 2 cm movement, and so on [see Figure 3-6(c)]. As will be explained the PMMC instrument can be used as a dc voltmeter, a dc ammeter, and an ohmmeter. When connected with rectifiers and transformers, it can also be employed to measure alternating voltage and current.

Example 3-1

A PMMC instrument with a 100-turn coil has a magnetic flux density in its air gaps of $B = 0.2$ T. The coil dimensions are $D = 1$ cm and $l = 1.5$ cm. Calculate the torque on the coil for a current of 1 mA.

Solution   *Equation 3-1,*

$$T_D = Bl\,IND$$

$$= 0.2\,\text{T} \times 1.5 \times 10^{-2} \times 1\,\text{mA} \times 100 \times 1 \times 10^{-2}$$

$$= 3 \times 10^{-6}\,\text{N} \cdot \text{m}$$

3-2 GALVANOMETER

A *galvanometer* is essentially a PMMC instrument designed to be sensitive to extremely low current levels. The simplest galvanometer is a very sensitive instrument with the type of center-zero scale illustrated in Figure 3-7(a). The deflection system is arranged so that the pointer can be deflected to either right or left of zero, depending on the direction of current through the moving coil. The scale may be calibrated in microamperes, or it may simply be a millimeter scale. In the latter case, the instrument *current sensitivity* (usually stated in μA/mm) is used to determine the current level that produces a measured deflection.

The torque equation for a galvanometer is exactly as discussed in Section 3-1. The deflecting torque is proportional to the number of coil turns, the coil dimensions, and the current flowing in the coil. The most sensitive moving-coil galvanometers use taut-band suspension, and the controlling torque is generated by the twist in the suspension ribbon. Eddy current damping may be provided, as in other PMMC instruments, by winding the coil on a nonmagnetic conducting coil former. Sometimes a nonconducting coil former is employed, and the damping currents are generated solely by the moving coil. In this case, the coil is shunted by a *damping resistor* which controls the level of eddy currents generated by the coil movements. Frequently, a *critical damping resistance* value is stated, which gives just sufficient damping to allow the pointer to settle down quickly with only a very small short-lived oscillation.

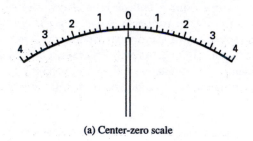

(a) Center-zero scale

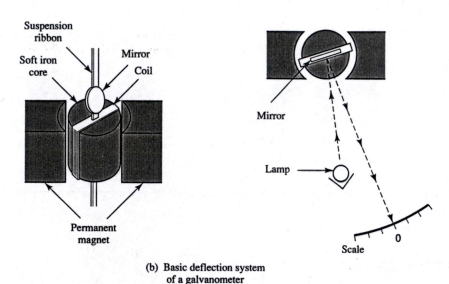

Suspension ribbon

Soft iron core

Mirror

Coil

Permanent magnet

Mirror

Lamp

Scale

0

(b) Basic deflection system of a galvanometer using a light beam

Figure 3-7 A galvanometer is simply an extremely sensitive PMMC instrument with a center-zero scale. For maximum sensitivity, the mass of the moving system is minimized by using a pointer that consists of a light beam reflected from a tiny mirror fastened to the coil.

With the moving-coil weight reduced to the lowest possible minimum for greatest sensitivity, the weight of the pointer can create a problem. This is solved in many instruments by mounting a small mirror on the moving coil instead of a pointer. The mirror reflects a beam of light on to a scale, as illustrated in Figure 3-7(b). The light beam behaves as a very long weightless pointer which can be substantially deflected by a very small coil current. This makes light-beam galvanometers sensitive to much lower current levels than pointer instruments.

Galvanometer *voltage sensitivity* is often expressed for a given value of critical damping resistance. This is usually stated in microvolts per millimeter. A *megohm sensitivity* is sometimes specified for galvanometers, and this is the value of resistance that must be connected in series with the instrument to restrict the deflection to one scale division when a potential difference of 1 V is applied across its terminals. Pointer galvanometers have current sensitivities ranging from 0.1 to 1 μA/mm. For light-beam instruments typical current sensitivities are 0.01 to 0.1 μA per scale division.

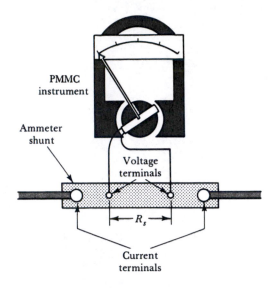

PMMC instrument

Ammeter shunt

Voltage terminals

R_s

Current terminals

(a) Construction of dc ammeter

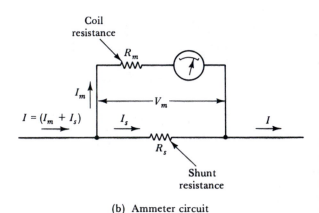

Coil resistance

R_m

I_m

V_m

$I = (I_m + I_s)$

I_s

I

R_s

Shunt resistance

(b) Ammeter circuit

Figure 3-9 A direct-current ammeter consists of a PMMC instrument and a low-resistance shunt. The meter current is directly proportional to the shunt current, so that the meter scale can be calibrated to indicate the total ammeter current.

maximum pointer deflection is produced by a very small current, and the coil is usually wound of thin wire that would be quickly destroyed by large currents. For larger currents, the instrument must be modified so that most of the current to be measured is shunted around the coil of the meter. Only a small portion of the current passes through the moving coil. Figure 3-9 illustrates how this is arranged.

A *shunt*, or very low resistance, is connected in parallel with the instrument coil [Figure 3-9(a)]. The shunt is sometimes referred to as a *four-terminal resistor*, because it has two sets of terminals identified as *voltage terminals* and *current terminals*. This is to ensure that the resistance in parallel with the coil (R_s) is accurately defined and the contact resistance of the current terminals is removed from R_s. Contact resistance can vary with change in current level and thus introduce errors.

In the circuit diagram in Figure 3-9(b), R_m is the meter resistance (or coil circuit resistance) and R_s is the resistance of the shunt. Suppose that the meter resistance is exactly 99 Ω and the shunt resistance is 1 Ω. The shunt current (I_s) will be 99 times the meter current (I_m). In this situation, if the meter gives FSD for a coil current of 0.1 mA, the scale should be calibrated to read 100×0.1 mA or 10 mA at full scale. The relationship between shunt current and coil current is further investigated in Examples 3-3 and 3-4.

Example 3-3

An ammeter (as in Figure 3-9) has a PMMC instrument with a coil resistance of $R_m = 99\ \Omega$ and FSD current of 0.1 mA. Shunt resistance $R_s = 1\ \Omega$. Determine the total current passing through the ammeter at (a) FSD, (b) 0.5 FSD, and (c) 0.25 FSD,

Solution

(a) *At FSD:*

$$meter\ voltage\ V_m = I_m R_m [\text{see Figure 3-9(b)}]$$

$$= 0.1\ mA \times 99\ \Omega$$

$$= 9.9\ mV$$

and

$$I_s R_s = V_m$$

$$I_s = \frac{V_m}{R_s} = \frac{9.9\ mV}{1\ \Omega} = 9.9\ mA$$

$$total\ current\ I = I_s + I_m = 9.9\ mA + 0.1\ mA$$

$$= 10\ mA$$

(b) *At 0.5 FSD:*

$$I_m = 0.5 \times 0.1\ mA = 0.05\ mA$$

$$V_m = I_m R_m = 0.05\ mA \times 99\ \Omega = 4.95\ mV$$

$$I_s = \frac{V_m}{R_s} = \frac{4.95\ mV}{1\ \Omega} = 4.95\ mA$$

$$total\ current\ I = I_s + I_m = 4.95\ mA + 0.5\ mA$$

$$= 5\ mA$$

(c) *At 0.25 FSD:*

$$I_m = 0.25 \times 0.1\ mA = 0.025\ mA$$

$$V_m = I_m R_m = 0.025\ mA \times 99\ \Omega$$

$$= 2.475\ mV$$

$$I_s = \frac{V_m}{R_s} = \frac{2.475\ mV}{1\ \Omega} = 2.475\ mA$$

$$\text{total current } I = I_s + I_m = 2.475 \text{ mA} + 0.025 \text{ mA}$$

$$= 2.5 \text{ mA}$$

Ammeter Scale

The total ammeter current in Example 3-3 is 10 mA when the moving-coil instrument indicates FSD. Therefore, the meter scale can be calibrated for FSD to indicate 10 mA. When the pointer indicates 0.5 FSD and 0.25 FSD, the current levels are 5 mA and 2.5 mA, respectively. Thus, the ammeter scale may be calibrated to linearly represent all current levels from zero to 10 mA. Figure 3-10 shows a panel meter (for mounting on a control panel) that has a direct current scale calibrated linearly from 0 mA to 50 μA.

Shunt Resistance

Refer again to Example 3-3. If a shunt having a smaller resistance is used, the shunt current and the total meter current will be larger than the levels calculated. In fact, shunt resistance values can be determined to convert a PMMC instrument into an ammeter for measuring virtually any desired level of current. Example 3-4 demonstrates how shunt resistances are calculated.

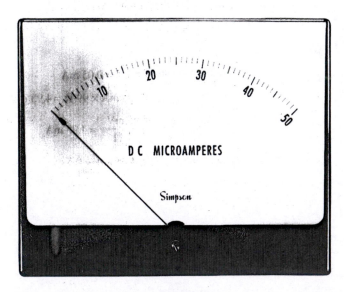

Figure 3-10 A dc ammeter made up of a PMMC instrument and a shunt has a linear current scale (Courtesy of bach-simpson limited.)

Example 3-4

A PMMC instrument has FSD of 100 μA and a coil resistance of 1 kΩ. Calculate the required shunt resistance value to convert the instrument into an ammeter with (a) FSD = 100 mA and (b) FSD = 1 A.

Solution

(a) $FSD = 100$ mA:

$$V_m = I_m R_m = 100 \ \mu A \times 1 \ k\Omega = 100 \ mV$$

$$I = I_s + I_m$$

$$I_s = I - I_m = 100 \ mA - 100 \ \mu A = 99.9 \ mA$$

$$R_s = \frac{V_m}{I_s} = \frac{100 \ mV}{99.9 \ mA} = 1.001 \ \Omega$$

(b) $FSD = 1$ A:

$$V_m = I_m R_m = 100 \ mV$$

$$I_s = I - I_m = 1 \ A - 100 \ \mu A = 999.9 \ mA$$

$$R_s = \frac{V_m}{I_s} = \frac{100 \ mV}{999.9 \ mA} = 0.10001 \ \Omega$$

Swamping Resistance

The moving coil in a PMMC instrument is wound with thin copper wire, and its resistance can change significantly when its temperature changes. The heating effect of the coil current may be enough to produce a resistance change. Any such change in coil resistance will introduce an error in ammeter current measurements. To minimize the effect of coil resistance variation, a *swamping resistance* made of *manganin* or *constantan* is connected in series with the coil, as illustrated in Figure 3-11. Manganin and constantan have resistance temperature coefficients very close to zero. If the swamping resistance is nine times the coil resistance, a 1% change in coil resistance would result in a total (swamping plus coil) resistance change of 0.1%.

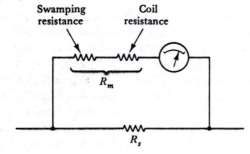

Figure 3-11   A swamping resistance made of a material with a near-zero temperature coefficient can be connected in series with the coil of a PMMC instrument to minimize temperature errors.

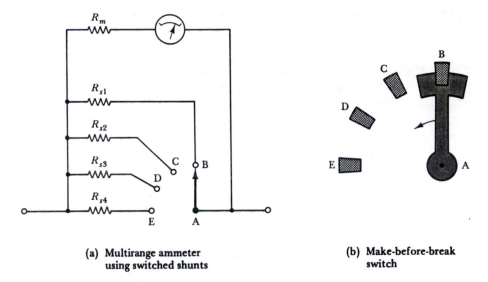

(a) Multirange ammeter
using switched shunts

(b) Make-before-break
switch

Figure 3-12 A multirange ammeter consists of a PMMC instrument, several shunts, and a switch that makes contact with the next shunt before losing contact with the previous one when range switching.

The ammeter shunt must also be made of manganin or constantan to avoid shunt resistance variations with temperature. As noted in Figure 3-11, the swamping resistance must be considered part of the meter resistance R_m when calculating shunt resistance values.

Multirange Ammeters

The circuit of a multirange ammeter is shown in Figure 3-12(a). As illustrated, a rotary switch is employed to select any one of several shunts having different resistance values. A *make-before-break switch* [Figure 3-12(b)] must be used so that the instrument is not left without a shunt in parallel with it even for a brief instant. If this occurred, the high resistance of the instrument would affect the current flowing in the circuit. More important, a current large enough to destroy the instrument might flow through its moving coil. When switching between shunts, the wide-ended moving contact of the make-before-break switch makes contact with the next terminal before it breaks contact with the previous terminal. Thus, during switching there are actually two shunts in parallel with the instrument.

Figure 3-13 shows another method of protecting the deflection instrument of an ammeter from excessive current flow when switching between shunts. Resistors R_1, R_2, and R_3 constitute an *Ayrton shunt*. In Figure 3-13(a) the switch is at contact B, and the total resistance in parallel with the instrument is $R_1 + R_2 + R_3$. The meter circuit resistance remains R_m. When the switch is at contact C [Figure 3-13(b)], the resistance R_3 is in series with the meter, and $R_1 + R_2$ is in parallel with $R_m + R_3$. Similarly, with the switch at contact D, R_1 is in parallel with $R_m + R_2 + R_3$. Because the shunts are permanently connected, and the switch makes contact with the shunt junctions, the deflection instrument is never left without a parallel-connected shunt (or shunts). In Ex-

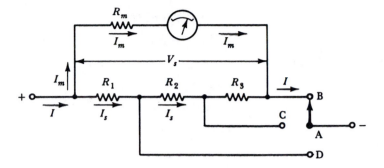

(a) $(R_1 + R_2 + R_3)$ in parallel with R_m

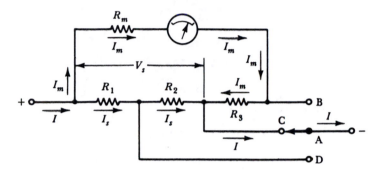

(b) $(R_1 + R_2)$ in parallel with $(R_m + R_3)$

Figure 3-13   An Ayrton shunt used with an ammeter consists of several series-connected resistors all connected in parallel with the PMMC instrument. Range change is effected by switching between the resistor junctions.

ample 3-5 ammeter current ranges are calculated for each switch position on an Ayrton shunt.

Example 3-5

A PMMC instrument has a three-resistor Ayrton shunt connected across it to make an ammeter, as in Figure 3-13. The resistance values are $R_1 = 0.05\ \Omega$, $R_2 = 0.45\ \Omega$, and $R_3 = 4.5\ \Omega$. The meter has $R_m = 1\ \text{k}\Omega$ and FSD $= 50\ \mu\text{A}$. Calculate the three ranges of the ammeter.

Solution   *Refer to Figure 3-13.*

Switch at contact B: $V_s = I_m R_m = 50\ \mu\text{A} \times 1\ \text{k}\Omega = 50\ \text{mV}$

$$I_s = \frac{V_s}{R_1 + R_2 + R_3}$$

$$= \frac{50\ \text{mV}}{0.05\ \Omega + 0.45\ \Omega + 4.5\ \Omega} = 10\ \text{mA}$$

$$I = I_m + I_s = 50\ \mu A + 10\ mA$$

$$= 10.05\ mA$$

Ammeter range $\simeq 10\ mA.$

Switch at contact C:
$$V_s = I_m(R_m + R_3)$$

$$= 50\ \mu A(1\ k\Omega + 4.5\ \Omega)$$

$$\simeq 50\ mV$$

$$I_s = \frac{V_s}{R_1 + R_2}$$

$$= \frac{50\ mV}{0.05\ \Omega + 0.45\ \Omega}$$

$$= 100\ mA$$

$$I = 50\ \mu A + 100\ mA$$

$$= 100.05\ mA$$

Ammeter range $\simeq 100\ mA.$

Switch at contact D:
$$V_s = I_m(R_m + R_3 + R_2)$$

$$= 50\ \mu A(1\ k\Omega + 4.5\ \Omega + 0.45\ \Omega)$$

$$\simeq 50\ mV$$

$$I_s = \frac{V_s}{R_1} = \frac{50\ mV}{0.05\ \Omega}$$

$$= 1\ A$$

$$I = 50\ \mu A + 1\ A$$

$$= 1.00005\ A$$

Ammeter range $\simeq 1\ A.$

3-4 DC VOLTMETER

Voltmeter Circuit

The deflection of a PMMC instrument is proportional to the current flowing through the moving coil. The coil current is directly proportional to the voltage across the coil. Therefore, the scale of the PMMC meter could be calibrated to indicate voltage. The coil resistance is normally quite small, and thus the coil voltage is also usually very small. Without any additional series resistance the PMMC instrument would only be able to measure very low voltage levels. The voltmeter range is easily increased by connecting a resistance in series with the instrument [see Figure 3-14(a)]. Because it increases the range of the voltmeter, the series resistance is termed a *multiplier resistance*. A multiplier resis-

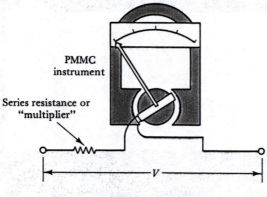

PMMC
instrument

Series resistance or
"multiplier"

V

(a) Construction of dc voltmeter

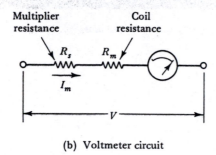

Multiplier
resistance

Coil
resistance

R_s R_m

I_m

V

(b) Voltmeter circuit

Figure 3-14 A dc voltmeter is made up of a PMMC instrument and a series *multiplier* resistor. The meter current is directly proportional to the applied voltage, so that the meter scale can be calibrated to indicate the voltage.

tance that is nine times the coil resistance will increase the voltmeter range by a factor of 10. Figure 3-14(b) shows that the total resistance of the voltmeter is (multiplier resistance) + (coil resistance).

Example 3-6

A PMMC instrument with FSD of 100 μA and a coil resistance of 1 kΩ is to be converted into a voltmeter. Determine the required multiplier resistance if the voltmeter is to measure 50 V at full scale (Figure 3-15). Also calculate the applied voltage when the instrument indicates 0.8, 0.5, and 0.2 of FSD.

Solution

$$V = I_m(R_s + R_m) \quad \text{[see Figure 3-14(b)]}$$

$$R_s + R_m = \frac{V}{I_m}$$

and
$$R_s = \frac{V}{I_m} - R_m$$

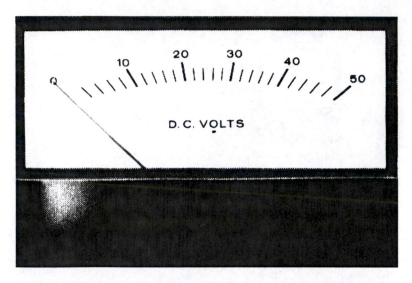

Figure 3-15 A dc voltmeter using a PMMC instrument has a linear voltage scale.

For V = 50 V FSD,	$I_m = 100 \ \mu A$
	$R_s = \dfrac{50 \ V}{100 \ \mu A} - 1 \ k\Omega$
	$= 499 \ k\Omega$
At 0.8 FSD:	$I_m = 0.8 \times 100 \ \mu A$
	$= 80 \ \mu A$
	$V = I_m(R_s + R_m)$
	$= 80 \ \mu A(499 \ k\Omega + 1 \ k\Omega)$
	$= 40 \ V$
At 0.5 FSD:	$I_m = 50 \ \mu A$
	$V = 50 \ \mu A(499 \ k\Omega + 1 \ k\Omega)$
	$= 25 \ V$
At 0.2 FSD:	$I_m = 20 \ \mu A$
	$V = 20 \ \mu A(499 \ k\Omega + 1 \ k\Omega)$
	$= 10 \ V$

Swamping Resistance

As in the case of the ammeter, the change in coil resistance (R_m) with temperature change can introduce errors in a PMMC voltmeter. However, the presence of the voltmeter multiplier resistor (R_s) tends to *swamp* coil resistance changes, except for low voltage ranges where R_s is not very much larger than R_m. R_s will also be temperature sensitive to some degree (not as much as the copper wire coil), and in some cases it might be necessary to construct the multiplier resistor of manganin or constantan.

Voltmeter Sensitivity

The voltmeter designed in Example 3-6 has a total resistance of

$$R_V = R_s + R_m = 500 \text{ k}\Omega$$

Since the instrument measures 50 V at full scale, its *resistance per volt* is

$$\frac{500 \text{ k}\Omega}{50 \text{ V}} = 10 \text{ k}\Omega/\text{V}$$

This quantity is also termed the *sensitivity* of the voltmeter. The sensitivity of a voltmeter is always specified by the manufacturer, and it is frequently printed on the scale of the instrument. If the sensitivity is known, the total voltmeter resistance is easily calculated as (sensitivity × range). [It is important to note that the total resistance is *not* (sensitivity × meter reading).] If the full-scale meter current is known, the sensitivity can be determined as the reciprocal of full-scale current.

Ideally, a voltmeter should have an extremely high resistance. A voltmeter is always connected across, or in parallel with, the points in a circuit at which the voltage is to be measured. If its resistance is too low, it can alter the circuit voltage. This is known as *voltmeter loading effect*.

Multirange Voltmeter

A multirange voltmeter consists of a deflection instrument, several multiplier resistors, and a rotary switch. Two possible circuits are illustrated in Figure 3-16. In Figure 3-16(a) only one of the three multiplier resistors is connected in series with the meter at any time. The range of this voltmeter is

$$V = I_m(R_m + R)$$

where R can be R_1, R_2, or R_3.

In Figure 3-16(b) the multiplier resistors are connected in series, and each junction is connected to one of the switch terminals. The range of this voltmeter can also be calculated from the equation $V = I_m(R_m + R)$, where R can now be R_1, $R_1 + R_2$, or $R_1 + R_2 + R_3$.

Of the two circuits, the one in Figure 3-16(b) is the least expensive to construct. This is because (as shown in Example 3-7) all of the multiplier resistors in Figure 3-16(a) must be special (nonstandard) values, while in Figure 3-16(b) only R_1 is a special resistor and all other multipliers are standard-value (precise) resistors.

Example 3-7

A PMMC instrument with FSD = 50 μA and R_m = 1700 Ω is to be employed as a voltmeter with ranges of 10 V, 50 V, and 100 V. Calculate the required values of multiplier resistors for the circuits of Figure 3-16(a) and (b).

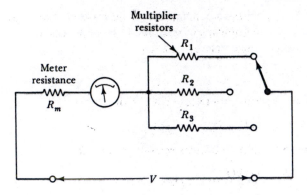

(a) Multirange voltmeter using switched
   multiplier resistors

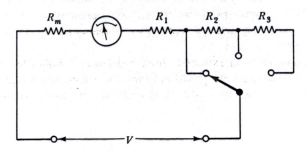

Figure 3-16 A multirange voltmeter consists of a PMMC instrument, several multiplier resistors, and a switch for range selection. Individual, or series-connected resistors may be used.

(b) Multirange voltmeter using series-connected
   multiplier resistors

Solution

Circuit as in Figure 3-16(a):

$$R_m + R_1 = \frac{V}{I_m}$$

$$R_1 = \frac{V}{I_m} - R_m$$

$$= \frac{10 \text{ V}}{50 \text{ μA}} - 1700 \text{ Ω}$$

$$= 198.3 \text{ kΩ}$$

$$R_2 = \frac{50 \text{ V}}{50 \text{ μA}} - 1700 \text{ Ω}$$

$$= 998.3 \text{ kΩ}$$

$$R_3 = \frac{100 \text{ V}}{50 \text{ μA}} - 1700 \text{ Ω}$$

$$= 1.9983 \text{ MΩ}$$

Circuit as in Figure 3-16(b):

$$R_m + R_1 = \frac{V_1}{I_m}$$

$$R_1 = \frac{V_1}{I_m} - R_m$$

$$= \frac{10\ V}{50\ \mu A} - 1700\ \Omega$$

$$= 198.3\ k\Omega$$

$$R_m + R_1 + R_2 = \frac{V_2}{I_m}$$

$$R_2 = \frac{V_2}{I_m} - R_1 - R_m$$

$$= \frac{50\ V}{50\ \mu A} - 198.3\ k\Omega - 1700\ \Omega$$

$$= 800\ k\Omega$$

$$R_m + R_1 + R_2 + R_3 = \frac{V_3}{I_m}$$

$$R_3 = \frac{V_3}{I_m} - R_2 - R_1 - R_m$$

$$= \frac{100\ V}{50\ \mu A} - 800\ k\Omega - 198.3\ k\Omega - 1700\ \Omega$$

$$= 1\ M\Omega$$

3-5 RECTIFIER VOLTMETER

PMMC Instrument on AC

As discussed earlier, the PMMC instrument is *polarized,* that is, its terminals are identified as + and −, and it must be connected correctly for positive (on-scale) deflection to occur. When an alternating current with a very low frequency is passed through a PMMC instrument, the pointer tends to follow the instantaneous level of the ac. As the current grows postively, the pointer deflection increases to a maximum at the peak of the ac. Then as the instantaneous current level falls, the pointer deflection decreases toward zero. When the ac goes negative, the pointer is deflected (off-scale) to the left of zero. This kind of pointer movement can occur only with ac having a frequency of perhaps 0.1 Hz or lower. With the normal 60 Hz or higher supply frequencies, the damping mechanism of the instrument and the inertia of the meter movement prevent the pointer from following the changing instantaneous levels. Instead, the instrument pointer settles at the average value of the current flowing through the moving coil. The average value of purely sinusoidal ac is zero. Therefore, a PMMC instrument connected directly to measure 60 Hz ac indicates zero. It is important to note that although a PMMC instrument connected to an ac supply may be indicating zero, there can actually be a very large rms current flowing in its coils.

Full-Wave Rectifier Voltmeter

Rectifier instruments use silicon or germanium diodes to convert alternating current to a series of unidirectional current pulses, which produce positive deflection when passed through a PMMC instrument. The full-wave bridge rectifier circuit in Figure 3-17 passes the positive half-cycles of the sinusoidal input waveform and inverts the negative half-cycles. When the input is positive, diodes D_1 and D_4 conduct, causing current to flow through the meter from top to bottom, as shown. When the input goes negative, D_2 and D_3 conduct, and current again flows through the meter from the positive terminal to the negative terminal. The resulting current waveform is a series of positive half-cycles without any intervening spaces (see Figure 3-17).

As in the case of a dc voltmeter, the rectifier voltmeter circuit in Figure 3-17 uses a series-connected multiplier resistor to limit the current flow through the PMMC instrument. The meter deflection is proportional to the average current, which is 0.637 × peak current. But the actual current (or voltage) to be indicated in ac measurements is normally the rms quantity, which is 0.707 of the peak value, or 1.11 times the average value. Since there are direct relationships between rms, peak, and average values, the meter scale can be calibrated to indicate rms volts.

A rectifier voltmeter as discussed above is for use only on pure sine-wave voltages. When other than pure sine waves are applied, the voltmeter will *not* indicate the rms voltage.

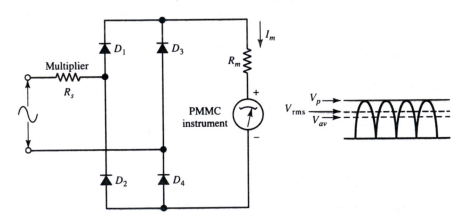

Figure 3-17 An ac voltmeter may be constructed of a PMMC instrument, a multiplier resistor, and a full-wave bridge rectifier. The instrument scale is correct only for pure sine waves.

Example 3-8

A PMMC instrument with FSD = 100 μA and $R_m = 1$ kΩ is to be employed as an ac voltmeter with FSD = 100 V (rms). Silicon diodes are used in the bridge rectifier circuit of Figure 3-17. Calculate the multiplier resistance value required.

Solution   *At FSD, the average current flowing through the PMMC instrument is*

$$I_{av} = 100 \text{ μA}$$

$$peak \; current \; I_m = \frac{I_{av}}{0.637} = \frac{100 \text{ μA}}{0.637} \simeq 157 \text{ μA}$$

$$I_m = \frac{(applied \; peak \; voltage) - (rectifier \; volt \; drop)}{total \; circuit \; resistance}$$

rectifier volt drops $= 2V_F$ (for D_1 and D_4 or D_2 and D_3)

applied peak voltage $= 1.414 V_{rms}$

total circuit resistance $= R_s + R_m$

$$I_m = \frac{1.414 V_{rms} - 2V_F}{R_s + R_m}$$

$$R_s = \frac{1.414 V_{rms} - 2V_F}{I_m} - R_m$$

$$= \frac{(1.414 \times 100 \text{ V}) - (2 \times 0.7 \text{ V})}{157 \text{ μA}} - 1 \text{ kΩ}$$

$$= 890.7 \text{ kΩ}$$

Example 3-9

Calculate the pointer indications for the voltmeter in Example 3-8, when the rms input voltage is (a) 75 V and (b) 50 V.

Solution

(a) $$I_{av} = 0.637 I_m = 0.637 \left(\frac{1.414 V_{rms} - 2V_F}{R_s + R_m} \right)$$

$$= 0.637 \left[\frac{(1.414 \times 75 \text{ V}) - (2 \times 0.7 \text{ V})}{890.7 \text{ kΩ} + 1 \text{ kΩ}} \right]$$

$$\simeq 75 \text{ μA} = 0.75 \text{ FSD}$$

$$\text{(b)} \qquad I_{av} = 0.637 \left[\frac{(1.414 \times 50 \text{ V}) - (2 \times 0.7 \text{ V})}{890.7 \text{ k}\Omega + 1 \text{ k}\Omega} \right]$$

$$\simeq 50 \ \mu\text{A} = 0.5 \text{ FSD}$$

Example 3-10

Calculate the sensitivity of the voltmeter in Example 3-8.

Solution

$$I_m = 157 \ \mu\text{A}$$

$$I_{rms} = 0.707 I_m = 0.707 \times 157 \ \mu\text{A}$$

$$\simeq 111 \ \mu\text{A (at FSD)}$$

$$V_{rms} = 100 \text{ V (at FSD)}$$

$$\text{total } R = \frac{100 \text{ V}}{111 \ \mu\text{A}} = 900.9 \text{ k}\Omega$$

$$\text{sensitivity} = \frac{R}{V} = \frac{900.9 \text{ k}\Omega}{100 \text{ V}} \ \Omega/\text{V}$$

$$= 9.009 \text{ k}\Omega/\text{V}$$

$$\simeq 9 \text{ k}\Omega/\text{V}$$

Examples 3-8 and 3-9 demonstrate that the rectifier voltmeter designed to indicate 100 V rms at full scale also indicates 0.75 FSD when 75 V rms is applied, and 0.5 FSD for 50 V rms. Therefore, the instrument has a linear scale. At low levels of input voltage the rectifier current is also low, and this can result in errors due to variations in the diode voltage drop. The effect can be countered by using a shunt resistor across the meter, as discussed next for a half-wave rectifier voltmeter.

Half-Wave Rectifier Voltmeter

Half-wave rectification is employed in the ac voltmeter circuit shown in Figure 3-18. R_{SH} shunting the meter is included to cause a relatively large current to flow through diode D_1 (larger than the meter current) when the diode is forward biased. This is to ensure that the diode is biased beyond the knee and well into the linear range of its characteristics. Diode D_2 conducts during the negative half-cycles of the input. When conducting, D_2 causes a small voltage drop (V_F) across D_1 and the meter, thus preventing the flow of any significant reverse leakage current through the meter via D_1. Diode D_2 also protects the meter against reverse voltages.

The waveform of voltage developed across the meter and R_{SH} is a series of positive half-cycles with intervening spaces, as illustrated. In half-wave rectification, $I_{av} = 0.5(0.637 I_m)$. This must be taken into account in the circuit design calculations.

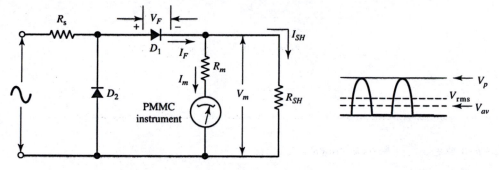

Figure 3-18 Half-wave rectification may be used with a PMMC instrument and a multiplier resistor for ac voltage measurements. A shunt resistor (R_{SH}) is included to ensure a satisfactory rectifier forward current level. The additional rectifier (D_2) minimizes reverse leakage current through D_1.

Example 3-11

A PMMC instrument with FSD = 50 μA and R_m = 1700 Ω is used in the half-wave rectifier voltmeter circuit illustrated in Figure 3-18. The silicon diode (D_1) must have a minimum (peak) forward current of 100 μA when the measured voltage is 20% of FSD. The voltmeter is to indicate 50 V rms at full scale. Calculate the values of R_s and R_{SH}.

Solution   At FSD, I_{av} = 50 μA.

Meter peak current,

$$I_m = \frac{I_{av}}{0.5 \times 0.637} = \frac{50 \text{ μA}}{0.5 \times 0.637} \approx 157 \text{ μA}$$

At 20% of FSD, diode peak current I_F must be at least 100 μA; therefore, at 100% of FSD,

$$I_{F(peak)} = \frac{100\%}{20\%} \times 100 \text{ μA} = 500 \text{ μA}$$

$$I_{F(peak)} = I_m + I_{SH}$$

$$I_{SH(peak)} = I_{F(peak)} - I_m$$

$$= 500 \text{ μA} - 157 \text{ μA} = 343 \text{ μA}$$

$$V_{m(peak)} = I_m R_m = 157 \text{ μA} \times 1700 \text{ Ω}$$

$$= 266.9 \text{ mV}$$

$$R_{SH} = \frac{V_{m(peak)}}{I_{SH(peak)}} = \frac{266.9 \text{ mV}}{343 \text{ μA}} = 778 \text{ Ω}$$

$$I_{F(peak)} = \frac{(\text{applied peak voltage}) - V_{m(peak)} - V_F}{R_s}$$

$$I_{F(peak)} = \frac{1.414 V_{rms} - V_{m(peak)} - V_F}{R_s}$$

$$R_s = \frac{1.414V_{\text{rms}} - V_{m(\text{peak})} - V_F}{I_{F(\text{peak})}}$$

$$= \frac{(1.414 \times 50 \text{ V}) - 266.9 \text{ mV} - 0.7 \text{ V}}{500 \text{ }\mu\text{A}}$$

$$= 139.5 \text{ k}\Omega$$

Half-Bridge Full-Wave Rectifier Voltmeter

The circuit in Figure 3-19 is that of an ac voltmeter employing a *half-bridge full-wave rectifier circuit.* The *half-bridge* name is applied because two diodes and two resistors are employed instead of the four diodes used in a full-wave bridge rectifier. This circuit passes full-wave rectified current through the meter, but as in the circuit of Figure 3-18, some of the current bypasses the meter.

During the positive half-cycle of the input, diode D_1 is forward biased and D_2 is reverse biased. Current flows from terminal 1 through D_1 and the meter (positive to negative), and then through R_2 to terminal 2. But R_1 is in parallel with the meter and R_2, which are connected in series. Therefore, much of the current flowing in D_1 passes through R_1, while only part of it flows through the meter and R_2. During the negative half-cycle of the input, D_2 is forward biased and D_1 is reverse biased. Current now flows from terminal 2 through R_1 and the meter, and through D_2 to terminal 1. Now, R_2 is in parallel with the series-connected meter and R_1. Once again, much of the diode current bypasses the meter by flowing through R_2. This arrangement forces the diodes to operate beyond the knee of their characteristics and helps to compensate for differences that might occur in the characteristics of D_1 and D_2.

3-6 RECTIFIER AMMETER

Like a dc ammeter, an ac ammeter must have a very low resistance because it is always connected in series with the circuit in which current is to be measured. This

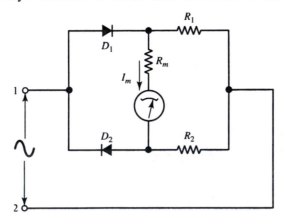

Figure 3-19 Circuit of an ac voltmeter using a half-bridge full-wave rectifier. The resistors in the half-bridge circuit appear in parallel with the PMMC instrument during alternate half-cycles to ensure a satisfactory rectifier forward current.

Electromechanical Instruments   Chap. 3

low-resistance requirement means that the voltage drop across the ammeter must be very small, typically not greater than 100 mV. However, the voltage drop across a diode is 0.3 to 0.7 V, depending on whether the diode is made from germanium or silicon. When a bridge rectifier circuit is employed, the total diode volt drop is 0.6 to 1.4 V. Clearly, a rectifier instrument is not suitable for direct application as an ac ammeter.

The use of a *current transformer* (Figure 3-20) gives the ammeter a low terminal resistance and low voltage drop. The transformer also steps up the input voltage (more secondary turns than primary turns) to provide sufficient voltage to operate the rectifiers, and at the same time it steps down the primary current to a level suitable for measurement by a PMMC meter. Since the transformer is used in an ammeter circuit, the current transformation ratio $I_p/I_s = N_s/N_p$ is very important.

A precise load resistor (R_L in Figure 3-20) is connected across the secondary winding of the transformer. This is selected to take the portion of secondary current not required by the meter. For example, suppose that the PMMC instrument requires 100 μA (average) for FSD, and the current transformer has $N_s = 2000$ and $N_p = 5$. If the rms primary current is 100 mA, the secondary rms current is

$$I_s = \frac{5}{2000} \times 100 \text{ mA} = 250 \text{ μA}$$

or an average of

$$I_{s(\text{av})} = \frac{1}{1.11} \times 250 \text{ μA} = 225.2 \text{ μA}$$

Since the meter requies 100 μA for FSD, the value of R_L is calculated to pass the remaining 125.2 μA.

The range of the instrument can be changed by switching-in different values of load resistance. Another method of range changing involves the use of additional terminals (or taps) on the primary winding to alter the number of primary turns, as shown in Figure 3-20.

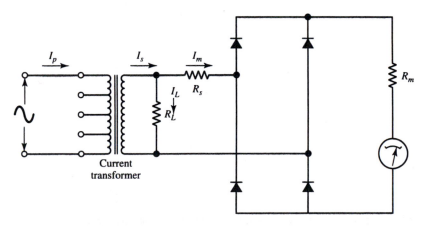

Figure 3-20   Ac ammeter circuit consisting of a current transformer, full-wave bridge rectifier, and a PMMC instrument.

Example 3 -12

A rectifier ammeter with the circuit shown in Figure 3-20 is to give FSD for a primary current of 250 mA. The PMMC meter has FSD = 1 mA and $R_m = 1700 \ \Omega$. The current transformer has $N_s = 500$ and $N_p = 4$. The diodes each have $V_F = 0.7$ V, and the series resistance is $R_s = 20 \ k\Omega$. Calculate the required value of R_L.

Solution

$$\text{Peak meter current } I_m = \frac{I_{av}}{0.637} = \frac{1 \text{ mA}}{0.637}$$

$$= 1.57 \text{ mA}$$

Transformer secondary peak voltage,

$$E_m = I_m(R_s + R_m) + 2V_F$$

$$= 1.57 \text{ mA } (20 \text{ k}\Omega + 1700 \ \Omega) + 1.4 \text{ V}$$

$$\approx 35.5 \text{ V}$$

or secondary voltage $\quad E_s = (0.707 \times 35.5 \text{ V}) \text{ rms}$

$$\approx 25.1 \text{ V}$$

and $\quad$ rms meter current $= 1.11 \ I_{av}$

$$= 1.11 \text{ mA}$$

Transformer rms secondary current,

$$I_s = I_p \frac{N_p}{N_s}$$

$$= 250 \text{ mA} \times \frac{4}{500} = 2 \text{ mA}$$

$$I_L = I_s - I_m$$

$$= 2 \text{ mA} - 1.11 \text{ mA} = 0.89 \text{ mA}$$

$$R_L = \frac{E_s}{I_L} = \frac{25.1 \text{ V}}{0.89 \text{ mA}}$$

$$= 28.2 \text{ k}\Omega$$

3-7 DEFLECTION INSTRUMENT ERRORS

Reading Errors

Some sources of error in measurements made by deflection instruments are: bearing friction, improperly adjusted zero, and incorrect reading of the pointer indication. Zero and friction errors can be minimized by carefully adjusting the mechanical zero of an instrument before use and by gently tapping the meter to relieve friction when zeroing and reading. Portable instruments should normally be used lying flat on their backs. Care in deciding the exact position of the pointer on the scale will reduce reading errors.

Even with an accurately marked scale and a sharp pointer, two observers may disagree about the exact scale reading. This occurs because of *parallax error:* the uncertainty about the eye of the observer being directly in line with the end of the pointer. Parallax error is eliminated in good instruments by the use of a knife-edge pointer and a mirror alongside the scale. When an observer lines up the pointer and the mirror image of the pointer, the observer's eye is exactly in the line with the pointer, and the scale can then be read accurately.

Specified Accuracy

High-quality instruments may have their accuracy specified as a percentage of the actual scale reading, or measured quantity. However, for most deflection instruments, manufacturers specify the accuracy as a percentage of FSD. This means, for example, that an instrument that gives FSD for a coil current of 100 μA, and which is specified as accurate to ±1%, has a ±1 μA accuracy at *all points* on its scale. Thus, as demonstrated in Example 3-13, the measurement error becomes progressively greater for low scale readings.

Example 3-13

An instrument that indicates 100 μA at FSD has a specified accuracy of ±1%. Calculate the upper and lower limits of measured current and the percentage error in the measurement for (a) FSD and (b) 0.5 FSD.

Solution

(a) *At FSD:*

$$indicated\ current = 100\ \mu A$$

$$error = \pm 1\%\ of\ 100\ \mu A$$

$$= \pm 1\ \mu A$$

$$actual\ measured\ current = 100 \pm 1\ \mu A$$

$$= 99\ to\ 101\ \mu A$$

$$error = \pm 1\%\ of\ measured\ current$$

(b) *At 0.5 FSD:*

$$indicated\ current = 0.5 \times 100\ \mu A$$

$$= 50\ \mu A$$

$$error = \pm 1\%\ of\ FSD$$

$$= \pm 1\ \mu A$$

$$actual\ measured\ current = 50\ \mu A \pm 1\ \mu A$$

$$= 49\ \mu A\ to\ 51\ \mu A$$

$$error = \frac{\pm 1 \ \mu A}{50 \ \mu A} \times 100\%$$

$$= \pm 2\% \ of \ measured \ current$$

3-8 SERIES OHMMETER

Basic Circuit

An *ohmmeter* (ohm-meter) is normally part of a *volt-ohm-milliammeter* (VOM), or *multi-function* meter. Ohmmeters do not usually exist as individual instruments. The simplest ohmmeter circuit consists of a voltage source connected in series with a pair of terminals, a standard resistance, and a low-current PMMC instrument. Such a circuit is shown in Figure 3-21(a). The resistance to be measured (R_x) is connected across terminals A and B.

The meter current indicated by the instrument in Figure 3-21(a) is (battery voltage)/(total series resistance):

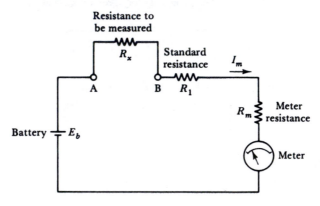

(a) Basic circuit of series ohmmeter

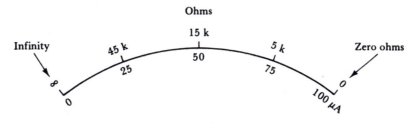

(b) Ohmmeter scale

Figure 3-21 Basic series ohmmeter circuit consisting of a PMMC instrument and a series-connected standard resistor (R_1). When the ohmmeter terminals are shorted ($R_X = 0$) meter full-scale deflection occurs. At half-scale deflection $R_X = R_1$, and at zero deflection the terminals are open-circuited.

$$I_m = \frac{E_b}{R_x + R_1 + R_m} \qquad (3\text{-}3)$$

When the external resistance is zero (i.e., terminals A and B short-circuited), Equation 3-3 becomes

$$I_m = \frac{E_b}{R_1 + R_m}$$

If R_1 and R_m are selected (or if R_1 is adjusted) to give FSD when A and B are short-circuited, FSD is marked as *zero ohms*. Thus, for $R_x = 0$, the pointer indicates 0 Ω [see Figure 3-21(b)]. When terminals A and B are open-circuited, the effective value of resistance R_x is infinity. No meter current flows, and the pointer indicates zero current. This point (zero current) is marked as *infinity* (∞) on the resistance scale [Figure 3-21(b)].

If a resistance R_x with a value between zero and infinity is connected across terminals A and B, the meter current is greater than zero but less than FSD. The pointer position on the scale now depends on the relationship between R_x and $R_1 + R_m$. This is demonstrated by Example 3-14.

Example 3-14

The series ohmmeter in Figure 3-21(a) is made up of a 1.5 V battery, a 100 μA meter, and a resistance R_1 which makes $(R_1 + R_m) = 15$ kΩ.
(a) Determine the instrument indication when $R_x = 0$.
(b) Determine how the resistance scale should be marked at 0.5 FSD, 0.25 FSD, and 0.75 FSD.

Solution

(a) *Equation 3-3,*

$$I_m = \frac{E_b}{R_x + R_1 + R_m} = \frac{1.5\ \text{V}}{0 + 15\ \text{k}\Omega}$$

$$= 100\ \mu\text{A (FSD)}$$

(b) *At 0.5 FSD:*

$$I_m = \frac{100\ \mu\text{A}}{2} = 50\ \mu\text{A}$$

From Equation 3-3,

$$R_x + R_1 + R_m = \frac{E_b}{I}$$

$$R_x = \frac{E_b}{I_m} - (R_1 + R_m)$$

$$= \frac{1.5\ \text{V}}{50\ \mu\text{A}} - 15\ \text{k}\Omega$$

$$= 15\ \text{k}\Omega$$

At 0.25 FSD:

$$I_m = \frac{100 \ \mu A}{4} = 25 \ \mu A$$

$$R_x = \frac{1.5 \ V}{25 \ \mu A} - 15 \ k\Omega$$

$$= 45 \ k\Omega$$

At 0.75 FSD:

$$I_m = 0.75 \times 100 \ \mu A = 75 \ \mu A$$

$$R_x = \frac{1.5 \ V}{75 \ \mu A} - 15 \ k\Omega$$

$$= 5 \ k\Omega$$

The ohmmeter scale is now marked as shown in Figure 3-21(b).

From Example 3-14, note that the measured resistance at center scale is equal to the internal resistance of the ohmmeter (i.e., $R_x = R_1 + R_m$). This makes sense because at FSD the total resistance is $R_1 + R_m$, and when the resistance is doubled, $R_x + R_1 + R_m = 2(R_1 + R_m)$, the circuit current is halved.

Ohmmeter with Zero Adjust

The simple ohmmeter described above will operate satisfactorily as long as the battery voltage remains exactly at 1.5 V. When the battery voltage falls (and the output voltage of all batteries fall with use), the instrument scale is no longer correct. Even if R_1 were adjusted to give FSD when terminals A and B are short-circuited, the scale would still be in error because now midscale would represent a resistance equal to the *new* value of $R_1 + R_m$. Falling battery voltage can be taken care of by an adjustable resistor connected in parallel with the meter (R_2 in Figure 3-22).

In Figure 3-22 the battery current I_b spls up into meter current I_m and resistor current I_2. With terminals A and B short-circuited, R_2 is adjusted to give FSD on the meter. At this time the total circuit resistance is $R_1 + R_2 \| R_m$. Since R_1 is always very much larger than $R_2 \| R_m$, the total circuit resistance can be assumed to equal R_1. When a resistance R_x equal to R_1 is connected across terminals A and B, the circuit resistance is doubled and the circuit current is halved. This causes both I_2 and I_m to be reduced to half of their previous levels (i.e., when A and B were short-circuited). Thus, the midscale measured resistance is again equal to the ohmmeter internal resistance R_1.

The equation for the battery current in Figure 3-22 is

$$I_b = \frac{E_b}{R_x + R_1 + R_2 \| R_m}$$

If $R_2 \| R_m \ll R_1$,

$$\boxed{I_b \simeq \frac{E_b}{R_x + R_1}} \qquad (3\text{-}4)$$

Also, the meter voltage is

$$V_m = I_b(R_2 \| R_m)$$

which gives meter current as

$$I_m = \frac{I_b(R_2 \| R_m)}{R_m} \qquad (3\text{-}5)$$

Each time the ohmmeter is used, terminals A and B are first short-circuited, and R_2 is adjusted for zero-ohm indication on the scale (i.e., for FSD). If this procedure is followed, then even when the battery voltage falls below its initial level, the scale remains correct. Example 3-15 demonstrates that this is so.

Example 3-15

The ohmmeter circuit in Figure 3-22 has $E_b = 1.5$ V, $R_1 = 15$ kΩ, $R_m = 50$ Ω, $R_2 = 50$ Ω, and meter FSD = 50 μA. Determine the ohmmeter scale reading at 0.5 FSD, and determine the new resistance value that R_2 must be adjusted to when E_b falls to 1.3 V. Also, recalculate the value of R_x at 0.5 FSD when $E_b = 1.3$ V.

Solution *At 0.5 FSD, with $E_b = 1.5$ V,*

$$V_m = I_m \times R_m = 25 \ \mu A \times 50 \ \Omega$$

$$= 1.25 \ \text{mV}$$

$$I_2 = \frac{V_m}{R_2} = \frac{1.25 \ \text{mV}}{50 \ \Omega}$$

$$= 25 \ \mu A$$

$$I_b = I_2 + I_m = 25 \ \mu A + 25 \ \mu A$$

$$= 50 \ \mu A$$

$$R_x + R_1 \approx \frac{E_b}{I_b} = \frac{1.5 \ \text{V}}{50 \ \mu A}$$

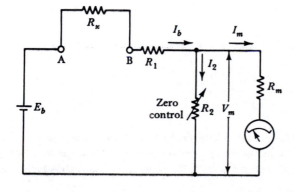

Figure 3-22 An adjustable resistor (R_2) connected in parallel with the meter provides an ohmmeter zero control. The ohmmeter terminals are initially short-circuited and the zero control is adjusted to give a zero-ohms reading. This eliminates errors due to variations in the battery voltage.

$$= 30 \text{ k}\Omega$$

$$R_x = 30 \text{ k}\Omega - R_1 = 30 \text{ k}\Omega - 15 \text{ k}\Omega$$

$$= 15 \text{ k}\Omega$$

With $R_x = 0$ and $E_b = 1.3$ V,

$$I_b \approx \frac{E_b}{R_x + R_1} = \frac{1.3 \text{ V}}{0 + 15 \text{ k}\Omega}$$

$$= 86.67 \ \mu\text{A}$$

$$I_2 = I_b - I_{m(FSD)} = 86.67 \ \mu\text{A} - 50 \ \mu\text{A}$$

$$= 36.67 \ \mu\text{A}$$

$$V_m = I_m R_m = 50 \ \mu\text{A} \times 50 \ \Omega$$

$$= 2.5 \text{ mV}$$

$$R_2 = \frac{V_m}{I_2} = \frac{2.5 \text{ mV}}{36.67 \ \mu\text{A}}$$

$$= 68.18 \ \Omega$$

At 0.5 FSD, with $E_b = 1.3$ V,

$$V_m = I_m \times R_m = 25 \ \mu\text{A} \times 50 \ \Omega$$

$$= 1.25 \text{ mV}$$

$$I_2 = \frac{V_m}{R_2} = \frac{1.25 \text{ mV}}{68.18 \ \Omega}$$

$$= 18.33 \ \mu\text{A}$$

$$I_b = I_2 + I_m = 18.33 \ \mu\text{A} + 25 \ \mu\text{A}$$

$$= 43.33 \ \mu\text{A}$$

$$R_x + R_1 \approx \frac{E_b}{I_b} = \frac{1.3 \text{ V}}{43.33 \ \mu\text{A}}$$

$$= 30 \text{ k}\Omega$$

$$R_x = 30 \text{ k}\Omega - R_1 = 30 \text{ k}\Omega - 15 \text{ k}\Omega$$

$$= 15 \text{ k}\Omega$$

3-9 SHUNT OHMMETER

Circuit and Scale

The series ohmmeter circuit discussed in Section 3-8 could be converted to a multirange ohmmeter by employing several values of standard resistor (R_1 in Figure 3-22) and a rotary switch. The major inconvenience of such a circuit is the fact that a large adjustment of the zero control (R_2 in Figure 3-22) would have to be made every time the resistance range is changed. In the *shunt ohmmeter* circuit, this adjustment is not necessary; once zeroed, the instrument can be switched between ranges with only minor zero adjustments.

Figure 3-23(a) shows the circuit of a typical multirange shunt ohmmeter as found in good-quality multifunction deflection instruments. The deflection meter used gives FSD when passing 37.5 μA, and its resistance (R_m) is 3.82 kΩ. The zero control is a 5 kΩ variable resistance, which is set to 2.875 kΩ when the battery voltages are at the normal levels. Two batteries are included in the circuit; a 1.5 V battery used on all ranges except the $R \times 10$ kΩ range, and a 15 V battery solely for use on the $R \times 10$ kΩ range. R_x, the resistance to be measured, is connected at the terminals of the circuit. The terminals are identified as + and − because the ohmmeter circuit is part of an instrument that also functions as an ammeter and as a voltmeter. It is important to note that the *negative* terminal of each battery is connected to the + terminal of the multifunction instrument.

The range switch in Figure 3-23(a) has a movable contact that may be step-rotated clockwise or counterclockwise. The battery terminals on the rotary switch are seen to be longer than any other terminals, so that they make contact with the largest part of the movable contact, while the other (short) terminals reach only to the tab of the moving contact. In the position shown, the $R \times 1$ k terminal is connected (via the movable contact) to the + terminal of the 1.5 V battery. If the movable contact is step-rotated clockwise, it will connect the 1.5 V battery in turn to $R \times 100$, $R \times 10$, and $R \times 1$ terminals. When rotated one step counterclockwise from the position shown, the movable contact is disconnected from the 1.5 V battery, and makes contact between the $R \times 10$ k terminal and the + terminal of the 15 V battery.

In Figure 3-23(b) the typical scale and controls for this type of ohmmeter are illustrated. When the range switch is set to $R \times 1$, the scale is read directly in ohms. On any other range the scale reading is multiplied by the range factor. On $R \times 100$, for example, the pointer position illustrated would be read as 30 Ω × 100 = 3 kΩ. The instrument must be *zeroed* before use to take care of battery voltage variation. This can be performed on any range, simply by short-circuiting the + and − terminals and adjusting the zero control until the pointer indicates exactly 0 Ω. When changing to or from the $R \times 10$ kΩ range, the ohmmeter zero must always be checked because the circuit supply is being switched between the 15 V and 1.5 V batteries.

The ohmmeter equivalent circuit for the $R \times 1$ range is shown in Figure 3-24. Current and resistance calculations are made in Example 3-16.

Example 3-16

Calculate the meter current and indicated resistance for the ohmmeter circuit of Figure 3-23(a) on its $R \times 1$ range when (a) $R_x = 0$ and (b) $R_x = 24$ Ω.

Solution

The equivalent circuit in Figure 3-24 is derived from Figure 3-23(a), for the $R \times 1$ range.

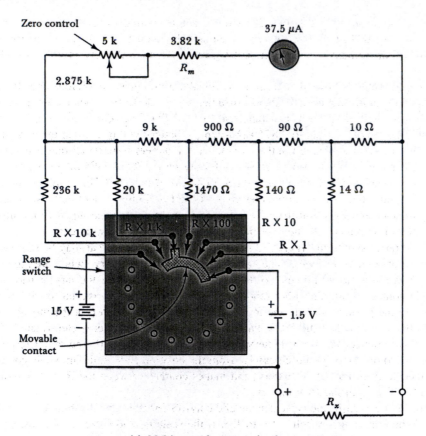

(a) Multirange ohmmeter circuit

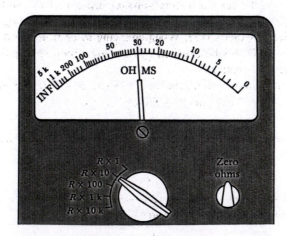

(b) Range switch and scale

Figure 3-23 Circuit, scale, and range switch for a typical multirange shunt ohmmeter, as used on a multifunction analog instrument. The 15 V battery is used only on the $R \times 10$ kΩ range, and the 1.5 V battery is the supply for all other ranges.

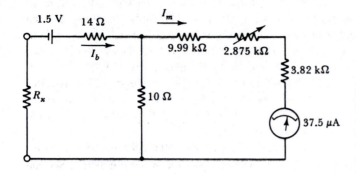

Figure 3-24 Equivalent circuit of the multirange shunt ohmmeter (Figure 3-23) on the $R \times 1$ range. The 9.99 kΩ resistance is the sum of the 9 kΩ, 900 Ω, and 90 Ω resistors. The 2.875 kΩ and the 3.82 kΩ resistors are the zero control and meter resistances, respectively.

(a) When $R_x = 0$:
battery current,

$$I_b = \frac{1.5 \text{ V}}{14 \text{ Ω} + [10 \text{ Ω}\|(9.99 \text{ kΩ} + 2.875 \text{ kΩ} + 3.82 \text{ kΩ})]}$$

$$= \frac{1.5 \text{ V}}{14 \text{ Ω} + [10 \text{ Ω}\|16.875 \text{ kΩ}]}$$

$$= 62.516 \text{ mA}$$

Using the current divider rule:

$$\text{meter current} \quad I_m = 62.516 \text{ mA} \times \frac{10 \text{ Ω}}{10 \text{ Ω} + 16.685 \text{ kΩ}}$$

$$= 37.5 \text{ μA}$$

$$= \text{full scale} = 0 \text{ Ω}$$

(b) When $R_x = 24$ Ω:

$$I_b = \frac{1.5 \text{ V}}{24 \text{ Ω} + 14 \text{ Ω} + (10 \text{ Ω}\|(16.685 \text{ kΩ})}$$

$$= 31.254 \text{ mA}$$

$$I_m = 31.254 \text{ mA} \times \frac{10 \text{ Ω}}{10 \text{ Ω} + 16.685 \text{ kΩ}}$$

$$= 18.72 \text{ μA}$$

$$\approx \text{half scale} = 24 \text{ Ω}$$

Ohmmeter Accuracy

Referring to Figure 3-23(b), it is clear that the ohmmeter scale is nonlinear. On the $R \times 1$ range the pointer indicates 24 Ω at 0.5 FSD. At 0.9 FSD, the indicated resistance is 2.6 Ω, and at 0.1 FSD the resistance measured is 216 Ω. (Although they are not marked on the scale, these resistance values can be calculated for 0.9 and 0.1 FSD.) Therefore, in the range 0.1 to 0.9 FSD, resistance values from 2.6 Ω to 216 Ω can be measured. But the portion of the scale from 0.1 FSD to zero deflection includes all resistance values from 216 Ω to infinity. Also, that part of the scale from 0.9 FSD to FSD covers all resistance values from 2.6 Ω to 0 Ω. Clearly, on this range of the ohmmeter, resistance values from 0 to 2.6 Ω and from 216 Ω to infinity cannot be measured or even roughly estimated. For example, at what points on the scale would 0.01 Ω and 200 kΩ be found? *The useful range of the ohmmeter scale is seen to be approximately from 10% to 90% of FSD.* Now consider the actual accuracy of the resistance measurement.

As already demonstrated, an ohmmeter indicates 0.5 FSD when the measured resistance R_x is equal to the ohmmeter internal resistance. Also, it was explained in Section 3-7 that the current meter accuracy is usually specified as a percentage of full scale. Now consider the errors that may occur in resistance measurement by an ohmmeter that uses an instrument with an accuracy of ±1%.

At 0.5 FSD, the accuracy of pointer deflection is ±1% of FSD, which, when used as a current meter, is ±2% of the indicated current. Also, at 0.5 FSD, (measured resistance R_x) = (ohmmeter internal resistance R_1) and

$$I_b = \frac{E_b}{R_1 + R_x}$$

Since the currrent meter accuracy (at 0.5 FSD) is ±2% of the indicated current, the accuracy of I_b is ±2%. Consequently, the accuracy of the total circuit resistance is ±2% (assuming that the ohmmeter was initially zeroed to suit the battery voltage). If R_1 is made up of precision resistors, virtually none of the ±2% resistance error can be assumed to reside in R_1. All of the resistance error must exist in R_x, the measured resistance.

The total resistance error is ±2% of $(R_1 + R_x)$. Since $R_1 = R_x$ at 0.5 FSD, the total error in R_x is = ±2% of $(2R_x)$ = ±4% of R_x. Thus, an ohmmeter that uses precision internal resistors and a current meter with an accuracy of ±1% of FSD measures resistance at 0.5 FSD with an accuracy of ±4%.

Example 3-17

Analyze the accuracy of the ohmmeter in Figure 3-21(a) when the pointer is at 0.8 FSD if the meter used has a 1% accuracy.

Solution

At 0.8 FSD:

$$R_x + R_1 = \frac{E_b}{0.8 I_{FSD}}$$

and

$$\frac{E_b}{I_{FSD}} = R_1$$

so

$$R_x + R_1 = \frac{R_1}{0.8} = 1.25 R_1$$

Electromechanical Instruments   Chap. 3

$$or \qquad 1.25R_1 - R_1 = R_x$$

$$0.25R_1 = R_x$$

$$R_1 = 4R_x$$

$$total\ error = 1\%\ of\ FSD$$

$$= \frac{1\%}{0.8}\ of\ pointer\ indication$$

$$= 1.25\%\ of\ pointer\ indication$$

$$total\ R_x\ error = 1.25\%\ of\ (R_1 + R_x)$$

$$= 1.25\%\ of\ (4R_x + R_x)$$

$$= 6.25\%\ of\ R_x$$

The analysis above demonstrates that when indicating half-scale deflection, the ohmmeter error is ±4 (current meter error). Also, at 0.8 FSD the ohmmeter error is ±6.25 (meter error). Similarly, at 0.2 FSD the ohmmeter error can be shown to be ±6.25 (meter error). It is seen that *for greatest accuracy the ohmmeter range should always be selected to give an indication as close as possible to 0.5 FSD.*

3-10 VOLT-OHM-MILLIAMMETER

As its name suggests, the *volt-ohm-milliammeter (VOM)* is a multifunction instrument that can be used to measure voltage resistance and current. All VOMs can measure resistance, dc voltage, dc current, and ac voltage. Some can also measure ac current, and some have decibel scales. A typical good-quality analog VOM (the *Simpson 250*) is illustrated in Figure 3-25.

Front Panel Controls

The left-hand and central knobs on the instrument in Figure 3-25 are used for function and range selections. The righ-hand knob is a *ZERO OHMS* control for the ohmmeter function. The mechanical zero control is just below the base of the pointer. The available selections for the left-hand knob are:

> *AC VOLTS:* for ac voltage measurements
> *−DC or +DC:* for dc current and voltage measurements
> *Musical note symbol:* for continuity testing

The central knob has the following function and range selections:

> *2.5 V to 1000 V:* for ac or dc voltage measurements

Figure 3-25 Typical deflection volt-ohm-milliammeter, or multimeter. (Courtesy of bach-simpson limited.)

1 mA to 500 mA: for dc current measurements (note that the instrument manufacturer uses *MA* as the symbol for milliamperes)

(R × 1) to (R × 10,000): for resistance measurements

Terminals

The terminals marked + *(plus)* and − *(minus)* (also identified as *COMMON*) are those normally employed for all voltage, current, and resistance measurements. With the central selector switch at the *250 V 500 V 1000 V* position, the maximum measured voltage is 250 V when using the *plus* and *common* terminals. A maximum of 500 V can be measured if the input is applied to the *common* and *500 V AC DC* terminals. Similarly, for 1000 V maximum, the input should be connected to the *common* and *1000 V AC DC* terminals. The *−10 A* and *+10 A* terminals are used for 1 A to 10 A dc measurements, and the *OUTPUT 350 VDC* terminal is for decibel measurements.

Overload Protection

The deflecting coil in a PMMC instrument is normally wound of fine copper wire with a plastic film or varnish-type insulation. If the coil current is too high, the insulation may be destroyed by overheating, and in extreme circumstances the copper wire might be melted. The instrument in Figure 3-25 is equipped with an overload protection circuit (as well as fuses). When the meter current exceeds a maximum safe level, the overload device open-circuits the instrument internally, and the *RESET* button pops up. The device is reset by pushing the button down.

Scales

The instrument shown in Figure 3-25 has a knife-edge pointer and a mirror scale to avoid parallax error (see Section 3-7). The top scale is used only for resistance measurements. When the *(R × 1)* range is selected, the resistance is read directly in ohms. On the *(R × 100)* and *(R × 10,000)* ranges, the scale reading must be multiplied by the appropriate factor. Thus, at center scale, the measured resistance might be 15 Ω, 1500 Ω, or 150 kΩ, depending on the range selected. The *0 to 250 V* scale is read as *0 to 2.5 V* when the 2.5 V range is selected, and as *0 to 25 V* for the 25 V range. For the 10 V and 50 V ranges, the appropriate scale markings are read. An *AC AMP CLAMP* scale is provided for use with a clamping-type high-current probe. (This is discussed in Section 4-7.) A decibel scale is also included for measuring audio power levels.

Accuracy

The specified accuracy for the Simpson 260 is ±2% of full scale for dc voltage and current, ±3% of full scale for ac voltage, and 2° to 25° ARC for resistance measurements. The most accurate voltage or current measurement is made on the range that gives the greatest on-scale deflection. Greatest accuracy of resistance measurement is made when the pointer is closest to half-scale (see Section 3-9). VOMs can typically be used for ac measurements up to a frequency of about 100 kHz.

Using a VOM as a dc Ammeter

1. Set the function and range switches to *+DC* and *500 MA* (the largest selectable range).
2. If necessary, adjust the mechanical zero control to set the pointer exactly at zero on the scale. Tap the instrument gently to relieve friction when zeroing.
3. Connect the instrument *in series* with the circuit or component in which the current is to be measured, with (conventional) current direction into the + terminal and out of the *COMMON* terminal. (If the current direction is reversed, positive pointer deflection can be obtained by switching from *+DC* to *−DC*.)
4. Adjust the range selection to give the greatest possible on-scale deflection.
5. Tap the instrument gently to relieve friction when reading the pointer position.

When using the instrument as an ammeter, there will be a voltage drop across the instrument, which might have some effect on the current being measured. The ammeter voltage drop for the Simpson 260 ranges from 250 mV to 500 mV.

Using a VOM as a dc Voltmeter

1. Set the function and range switches to *+DC* and *250 V* (the largest selectable range)
2. If necessary, adjust the mechanical zero control to set the pointer exactly at zero on the scale. Tap the instrument gently to relieve friction when zeroing.
3. Connect the instrument *in parallel* with the circuit or component that is to have its voltage measured. The + terminal should be connected to the most positive of the two points at which the voltage is to be measured. The *COMMON* terminal should be connected to the most negative of the two points. (If the voltage polarity is reversed, positive pointer deflection can be obtained by switching from *+DC* to *−DC*.)
4. Adjust the range selection to give the greatest possible on-scale deflection.
5. Tap the instrument gently to relieve friction when reading the pointer position.

When using the VOM as a voltmeter, the instrument resistance might affect the measured voltage. The voltmeter resistance can be calculated as (sensitivity × range). For the Simpson 260, the specified sensitivity is marked on the lower left-hand side of the scale as 20,000 Ω/V.

Using a VOM as an ac Voltmeter

1. Set the function and range switches to *AC VOLTS ONLY* and *250 V* (the largest selectable range).
2. Continue as for a dc voltmeter, with the exception that terminal polarity need not be observed.

For ac voltmeter applications, the sensitivity of the Simpson 260 is 5 kΩ/V.

VOM Probes

The range of a VOM can be extended by the use of high-voltage, high-current, and high-frequency probes. These are exactly as discussed in Section 4-7.

3-11 ELECTRODYNAMIC INSTRUMENT

Construction and Operation

The basic construction of an *electrodynamic* or *dynamometer* instrument is illustrated in Figure 3-26(a). When this is compared to the PMMC instrument in Figure 3-4, it is seen that the major difference is that two magnetic *field coils* are substituted in place of the permanent magnet. The magnetic field in which the moving coil is pivoted is generated by passing a current through the stationary field coils. When a current flows through the pivoted coil, the two fluxes interact (as in the PMMC instrument), causing the coil and pointer to be deflected. Spiral springs provide controlling force and connecting leads to the pivoted coil. Zero adjustment and moving system balance are also as in the PMMC instrument.

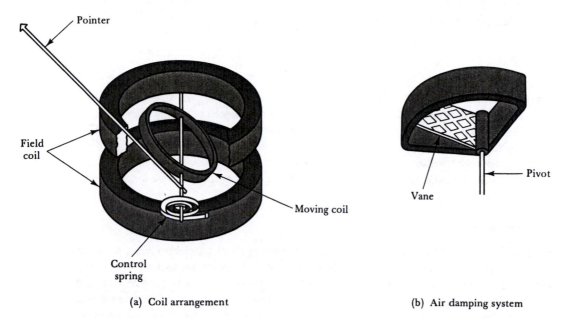

| (a) Coil arrangement | (b) Air damping system |

Figure 3-26 An electrodynamic instrument has a moving coil, as in a PMMC instrument, but the magnetic field is produced by two current-carrying field coils instead of a magnet. Damping is provided by an enclosed vane.

Another major difference from the PMMC instrument is that the electrodynamic instrument usualy has air damping. A lightweight vane pushes air around in an enclosure when the pivoted coil is in motion [see Figure 3-26(b)]. This damps out all rapid movements and oscillations of the moving system. As will be explained, electrodynamic instruments can be used on ac. The alternating current would induce unwanted eddy currents in a metallic coil former. Therefore, the damping method employed in a PMMC instrument would not be suitable for an electrodynamic instrument.

Normally, there is no iron core in an electrodynamic instrument, so the flux path is entirely an air path. Consequently, the field flux is much smaller than in a PMMC instrument. To produce a strong enough deflecting torque, the moving-coil current must be much larger than the small currents required in a PMMC instrument.

As in the case of the PMMC instrument, the deflecting torque of an electrodynamic instrument is dependent on field flux, coil current, coil dimensions, and number of coil turns. However, the field flux is directly proportional to the current through the field coils, and the moving-coil flux is directly proportional to the current through the moving coil. Consequently, the deflecting torque is proportional to the product of the two currents:

$$T_D \propto I_{\text{field coil}} I_{\text{moving coil}}$$

When the same current flows through field coils and pivoted coil, the deflecting torque is proportional to the square of the current:

$$T_D \propto I^2$$

This gives the deflection angle as

$$\boxed{\theta = CI^2} \tag{3-6}$$

where C is a constant. Because the deflection is proportional to I^2, the scale of the instrument is nonlinear: cramped at the low (left-hand) end and spaced out at the high end.

The major disadvantages of an electrodynamic instrument compared to a PMMC instrument are the lower sensitivity and the nonlinear scale. A major advantage of the electrodynamic instrument is that it is not polarized; that is, a positive deflection is obtained regardless of the direction of current in the coils. Thus the instrument can be used to measure ac or dc.

AC Operation

Consider Figure 3-27 in which the fixed and moving coils of an electrodynamic instrument are shown connected in series. In Figure 3-27(a) the current direction is such that the flux of the field coils sets up S poles at the top, and N poles at the bottom of each coil. The moving-coil flux produces an N pole at the right-hand side of the coil, and an S pole at the left-hand side. The N pole of the moving coil is adjacent to the N pole of the upper field coil, and the S pole of the moving coil is adjacent to the S pole of the lower field coil. Since like poles repel, the moving coil rotates in a clockwise direction, causing the poiner to move to the right from its zero position on the scale.

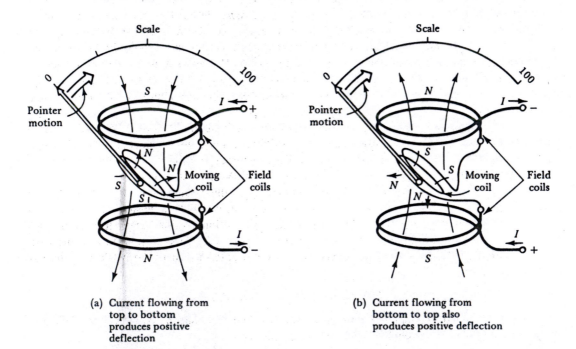

(a) Current flowing from top to bottom produces positive deflection

(b) Current flowing from bottom to top also produces positive deflection

Figure 3-27 Positive deflection of pointer occurs in an electrodynamic instrument regardless of current direction. The instrument can be used directly for both ac and dc measurements.

Electromechanical Instruments   Chap. 3

Now consider what occurs when the current through all three coils is reversed. Figure 3-27(b) shows that the reversed current causes the field coils to set up N poles at the top and S poles at the bottom of each coil. The moving-coil flux is also reversed so that it has an S pole at the right-hand side and an N pole at the left. Once again similar poles are adjacent, and repulsion produces clockwise rotation of the coil and pointer.

It is seen that the electrodynamic instrument has a positive deflection, regardless of the direction of current through the meter. Consequently, the terminals are *not* marked + and − (i.e., the instrument is *not* polarized).

As already explained, the electrodynamic instrument deflection is proportional to I^2 (i.e., when the same current flows in the moving coil and field coils). When used on ac, the deflection settles down to a position proportional to the average value of I^2. Thus, the deflection is proportional to the *mean-squared value* of the current. Since the scale of the meter is calibrated to indicate I, rather than I^2, the meter indicates *root-mean-squared current,* or the rms value. The rms value has the same effect as a numerically equivalent dc value. Therefore, the scale of the instrument can be read as either dc or rms ac. This is the characteristic of a *transfer instrument,* which can be calibrated on dc and then used to measure ac. (Voltmeters are available that operate on an electrostatic principle. These are also ac/dc transfer instruments.)

Because the reactance of the coils increase rapidly with increasing frequency, electrodynamic instruments are useful only at low frequencies. Electrodynamic wattmeters, in particular, perform very satisfactorily at domestic and industrial power frequencies.

Electrodynamic Voltmeter and Ammeter

Figure 3-28(a) shows the usual circuit arrangement for an electrodynamic voltmeter. Since a voltmeter must have a high resistance, all three coils are connected in series, and a multiplier resistor (made of manganin or constantan) is included. When the total resistance of the coils, and the required current for FSD are known, the multiplier resistance is calculated exactly as for dc voltmeters. The instrument scale can be read either as dc voltage or rms ac voltage.

Because electrodynamic instruments usually require at least 100 mA for FSD, an electrodynamic voltmeter has a much lower sensitivity than a PMMC voltmeter. At 100 mA FSD, the sensitivity is 1/100 mA = 10 Ω/V. For a 100 V instrument, this sensitivity gives a total resistance of only 1 kΩ. Therefore, an electrodynamic voltmeter is not suitable for measuring voltages in electronic circuits because of the loading effect.

In an electrodynamic ammeter, the moving coil and its series-connected swamping resistance are connected in parallel with the ammeter shunt. This is illustrated in Figure 3-28(b). The two field coils should be connected in series with the parallel arrangement of shunt and moving coil, as shown.

Because the field coils are always passing the actual current to be measured, resistance changes in the coils with temperature variations have no effect on the instrument performance. However, as in PMMC ammeters, the moving coil must have a manganin or constantan swamping resistance connected in series. Also, the shunt resistor must be made of manganin.

The scale of the electrodynamic ammeter can be read either as dc levels or rms ac values. Like the electrodynamic voltmeter, this instrument can be calibrated on dc and then used to measure either dc or ac.

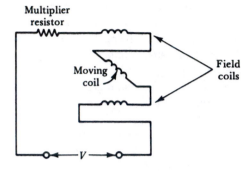

(a) Electrodynamic voltmeter

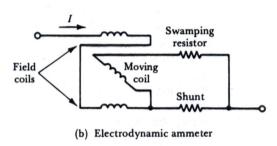

(b) Electrodynamic ammeter

Figure 3-28 For use as a voltmeter, an electrodynamic instrument has the field coils, moving coil, and multiplier resistor all connected in series. For use as an ammeter, the field coils are seriesed with the parallel-connected shunt and moving-coil circuit.

Electrodynamic Wattmeter

For both dc and ac applications, the most important use of the electrodynamic instrument is as a wattmeter. The coil connections for power measurement are illustrated in Figure 3-29(a). The field coils are connected in series with the load in which power is to be measured. The moving coil and a multiplier resistor are connected in parallel with the load. Thus, the field coils carry the load current, and the moving-coil current is proportional to the load voltage. Since the instrument deflection is proportional to the product of the two currents, deflection = $C \times EI$, where C is a constant, or meter indication = EI watts. In Figure 3-29(b) the electrodynamic wattmeter is shown in a slightly less complicated form than in Figure 3-29(a). A single-coil symbol is used to represent the two series-connected field coils.

Suppose that the instrument is correctly connected and giving a positive deflection. If the supply voltage polarity were reversed, the fluxes would reverse in both the field coils and the moving coil. As already explained, the instrument would still have a positive deflection. In ac circuits where the supply polarity is reversing continuously, the electrodynamic wattmeter gives a positive indication proportional to $E_{rms}I_{rms}$. Like electrodynamic ammeters and voltmeters, the wattmeter can be calibrated on dc and then used to measure power in either dc or ac circuits.

In ac circuits the load current may lead or lag the load voltage by a phase angle ϕ. The wattmeter deflection is proportional to the in-phase components of the current and voltage. As shown in Figure 3-29(c), the instrument deflection is proportional to EI cos

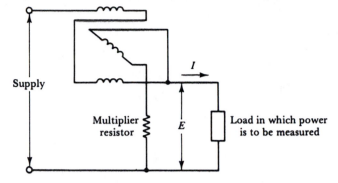

(a) Electrodynamic wattmeter circuit

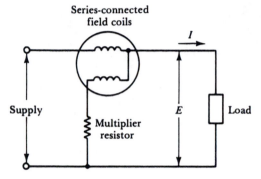

(b) Another way to show the wattmeter circuit

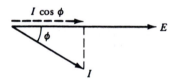

(c) Wattmeter measures $EI \cos \phi$

Figure 3-29 An electrodynamic wattmeter has the moving coil and multiplier resistor connected in parallel with the load, and the field coils in series with the load. Instrument deflection is proportional to $EI \cos \phi$.

ϕ. Since the true power dissipated in a load with an ac supply is $EI \cos \phi$, the electrodynamic wattmeter measures true power.

An important source of error in the wattmeter is illustrated in Figure 3-30(a) and (b). Figure 3-30(a) shows that if the moving coil (or voltage coil) circuit is connected in parallel with the load, the field coils pass a current $(I + I_v)$, the sum of the load current and the moving-coil current. This results in the wattmeter indicating the load power (EI), plus a small additional quantity (EI_v). Where the load current is very much larger than I_v, this error may be negligible. In low-load-current situations, the error may be quite significant.

Sec. 3-11   Electrodynamic Instrument

77

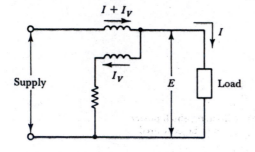

Deflection $\alpha\ E(I + I_V)$

$\alpha\ EI + \underbrace{EI_V}_{\text{Error}}$

(a) Error due to moving-coil current

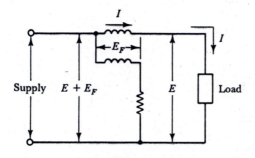

Deflection $\alpha\ (E + E_F)I$

$\alpha\ EI + \underbrace{E_FI}_{\text{Error}}$

(b) Error due to field coils voltage drop

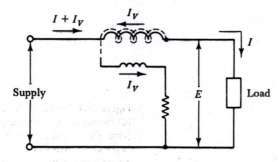

Deflection $\alpha\ E(I + I_V - I_V)$

$\alpha\ EI$

(c) Compensated wattmeter using an
additional coil wound alongside
the field coils

Figure 3-30 Wattmeter coil connections can result in significant reading errors. The error in (a) is not important if $I_V \ll I$, and that in (b) is not important when $E_F \ll E$. The I_V error is eliminated by the compensated wattmeter connection in (c).

In Figure 3-30(b) the voltage coil is connected to the supply side of the field coils so that only the load current flows through the field coils. However, the voltage applied to the series-connected moving coil and multiplier is $E + E_F$ (the load voltage plus the voltage drop across the field coils). Now the wattmeter indicates load power (EI) plus an additional quantity (E_FI). In high-voltage circuits, where the load voltage is very much larg-

Electromechanical Instruments   Chap. 3

er than the voltage drop across the field coils, the error may be insignificant. In low-voltage conditions, this error may be serious.

The *compensated wattmeter* illustrated in Figure 3-30(c) eliminates the errors described above. Since the field coils carry the load current, they must be wound of thick copper wire. In the compensated wattmeter, an additional thin conductor is wound right alongside every turn on the field coils. This additional coil, shown dashed in Figure 3-30(c), becomes part of the voltage coil circuit. The voltage coil circuit is seen to be connected directly across the load, so that the moving-coil current is always proportional to load voltage. The current through the field coils in $I + I_v$, so that a field coil flux is set up proportional to $I + I_v$. But the additional winding on the field coils carries the moving-coil current I_v, and this sets up a flux in opposition to the main flux of the field coils. The resulting flux in the field coils is proportional to $[(I + I_v) - I_v] \propto I$. Thus, the additional winding cancels the field flux due to I_v, and the wattmeter deflection is now directly proportional to EI.

The range of voltages that may be applied to the moving-coil circuit of a wattmeter can be changed by switching different values of multiplier resistors into or out of the circuit, exactly as in the case of a voltmeter. Current ranges can most easily be changed by switching the two field coils from series connection to parallel connection. Figure 3-31 illustrates the circuitry, controls, and scale for a typical multirange wattmeter.

In the circuit shown in Figure 3-31(a), the series-connected multiplier resistors give three possible voltage range selections: 60 V, 120 V, and 240 V. The current range switch connects the field coils in series when set to the right, and in parallel when switched left. The wattmeter scale and controls illustrated in Figure 3-31(b) relate to the circuitry in Figure 3-31(a). With the range switches set at 0.5 A and 240 V, the instrument scale reads directly in watts, and FSD indicates 120 W. Similarly, with the 1 A and 120 V ranges selected, the scale may again be read directly in watts. When the range selections are 120 V and 0.5 A,

$$FSD = 120 \text{ V} \times 0.5 \text{ A} = 60 \text{ W}$$

Also, for a range selection of 1 A and 60 V,

$$FSD = 60 \text{ V} \times 1 \text{ A} = 60 \text{ W}$$

and for the switch at 0.5 A and 60 V, maximum deflection indicates $0.5 \text{ A} \times 60 \text{ V} = 30 \text{ W}$.

It is seen that to read the wattmeter correctly, the selected voltage and current ranges must be multiplied together to find the FSD power. In using a wattmeter it is possible to obtain a reasonable on-scale deflection, while actually overloading either the current or voltage coils. For example, suppose that the wattmeter voltage range is set to 60 V and the current range to 1 A. The instrument will have FSD = 60 V × 1 A = 60 W. Now suppose that the actual load current is 0.5 A, and the actual supply voltage is 120 V. The indicated power is

$$P = 120 \text{ V} \times 0.5 \text{ A} = 60 \text{ W}$$

Thus, the instrument would indicate 60 W at full scale, and there is no obvious problem. However, because the voltage circuit has 120 V applied to it, while set at a 60 V range, the moving coil is actually passing twice as much current as it is designed to take. This could cause overheating, which may destroy the insulation on the moving coil.

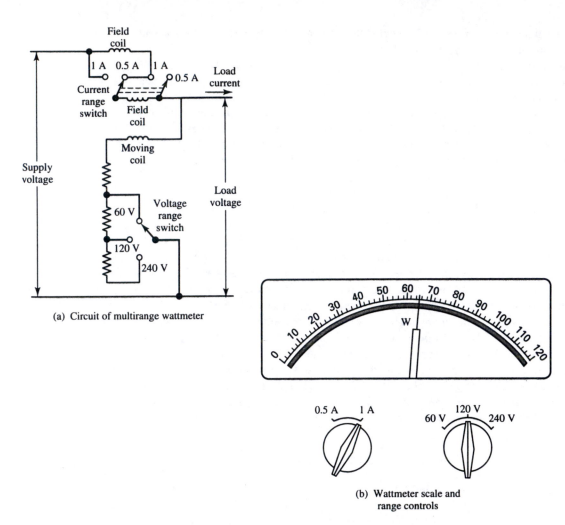

(a) Circuit of multirange wattmeter

(b) Wattmeter scale and range controls

Figure 3-31 In a multirange wattmeter, the field coils may be switched between series connection and parallel connection, and the moving-coil multiplier resistor is selectable. The scale illustrated reads directly in watts only for the 1 A × 120 V and 0.5 A × 240 V ranges.

With the electrodynamic wattmeter, the moving coil and field coils are supplied independently. Usually, a load in which power is to be measured has a constant level of supply voltage. When the load current changes, the supply voltage does not change. In this situation, the moving coil carries a constant current proportional to the supply voltage. The instrument deflection is now directly proportional to the load current, and the scale can be calibrated linearly.

Using Wattmeters

Before connecting a wattmeter into a circuit, check the mechanical zero of the instrument and adjust it if necessary. While zeroing, tap the instrument gently to relieve bearing friction. The current circuit of a wattmeter must be connected in series with the load in which power is to be measured. The voltage circuit must be connected in parallel with the load. If the pointer deflects to the left of zero, either the current terminals or voltage terminals must be reversed.

Before connecting a multirange wattmeter into a circuit, select a voltage range equal to or higher than the supply voltage. Select the highest current range. Then, switch down to the current range that gives the greatest on-scale deflection. Do not adjust the voltage range below the level of the supply voltage. This step ensures that the (low-current) voltage coil does not have an excessive current flow. However, it is still possible that excessive current may be passing through the current coils, although the meter is indicating less than full scale. This should also be avoided, but it is less damaging than excessive voltage coil current.

Electrodynamic wattmeters are useful for measurement on supply frequencies up to a maximum of 500 Hz. Thus, they are not suitable for high-frequency power measurements.

REVIEW QUESTIONS

3-1 List the three forces involved in the moving system of a deflection instrument. Explain the function of each force and how it is typically produced. Illustrate the explanations with suitable sketches.

3-2 Describe jeweled-bearing suspension and taut-band suspension as used in deflection instruments. Discuss the merits of each.

3-3 Sketch the basic construction of a typical PMMC instrument. Identify each part of the instrument and explain its operation.

3-4 Sketch the construction of a core-magnet type of PMMC instrument. Explain.

3-5 Develop the torque equation for a PMMC instrument and show that its scale is linear.

3-6 Sketch the basic construction of a light-beam galvanometer. Explain its operation.

3-7 For a galvanometer define current sensitivity, critical damping resistance, voltage sensitivity, and megohm sensitivity.

3-8 Discuss galvanometer applications. Show how a variable shunt should be used for galvanometer protection. Explain.

3-9 Sketch a circuit diagram to show how a PMMC instrument can be used as a dc ammeter. Explain the circuit operation.

3-10 Explain the following terms: four-terminal resistor, ammeter swamping resistance, make-before-break switch, Ayrton shunt.

3-11 Sketch the circuit diagram for a multirange ammeter using (a) several individual shunts and (b) an Ayrton shunt. Explain.

3-12 Sketch a circuit diagram to show how a PMMC instrument can be used as a dc voltmeter. Explain the circuit operation.

3-13 Sketch the circuit diagram for a multirange voltmeter using (a) individual multiplier resistors and (b) series-connected multiplier resistors. Explain the circuit operation in each case.

3-14 Explain voltmeter sensitivity, voltmeter loading effect, and voltmeter swamping resistance.

3-15 Discuss the response of a PMMC instrument to alternating current and to rectified ac.

3-16 Sketch the circuit and waveforms for an ac voltmeter using a PMMC instrument and a bridge rectifier. Explain the circuit operation.

3-17 Sketch the circuit and waveforms for an ac voltmetmer using a PMMC instrument and a half-wave rectifier. Explain the circuit operation.

3-18 Sketch the circuit and waveforms for an ac voltmeter using a PMMC instrument and a half-bridge full-wave rectifier. Explain the circuit operation.

3-19 Sketch the circuit of a rectifier ammeter and explain its operation.

3-20 Explain parallax error as it applies to a deflection instrument, and show how it can be minimized.

3-21 Discuss the measurement accuracy of a PMMC instrument.

3-22 Sketch the circuit of a series ohmmeter with a zero control. Explain the circuit operation.

3-23 Sketch a typical ohmmeter scale. Explain.

3-24 Sketch the circuit of a multirange shunt ohmmeter. Also, sketch typic range controls and scale. Explain the circuit operation, and discuss the scale readings on each range.

3-25 Discuss ohmmeter accuracy, and explain which part of the scale gives the most accurate resistance measurement.

3-26 List the procedure for using the VOM in Figure 3-25 as a dc ammeter.

3-27 List the procedure for using the VOM in Figure 3-25 as **(a)** a dc voltmeter and **(b)** an ac voltmeter.

3-28 Sketch the construction of an electrodynamic instrument. Identify each part of the instrument, and explain its operation.

3-29 Write the torque equation for an electrodynamic instrument. List the advantages and disadvantages of an electrodynamic instrument compared to a PMMC instrument.

3-30 Sketch the arrangement of coil connections in an electrodynamic instrument, and explain the instrument ac operation.

3-31 Sketch the circuit of an electrodynamic instrument employed as **(a)** a voltmeter and **(b)** an ammeter. Explain each circuit.

3-32 Sketch the circuit of an electrodynamic instrument employed as a wattmeter. Explain why the instrument measures dc power and true ac power.

3-33 Sketch the circuit of a compensated wattmeter, and explain how it eliminates measurement errors.

3-34 Sketch the circuit of a multirange electrodynamic wattmeter. Explain its operation and discuss the precautions that should be observed when using the instrument.

PROBLEMS

3-1 A PMMC instrument with a 300-turn coil has a 0.15 T magnetic flux density in its air gaps. The coil dimensions are $D = 1.25$ cm and $l = 2$ cm. Calculate the torque when the coil current is 500 μA.

3-2 A PMMC instrument has a 0.12 T magnetic flux density in its air gaps. The coil dimensions are $D = 1.5$ cm and $l = 2.25$ cm. Determine the number of coil turns required to give a torque of 4.5 μN · m when the coil is current is 100 μA.

3-3 A galvanometer has a current sensitvity of 500 nA/mm and a 3 kΩ critical damping resistance. Calculate its voltage sensitivity and megohm sensitivity.

3-4 A galvanometer has a 300 μV/mm voltage sensitivity and a megohm sensitivity of 1.5 MΩ. Determine its critical damping resistance.

3-5 Determine the current sensitivity and megohm sensitivity for a galvanometer that deflects by 5 cm when the coil current is 20 μA.

3-6 A PMMC instrument with a 750 Ω coil resistance gives FSD with a 500 μA coil current. Determine the required shunt resistance to convert the instrument into a dc ammeter with an FSD of **(a)** 50 mA and **(b)** 30 mA.

3-7 A dc ammeter is constructed of a 133.3 Ω resistance in parallel with a PMMC instrument. If the instrument has a 1.2 kΩ coil resistance and 30 μA FSD, determine the measured current at FSD, 0.5 FSD, and 0.33 FSD.

3-8 A dc ammeter consists of an Ayrton shunt in parallel with a PMMC instrument that has a 1.2 kΩ coil resistance and 100 μA FSD. The Ayrton shunt is made up of four 0.1 Ω series-connected resistors. Calculate the ammeter range at each setting of the shunt.

3-9 A 12 V source supplies 25 A to a load. Calculate the load current that would be measured when using an ammeter with a resistance of **(a)** 0.12 Ω, **(b)** 0.52 Ω, and **(c)** 0.002 Ω.

3-10 An ammeter measures the current in a 10 Ω load supplied from a 10 V source. Calculate the measured load current when the ammeter resistance is **(a)** 0.1 Ω and **(b)** 1 Ω.

3-11 A PMMC instrument with a 900 Ω coil resistance and an FSD of 75 μA is to be used as a dc voltmeter. Calculate the individual multiplier resistance to give an FSD of **(a)** 100 V, **(b)** 30 V, and **(c)** 5 V. Also, determine the voltmeter sensitivity.

3-12 Calculate the multiplier resistance values required for the voltmeter in Problem 3-11 when series-connected multipliers are used.

3-13 A PMMC instrument with $R_m = 1.3$ kΩ and FSD = 500 μA is used in a multirange dc voltmeter. The series-connected multiplier resistors are $R_1 = 38.7$ kΩ, $R_2 = 40$ kΩ, and $R_3 = 40$ kΩ. Calculate the three voltage ranges and determine the voltmeter sensitivity.

3-14 Two resistors, $R_1 = 47$ kΩ and $R_2 = 82$ kΩ, are connected in series across a 15 V supply. A voltmeter on a 10 V range is connected to measure the voltage across R_2. The voltmeter sensitivity is 10 kΩ/V. Calculate V_{R2} **(a)** with the voltmeter connected and **(b)** with the voltmeter disconnected.

3-15 A 100 kΩ potentiometer and a 33 kΩ resistor are connected in series across a 9 V supply. Calculate the maximum voltage that can be measured across the potentiometer using a voltmeter with **(a)** a 20 kΩ/V sensitivity and a 15 V range and **(b)** a 100 kΩ/V sensitivity and a 10 V range.

3-16 Two resistors, $R_1 = 70$ kΩ and $R_2 = 50$ kΩ, are connected in series across a 12 V supply. A voltmeter on a 5 V range is connected to measure the voltage across R_2. Calculate V_{R2} **(a)** with the voltmeter disconnected, **(b)** with a voltmeter having a sensitivity of 20 kΩ/V, and **(c)** with a voltmeter that has a sensitivity of 200 kΩ/V.

3-17 An ac voltmeter uses a bridge rectifier with silicon diodes and a PMMC instrument with FSD = 75 μA. If the meter coil resistance is 900 Ω and the multiplier resistor is 708 kΩ, calculate the applied rms voltage when the voltmeter indicates FSD.

3-18 Determine the new multiplier resistance required for the voltmeter in Problem 3-17 to change its range to 300 V FSD.

3-19 Determine the pointer position on the voltmeter in Problem 3-18 when the applied rms voltage is **(a)** 30 V and **(b)** 10 V.

3-20 A PPMC instrument with a 900 Ω coil resistance and an FSD of 75 μA is to be used with a half-wave rectifier circuit as an ac voltmeter. Silicon diodes are used, and the minimum diode forward current is to be 80 μA when the instrument indicates 0.25 FSD. Calculate the shunt and multiplier resistance values required to give 200 V FSD.

3-21 Calculate the sensitivity of the ac voltmeter in Problem 3-17.

3-22 Calculate the sensitivity of the ac voltmeter in Problem 3-20.

3-23 A rectifier ammeter is to indicate full scale for a 1 A rms current. The PMMC instrument used has a 1200 Ω coil resistance and 500 μA FSD, and the current transformer has $N_s = 7000$ and $N_p = 10$. Silicon diodes are used and the meter series resistance is $R_s = 150$ kΩ. Determine the required secondary shunt resistance value.

3-24 A rectifier ammeter has the following components: PMMC instrument with FSD = 200 μA and $R_m = 900$ Ω; current transformer with $N_s = 600$ and $N_p = 5$; diodes with $V_F \approx 0.3$ V; meter series resistance $R_s = 270$ kΩ; transformer shunt secondary resistance $R_L = 98.7$ kΩ. Calculate the level of transformer primary current for instrument FSD.

3-25 Calculate the sensitivity of the ac voltmeter in Example 3-11 when diode D_2 is **(a)** included in the circuit and **(b)** omitted from the circuit.

3-26 A PMMC instrument with 250 μA FSD has a specified accuracy of ±2%. Calculate the measurement accuracy at currents of 200 μA and 100 μA.

3-27 A deflection instrument with 100 μA FSD has a ±3% specified accuracy. Calculate the possible error when the meter indication is **(a)** 50 μA and **(b)** 10 μA.

3-28 A 25 μA current is measured on an instrument with 37.5 μA FSD. If the measurement is to be accurate to within ±5%, determine the required instrument accuracy.

3-29 A series ohmmeter is made up of the following components: supply voltage $E_B = 3$ V, series resistor $R_1 = 30$ kΩ, meter shunt resistor $R_2 = 50$ Ω, meter FSD = 50 μA, and meter resistance $R_m = 50$ Ω. Determine the resistance measured at 0, 0.25, 0.5, and 0.75 of full-scale deflection.

3-30 A series ohmmeter that has a standard internal resistance of $R_1 = 50$ kΩ uses a meter with FSD = 75 μA and $R_m = 100$ Ω. The meter shunt resistance is $R_2 = 300$ Ω, and the battery voltage is $E_B = 5$ V. Determine the resistance measured at 0, 25%, 50%, 75%, and 100% of full-scale deflection.

3-31 For the ohmmeter circuit in Problem 3-29, determine the new resistance to which R_2 must be adjusted when E_B falls to 2.5 V. Also, determine the new resistances measured at 0.5 and 0.75 of full-scale deflection.

3-32 Calculate the accuracy of resistance measurement for the ohmmeter in Problem 3-29 at 0.5 and 0.75 of FSD if the meter used has a specified accuracy of ±2%.

3-33 Using a 4.5 V battery together with a meter that has 100 μA FSD and a coil resistance of 100 Ω, design a series ohmmeter to have a range of 1 kΩ to 100 kΩ.

3-34 Calculate the meter current for the ohmmeter circuit in Figure 3-23(a) on its $R \times 10$ range when R_x is **(a)** 0 Ω, **(b)** 500 Ω, and **(c)** 70 Ω.

3-35 Determine the resistance measured at 0.75 FSD with the ohmmeter in Example 3-15 when E_B is **(a)** 1.5 V and **(b)** 1.3 V.

3-36 Calculate the meter current for the ohmmeter circuit in Figure 3-23(a) on the $(R \times 100)$ and $(R \times 10$ k) ranges when $R_x = 0$.

3-37 Determine the accuracy of an ohmmeter at 20% of FSD if the meter used is accurate to ±1%.

3-38 Calculate the resistance of the VOM in Figure 3-25 on its 50 V dc range and on its 250 V ac range.

Analog Electronic
Volt-Ohm-Milliammeters

4

Objectives

You will be able to:

1. Sketch various transistor analog voltmeter circuits, and explain the operation of each circuit. Calculate circuit currents, voltages, and input resistance.

2. Sketch an input attenuator circuit as used with an electronic voltmeter. Explain its operation, and define the circuit input resistance.

3. Using illustrations, explain the problems that can occur with electronic voltmeter ground terminals when measuring voltages in a circuit.

4. Draw the circuit diagrams of various op-amp analog voltmeters. Explain the operation of each circuit.

5. Draw series, shunt, and linear ohmmeter circuits as used in electronic instruments, and explain the operation of each circuit.

6. Sketch the circuit diagrams of various ac electronic voltmeters, and explain their operation.

7. Sketch a circuit to show how current is measured by an electronic voltmeter. Explain the circuit operation.

8. Draw the front panel of a typical analog electronic voltmeter showing the various controls and meter scales. State typical performance specifications for the instrument, and discuss its applications.

Introduction

Voltmeters constructed of moving-coil instruments and multiplier resistors (see Chapter 3) have some important limitations. They cannot measure very low voltages, and their resistance is too low for measurements in high-impedance circuitry. These restrictions are overcome by the use of electronic circuits that offer high input resistance, and which amplify low voltages to measurable levels. When such circuits are used, the instrument becomes an *electronic voltmeter.*

Electronic voltmeters can be *analog* instruments, in which the measurement is indicated by a pointer moving over a calibrated scale, or *digital* instruments, which display the measurement in numerical form (see Chapter 5). As well as amplification, transistor and operational amplifier circuits offer advantages in the measurement of resistance, direct current, and alternating current.

4-1 TRANSISTOR VOLTMETER CIRCUITS

Emitter-Follower Voltmeters

Voltmeter loading (see Section 3-4) can be greatly reduced by using an *emitter follower.* An emitter follower offers a high input resistance to voltages being measured, and provides a low output resistance to drive current through the coil of a deflection meter. The basic emitter-follower voltmeter circuit illustrated in Figure 4-1 shows a PMMC instrument and a multiplier resistance (R_s) connected in series with the transistor emitter. The dc supply is connected—positive to the transistor collector and the negative to the deflection meter. The positive terminal of voltage E (to be measured) is supplied to the transistor base, and its negative is connected to the same terminal as the power supply negative.

The transistor base current in Figure 4-1 is substantially lower than the meter current.

$$I_B \approx \frac{I_m}{h_{FE}}$$

where h_{FE} is the transistor current gain. Thus, the circuit input resistance is

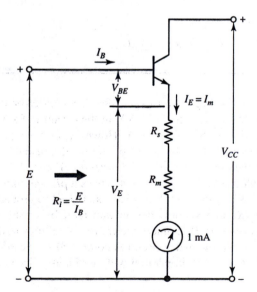

Figure 4-1 An emitter follower offers a high input resistance to a measured voltage, and a low output resistance to a deflection voltmeter circuit. V_{BE} introduces an error in the measurement.

$$R_i \approx \frac{E}{I_B}$$

which is much larger than the meter circuit resistance ($R_s + R_m$).

Example 4-1

The simple emitter-follower voltmeter circuit in Figure 4-1 has $V_{CC} = 20$ V, $R_s + R_m = 9.3$ kΩ, $I_m = 1$ mA at full scale, and transistor $h_{FE} = 100$.
(a) Calculate the meter current when $E = 10$ V,
(b) Determine the voltmeter input resistance with and without the transistor.

Solution

(a)
$$V_E = E - V_{B1} = 10 \text{ V} - 0.7 \text{ V}$$
$$= 9.3 \text{ V}$$

$$I_m = \frac{V_E}{R_s + R_m} \approx \frac{9.3 \text{ V}}{9.3 \text{ k}\Omega}$$
$$= 1 \text{ mA}$$

(b) *With the transistor,*
$$I_B \approx \frac{I_m}{h_{FE}} \approx \frac{1 \text{ mA}}{100}$$
$$= 10 \text{ }\mu\text{A}$$

$$R_i \approx \frac{E}{I_B} \approx \frac{10 \text{ V}}{10 \text{ }\mu\text{A}}$$
$$= 1 \text{ M}\Omega$$

Without the transistor,

$$R_i = R_s + R_m = 9.3 \text{ k}\Omega$$

The transistor base–emitter voltage drop (V_{BE}) introduces an error in the simple emitter-follower voltmeter. For example, when E is 5 V in the circuit in Example 4-1, the meter should read half of full-scale, that is, 0.5 mA. However, as a simple calculation shows, the meter current is actually 0.46 mA. The error can be eliminated by using a potential divider and an additional emitter follower, as illustrated in Figure 4-2.

The practical emitter-follower circuit in Figure 4-2 uses a *plus-and-minus,* or *dual-polarity* supply (typically, ±12 V). Transistor Q_1 has its base biased to ground via resistor R_1, and a potential divider (R_4, R_5, and R_6) provides an adjustable bias voltage (V_p) to the base of transistor Q_2. Resistors R_2 and R_3 connect the transistor emitter terminals to the negative supply voltage ($-V_{EE}$), and the meter circuit is connected between the transistor emitters. The circuit input resistance is R_1 in parallel with the input resistance at the transistor base.

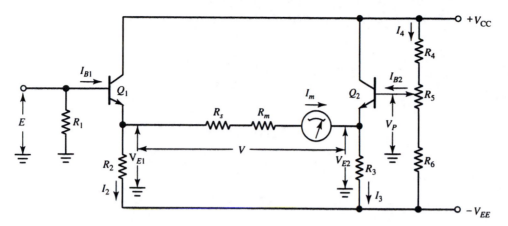

Figure 4-2 Practical emitter-follower voltmeter circuit using a second transistor (Q_2) and a potential divider (R_4, R_5, and R_6) to eliminate the V_{BE} error produced by Q_1.

When no input voltage is applied ($E = 0$ V), the base voltage of Q_2 is adjusted to give zero meter current. This makes $V_p = 0$ V, $V_{E1} = V_{E2} = -0.7$ V, and (meter circuit voltage) $V = 0$ V. Now suppose that a 5 V input is applied to the Q_1 base. The meter voltage is

$$V = V_{E1} - V_{E2}$$
$$= (E - V_{BE1}) - V_{E2}$$
$$= (5\text{ V} - 0.7\text{ V}) - (-0.7\text{ V})$$
$$= 5\text{ V}$$

Thus, unlike the case of the simple emitter-follower voltmeter, all of the voltage to be measured appears across the meter circuit; no part of it is lost as transistor V_{BE}.

Example 4-2

An emitter-follower voltmeter circuit such as that in Figure 4-2 has $R_2 = R_3 = 3.9$ kΩ and $V_{CC} = \pm 12$ V.
(a) Determine I_2 and I_3 when $E = 0$ V.
(b) Calculate the meter circuit voltage when $E = 1$ V and when $E = 0.5$ V.

Solution

(a)
$$V_{R2} = V_{R3} = 0\text{ V} - V_{BE} - V_{EE}$$
$$= 0\text{ V} - 0.7\text{ V} - (-12\text{ V})$$
$$= 11.3\text{ V}$$

$$I_2 = I_3 = \frac{V_{R2}}{R_2} = \frac{11.3\text{ V}}{3.9\text{ k}\Omega}$$
$$\approx 2.9\text{ mA}$$

(b) *When E = 1 V,*

$$V_{E1} = E - V_{BE} = 1\ V - 0.7\ V$$

$$= 0.3\ V$$

$$V_{E2} = V_p - V_{BE} = 0\ V - 0.7\ V$$

$$= -0.7\ V$$

$$V = V_{E1} - V_{E2} = 0.3\ V - (-0.7\ V)$$

$$= 1\ V$$

When E = 0.5 V,

$$V_{E1} = E - V_{BE} = 0.5\ V - 0.7\ V$$

$$= -0.2\ V$$

$$V_{E2} = V_p - V_{BE} = 0\ V - 0.7\ V$$

$$= -0.7\ V$$

$$V = V_{E1} - V_{E2} = -0.2\ V - (-0.7\ V)$$

$$= 0.5\ V$$

Ground Terminals and Floating Power Supplies

The circuit in Figure 4-2 shows the input voltage E as being measured with respect to ground. However, this may not always be convenient. For example, suppose that the voltage across resistor R_b in Figure 4-3(a) were to be measured by a voltmeter with its negative terminals grounded. The voltmeter ground would short-circuit resistor R_C and seriously affect the voltage and current conditions in the resistor circuit. Clearly, the voltmeter should *not* have one of its terminals grounded.

For the circuit in Figure 4-2 to function correctly, the lower end of R_1 must be at zero volts with respect to $+V_{CC}$ and $-V_{EE}$. The + and − supply voltage may be derived from two batteries [Figure 4-3(b)] or from two dc power supply circuits [Figure 4-3(c)]. In both cases, the negative terminal of the positive supply is connected to the positive terminal of the negative supply. For ±9 V supplies, V_{CC} is +9 V with respect to the common terminal, and V_{EE} is −9 V with respect to the common terminal. In many electronic circuits, the power supply common terminal is grounded. In electronic voltmeter circuits, this terminal is not grounded, simply to avoid the kind of problem already discussed. When left without any grounded terminal, the voltmeter supply voltages are said to be *floating*. This means that the common terminal assumes the absolute voltage (with respect to ground) of any terminal to which it may be connected. An inverted triangular symbol is employed to identify the common terminal or *zero voltage terminal* in a circuit [see Figure 4-3(b),(c)].

Although the electronic voltmeter supply voltages are allowed to float, some instruments have their common terminal connected to ground via a capacitor, usually 0.1 μF. If batteries are used as supply, the capacitor is connected to the chassis. Where a

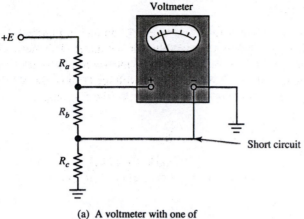

(a) A voltmeter with one of it's terminals grounded can short-circuit a component in a circuit in which voltage is being measured.

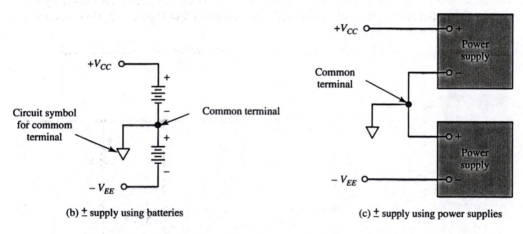

(b) ± supply using batteries

(c) ± supply using power supplies

Figure 4-3 Serious measurement errors can result when a grounded voltmeter terminal is incorrectly connected to a circuit. When a circuit has a plus-and-minus supply voltage, the voltmeter common terminal should always be connected to the common terminal of the supply.

115 V power supply is included in the voltmeter, the chassis and the capacitor are grounded. Thus, when measuring voltage levels in a transistor circuit, for example, the common terminal introduces a capacitance to ground wherever it is connected in the circuit. To avoid any effect on conditions within the circuit (oscillations or phase shifts), the voltmeter common terminal should always be connected to the transistor circuit ground or zero voltage terminal. All voltages are then measured with respect to this point.

Voltmeter Range Changing

The potential divider constituted by resistors R_a, R_b, R_c, and R_d in Figure 4-4 allows large input voltages to be measured on an emitter-follower voltmeter. This network, called an *input attenuator,* accurately divides the voltage to be measured before it is applied to the input transistor. Calculation shows that the Q_3 input voltage (E_G) is always 1 V when the maximum input is applied on any range. For example, on the 5 V range,

$$E_G = 5 \text{ V} \times \frac{R_b + R_c + R_d}{R_a + R_b + R_c + R_d}$$

$$= 5 \text{ V} \times \frac{100 \text{ k}\Omega + 60 \text{ k}\Omega + 40 \text{ k}\Omega}{800 \text{ k}\Omega + 100 \text{ k}\Omega + 60 \text{ k}\Omega + 40 \text{ k}\Omega}$$

$$= 1 \text{ V}$$

The input resistance offered by this circuit to a voltage being measured is the total resistance of the attenuator, which is 1 MΩ. A 9 MΩ resistor could be included in series with the input terminal to raise the input resistance to 10 MΩ. This would further divide the input voltage by a factor of 10 before it is applied to the gate terminal of Q_3.

FET-Input Voltmeter

The input resistance of the transistor voltmeter circuit can be increased further by using an additional emitter follower connected at the base of Q_1 in Figure 4-2. However, the use

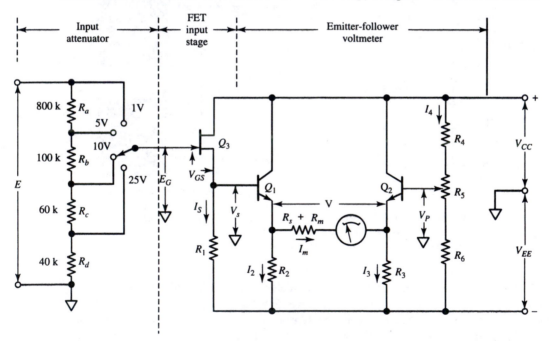

Figure 4-4 A voltmeter input attenuator is simply a potential divider that accurately divides the voltage to be measured. The FET input stage (Q_3) gives the emitter follower a very high input resistance.

Analog Electronic Volt-Ohm-Millimeters   Chap. 4

of a FET *source follower* (Q_3), as illustrated in Figure 4-4 gives a higher input resistance than can be achieved with a bipolar transistor. The FET source terminal is able to supply all of the base current required by Q_1, while the input resistance at the FET gate is typically in excess of 1 MΩ.

Consider the voltage levels in the circuit of Figure 4-4. When $E = 0$ V, the FET gate is at the zero voltage level. But the gate of an *n*-channel FET must always be negative with respect to its source terminal. This is the same as stating that the source must be positive with respect to the gate. If V_{GS} is to be −5 V, and $E_G = 0$ V, the source terminal voltage must be +5 V. This means that the base terminal of Q_1 is at +5 V, and, since Q_2 base voltage must be equal to Q_1 base voltage, Q_2 base must also be at +5 V. As in the circuit of Figure 4-2, R_5 in Figure 4-4 is used to zero the meter when the input voltage is 0 V.

Now consider what occurs when a voltage to be measured is applied to the circuit input. With the attenuator shown, E_G will be a maximum of 1 V. This causes the FET source terminal to increase until V_{GS} is again −5 V. That is, V_S goes from +5 to +6 V to maintain V_{GS} equal to −5 V. The V_S increase of 1 V is also a 1 V increase in the base voltage of Q_1. As already explained, all of this (1 V) increase appears across the meter circuit.

Example 4-3

Determine the meter reading for the circuit in Figure 4-4 when $E = 7.5$ V and the meter is set to its 10 V range. The FET gate–source voltage is −5 V, $V_p = +5$ V, $R_s + R_m = 1$ kΩ, and $I_m = 1$ mA at full scale.

Solution   *On the 10 V range:*

$$E_G = E \frac{R_c + R_d}{R_a + R_b + R_c + R_d}$$

$$= 7.5 \text{ V} \times \frac{60 \text{ k}\Omega + 40 \text{ k}\Omega}{800 \text{ k}\Omega + 100 \text{ k}\Omega + 60 \text{ k}\Omega + 40 \text{ k}\Omega}$$

$$= 0.75 \text{ V}$$

$$V_S = E_G - V_{GS} = 0.75 \text{ V} - (-5 \text{ V})$$

$$= 5.75 \text{ V}$$

$$V_{E1} = V_S - V_{BE} = 5.75 \text{ V} - 0.7 \text{ V} = 5.05 \text{ V}$$

$$V_{E2} = V_P - V_{BE} = 5 \text{ V} - 0.7 \text{ V}$$

$$= 4.3 \text{ V}$$

$$V = V_{E1} - V_{E2} = 5.05 \text{ V} - 4.3 \text{ V}$$

$$= 0.75 \text{ V} = E_G$$

$$I_m = \frac{V}{R_S + R_m} = \frac{0.75 \text{ V}}{1 \text{ k}\Omega}$$

$$= 0.75 \text{ mA} \ (75\% \text{ of full scale})$$

On the 10 V range, full scale represents 10 V, and 75% of full scale would be read as 7.5 V.

Difference Amplifier Voltmeter

The instruments discussed so far can measure a maximum of around 25 V. This could be extended further, of course, simply by modifying the input attenuator. The *minimum* (full-scale) voltage measurable by the electronic voltmeter circuits already considered is 1 V. This too can be altered to perhaps a minimum of 100 mV by selection of a meter that will give FSD when 100 mV appears across $R_s + R_m$. However, for accurate measurement of low voltage levels, the voltage must be amplified before it is applied to the meter.

Transistors Q_1 and Q_2 together with R_{L1}, R_{L2}, and R_E in Figure 4-5(a) constitute a *differential amplifier,* or *emitter-coupled amplifier.* The circuit as a whole is known as a *difference amplifier voltmeter.* This is because when the voltage at the base of Q_2 is zero, and an input voltage (E) is applied to the Q_1 base, the difference between the two base voltages is amplified and applied to the meter circuit.

When a small positive voltage is applied to the base of Q_1 in Figure 4-5, the current through Q_1 is increased, and that through Q_2 is decreased. An increase in I_{C1} causes $I_{C1}R_{L1}$ to increase and thus produces a fall in voltage V_{C1}. Similarly, a decrease in I_{C2} produces a rise in V_{C2}. The consequence of this is that the voltage across the meter circuit increases positively at the right-hand side and negatively at the left. This meter voltage (V) is directly proportional to the input voltage (E).

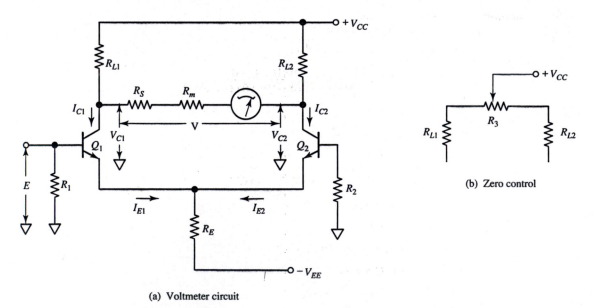

(a) Voltmeter circuit

(b) Zero control

Figure 4-5 A difference amplifier voltmeter amplifies low-level input voltages for measurement on the deflection voltmeter circuit.

Potentiometer R_3 in Figure 4-5(b) is an alternative method of providing meter-zero adjustment. Q_2 base control, as in Figure 4-4, could also be used in the circuit of Figure 4-5. When the movable contact of R_3 is adjusted to the right, the portion of R_3 added to R_{L1} is increased and the portion of R_3 added to R_{L2} is reduced. When the contact is moved left, the reverse is true. Thus, V_{C1} and V_{C2} can be adjusted differentially by means of R_3, and the meter voltage can be set to zero.

4-2 OPERATIONAL AMPLIFIER VOLTMETER CIRCUITS

Op-Amp Voltage-Follower Voltmeter

The operational amplifier voltage-follower voltmeter in Figure 4-6 is comparable to the simple emitter-follower circuit. However, unlike the emitter-follower, there is no base-emitter voltage drop from input to output. The voltage-follower also has a much higher input resistance and lower output resistance than the emitter-follower. The voltage-follower input (E_B) is applied to the op-amp noninverting input terminal, and the feedback from the output goes to the inverting input. The very high internal voltage gain of the operational amplifier, combined with the negative feedback, tends to keep the inverting terminal voltage exactly equal to that at the noninverting terminal. Thus, the output voltage (V_o) exactly follows the input. As discussed earlier, the attenuator selects the voltmeter range.

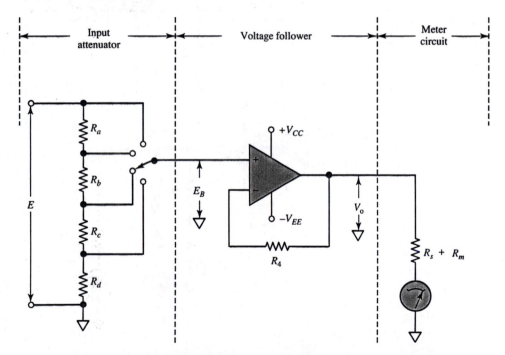

Figure 4-6 An IC operational amplifier voltage-follower voltmeter is similar to the emitter-follower voltmeter, except that the voltage-follower input resistance is much higher than that of the emitter follower, and there is no base–emitter voltage drop.

Op-Amp Amplifier Voltmeter

Like a transistor amplifier, an IC operational amplifier circuit can be used to amplify low voltages to levels measurable by a deflection instrument. Figure 4-7 shows a suitable op-amp circuit for this purpose. Input voltage E is applied to the op-amp noninverting input, the output voltage is divided across resistors R_3 and R_4, and V_{R3} is fed back to the op-amp inverting input terminal. The internal voltage gain of the op-amp causes V_{R3} to always equal E. Consequently, the output voltage is

$$V_o = E \frac{R_3 + R_4}{R_3}$$

(4-1)

The circuit is known as a *noninverting amplifier*, because its output is positive when a positive input voltage is applied, and negative when the input is a negative quantity. The noninverting amplifier has a very high input resistance, very low output resistance, and a voltage gain of

$$A_v = \frac{R_3 + R_4}{R_3}$$

(4-2)

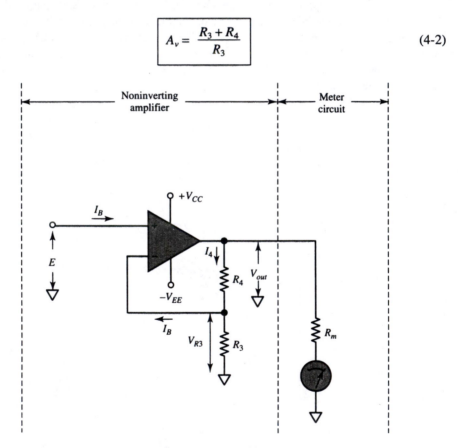

Figure 4-7 An operational amplifier noninverting amplifier can be used to amplify low input voltages to a level suitable for the deflection meter circuit. The voltmeter gain is $(R_3 + R_4)/R_3$.

An op-amp noninverting amplifier voltmeter is very easily designed. Current I_4 through R_3 and R_4 is first selected very much larger than the op-amp input bias current (I_B). Then the resistors are calculated as

$$R_3 = \frac{E}{I_4} \quad \text{and} \quad R_4 = \frac{V_o - E}{I_4}$$

Example 4-4

An op-amp voltmeter circuit as in Figure 4-7 is required to measure a maximum input of 20 mV. The op-amp input current is 0.2 μA, and the meter circuit has $I_m = 100$ μA FSD and $R_m = 10$ kΩ. Determine suitable resistance values for R_3 and R_4.

Solution

$$I_4 \gg I_B$$

Select

$$I_4 = 1000 \times I_B = 1000 \times 0.2\ \mu A$$
$$= 0.2\ mA$$

At full scale,

$$I_m = 100\ \mu A$$

and

$$V_{out} = I_m \times R_m = 100\ \mu A \times 10\ k\Omega$$
$$= 1\ V$$

$$R_3 = \frac{E}{I_4} = \frac{20\ mV}{0.2\ mA}$$
$$= 100\ \Omega$$

$$R_4 = \frac{V_o - E}{I_4} = \frac{1\ V - 20\ mV}{0.2\ mA}$$
$$= 4.9\ k\Omega$$

Voltage-to-Current Converter

The circuit shown in Figure 4-8 is essentially a noninverting amplifier, as in Figure 4-7. However, instead of connecting the meter between the op-amp output and ground, it is substituted in place of resistor R_4 (in Figure 4-7). Once again, V_{R3} remains equal to the input voltage, and as long as I_{R3} is very much greater than I_B, the meter current is

$$\boxed{I_m = I_{R3} = \frac{E}{R_3}} \tag{4-3}$$

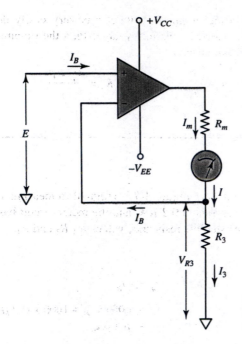

Figure 4-8 Voltmeter circuit using an op-amp voltage-to-current converter. The meter current is E/R_3.

Example 4-5

Calculate the value of R_3 for the circuit in Figure 4-8 if $E = 1$ V is to give FSD on the meter. The moving-coil meter has $I = 1$ mA at full scale and $R_m = 100$ Ω. Also determine the maximum voltage at the operational amplifier output terminal.

Solution   From Equation 4-3,

$$R_3 = \frac{E}{I_{(FSD)}} = \frac{1 \text{ V}}{1 \text{ mA}} = 1 \text{ k}\Omega$$

$$V_o = I(R_3 + R_m)$$

$$= 1 \text{ mA}(1 \text{ k}\Omega + 100 \text{ }\Omega)$$

$$= 1.1 \text{ V}$$

4-3 OHMMETER FUNCTION IN ELECTRONIC INSTRUMENTS

Series Ohmmeter

Since analog electronic instruments contain a moving-coil deflection meter, there is no reason why they cannot be made to function as an ohmmeter in the same way as described in Chapter 3. The circuit in Figure 4-9(a) shows a series ohmmeter circuit that

Analog Electronic Volt-Ohm-Millimeters   Chap. 4

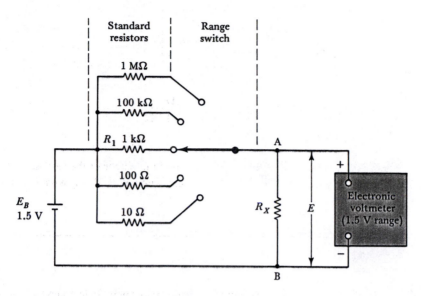

(a) Series ohmmeter circuit for electronic instrument

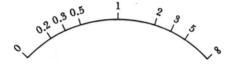

(b) Ohmmeter scale for electronic instrument

Figure 4-9 A series ohmmeter used in an electronic voltmeter is simply a potential divider constituted by a selected standard resistor (R_1) and the unknown resistance (R_x). The voltmeter measures the voltage drop across R_x, and the pointer deflection indicates the ratio of R_x to R_1.

uses the electronic voltmeter on a 1.5 V range. A 1.5 V battery and several standard resistors are included, as illustrated. The unknown resistance (R_x) is connected across terminals A and B, so that the voltmeter input (E) is the voltage drop across R_x. This circuit is similar to the series ohmmeter in Figure 3-22, except that the voltage across R_x is measured instead of its current.

Suppose that the range switch is set, as shown, to the 1 kΩ standard resistor (R_1). With terminals A and B open-circuited (R_x not connected), the voltmeter indicates full scale (1.5 V). Therefore, FSD (right-hand side of the scale) represents $R_x = \infty$ [see Figure 4-9(b)]. If terminals A and B are short-circuited, E becomes zero, and the pointer is at the left-hand side of the scale. Thus, the left-hand side represents $R_x = 0\ \Omega$. Now suppose that an unknown resistance greater than zero but less than infinity is connected to terminals A and B. The battery voltage (E_B) is potentially divided across R_1 and R_x, giving

$$E = E_B \frac{R_x}{R_1 + R_x} \qquad (4\text{-}4)$$

When $R_x = R_1 = 1 \text{ k}\Omega$,

$$E = 1.5 \text{ V} \times \frac{1 \text{ k}\Omega}{1 \text{ k}\Omega + 1 \text{ k}\Omega}$$

$$= 0.75 \text{ V}$$

So the meter indicates one-half of full scale, and the center of the resistance scale is marked 1. The meter will always indicate half scale when $R_x = R_1$ on whatever range is selected.

Example 4-6

For the circuit shown in Figure 4-9, determine the resistance scale markings at $\frac{1}{3}$ and $\frac{2}{3}$ of full scale.

Solution

From Equation 4-4,

$$R_x = \frac{R_1}{\dfrac{E_B}{E} - 1}$$

At $\frac{1}{3}$ FSD $\qquad E = \dfrac{E_B}{3}$

$$R_x = \frac{R_1}{\left[\dfrac{E_B \times 3}{E_B} - 1 \right]} = \frac{R_1}{2}$$

$$= 0.5 \, R_1$$

At $\frac{2}{3}$ FSD $\qquad E = \dfrac{2 \, E_B}{3}$

$$R_x = \frac{R_1}{\left[\dfrac{E_B \times 3}{2 \, E_B} - 1 \right]}$$

$$= 2 \, R_1$$

The two points calculated in Example 4-6 are shown on the resistance scale in Figure 4-9(b). Further calculations demonstrate that the scale becomes progressively cramped at both extremities. Thus, as in the case of the nonelectronic ohmmeter, this instrument measures resistance most accurately when indicating close to half-scale deflection.

The discussion above assumes that the electronic voltmeter is operating on its 1.5 V range, and that the battery voltage (E_B) is precisely 1.5 V. A battery voltage slightly larger or slightly smaller than this is easily taken care of by including an adjustable resistance (R_s) in series with the deflection meter. If $E_B = 1.4$ V, R_s is adjusted to give FSD for 1.4 V when terminals A and B are open-circuited. Then when $R_x = R_1$, $E = 0.5 E_B$, as before, and the pointer once again indicates half scale. All points on the scale are correct once the meter has been adjusted for full scale with terminal A and B *open-circuited.*

It should be noted that the resistance measuring system described above requires two adjustments before use. First, the voltmeter must be zeroed electrically when the terminals are short-circuited. After that, the calibration control must be adjusted to give FSD when the terminals are open-circuited. The unknown resistance can be measured only after both adjustments have been made.

Shunt Ohmmeter

A *shunt-type ohmmeter* circuit for use with an electronic instrument is illustrated in Figure 4-10. The precision resistors are connected in shunt with the supply, instead of in series. In this case a regulated power supply is used to provide a stable 6 V supply, instead of employing a battery. This eliminates the need for a voltmeter calibration control, because, unlike a battery, the regulated power supply voltage does not drop.

With terminals A and B in Figure 4-10 open-circuited, $R_x = \infty$ and

$$E = E_B \frac{R_2}{R_1 + R_2}$$

$$= 6 \text{ V} \times \frac{1.33 \text{ k}\Omega}{4 \text{ k}\Omega + 1.33 \text{ k}\Omega}$$

$$= 1.5 \text{ V}$$

Therefore, in this circuit the voltmeter is once again on a 1.5 V range to give FSD when $R_x = \infty$.

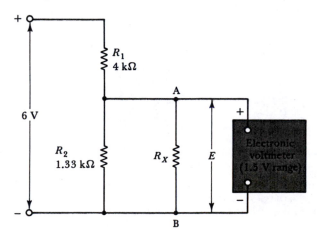

Figure 4-10   A shunt ohmmeter circuit used in an electronic voltmeter employs a regulated power supply and standard-value resistors connected as a potential divider. Half-scale deflection is obtained when $R_x = R_1 \| R_2$.

When $R_x = 0\ \Omega$ (A and B short-circuited), $E = 0$ V. Here again the pointer is at the left-hand side for $R_x = 0\ \Omega$. At any value of R_x:

$$E = E_B \frac{R_2 \| R_x}{R_1 + R_2 \| R_x} \tag{4-5}$$

The meter indicates half-scale when $R_x = R_1 \| R_2$. The values of R_1 and R_2 used in Figure 4-10 gives the instruments a 1 kΩ range. Resistance values 10 times larger would give a 10 kΩ range. Similarly, resistances 10 times smaller (than 4 kΩ and 1.33 kΩ) give a 100 Ω range. The scale on the instrument is exactly the same as that in Figure 4-9(b).

The major advantage of this system over the series ohmmeter described previously is that (because of the stable supply voltage) only one adjustment is required before resistance measurements are made; the ohmmeter terminals are short-circuited, and the instrument is electrically zeroed.

Linear Ohmmeter

In the circuit of Figure 4-11, transistor Q_1 together with resistors R_1, R_2, and R_E operates as a *constant-current circuit*. Resistors R_1 and R_2 potentially divide the supply voltage to give 5.7 V across R_1. When applied to the base of *pnp* transistor Q_1, this gives 5.7 V − $V_{BE} = 5$ V across resistor R_E. So the current I_E is 5 V/R_E, and this is a constant quantity. Since $I_C \simeq I_E$, the collector current is also a constant quantity. This constant current is

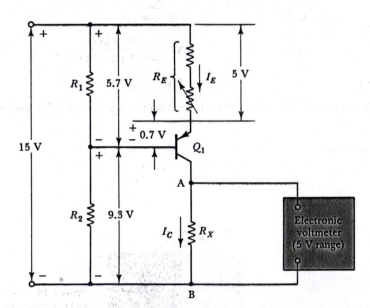

Figure 4-11   A linear ohmmeter circuit uses a constant-current circuit (R_1, R_2, R_E, and Q_1) to pass a fixed current level through the unknown resistance (R_x). The voltage drop across R_x is then directly proportional to R_x, thus giving a linear resistance scale.

passed through the unknown resistance R_x, and the voltage across R_x is measured by the voltmeter. The voltmeter scale can now be multiplied by an appropriate factor and used directly as a resistance scale.

Suppose that R_E is adjusted to give a collector current of exactly 1 mA. If the voltmeter indicates exactly 5 V, then

$$R_x = \frac{5 \text{ V}}{1 \text{ mA}} = 5 \text{ k}\Omega$$

When the voltmeter indicates 3 V, R_x is exactly 3 kΩ. Thus, the 0 to 5 V scale is a *linear* 0 to 5 kΩ resistance scale.

Other standard resistors substituted in place of R_E can reduce I_C to 100 μA, or increase it to 10 mA. When $I_C = 100$ μA, the maximum resistance measurable on the 5 V range is now

$$R_x = \frac{5 \text{ V}}{100 \text{ }\mu\text{A}} = 50 \text{ k}\Omega$$

With $I_C = 10$ mA, the maximum value of R_x becomes 500 Ω.

For multirange operations, a switching arrangement must be provided for selection of R_E from several standard resistor values. Some series adjustment for calibration of each range is necessary. Calibration is easily effected by connecting a known standard resistor in place of R_x and adjusting R_E for the appropriate deflection.

4-4 AC ELECTRONIC VOLTMETERS

The IC op-amp voltage-follower voltmeter described in Section 4-2 is a dc instrument. Connecting a rectifier in series with the meter circuit of this instrument, as shown in Figure 4-12(a), converts it into a half-wave rectifier voltmeter. The output from the voltage follower is exactly the same as the input. So the voltage fed to the meter circuit is simply a half-wave-rectified version of the input voltage E_B from the attenuator. The operation of this circuit and the design calculations are just as described in Section 3-5 for the nonelectronic half-wave rectifier voltmeter. The difference between the nonelectronic instrument and the electronic ac voltmeter circuit in Figure 4-12(a) is, of course, that the electronic instrument has a very high input impedance. Note the coupling capacitor (C_1) in Figure 4-12(a). This is usually provided at the input of an ac voltmeter to block unwanted dc voltages.

The voltage drop (V_F) across the rectifier is a source of error in the circuit in Figure 4-12(a). V_F can be taken into account in design calculations when the instrument is indicating full scale. However, at other points on the scale an error occurs due to V_F. Also, the rectifier voltage drop is not always exactly 0.7 V, as usually assumed for a silicon diode, and it varies with temperature change. To avoid these errors, the voltage follower feedback connection to the inverting terminal is taken from the cathode of rectifier D_1 instead of from the amplifier output [see Figure 4-12(b)]. The result is that the half-wave-rectified output precisely follows the positive half-cycle of the input voltage.

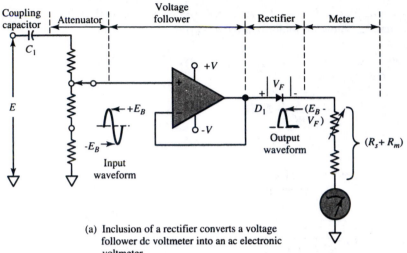

(a) Inclusion of a rectifier converts a voltage follower dc voltmeter into an ac electronic voltmeter.

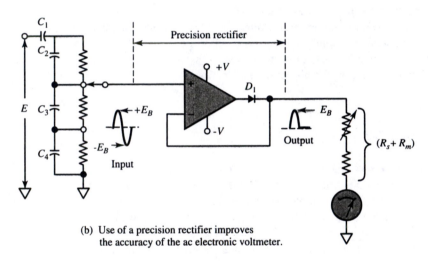

(b) Use of a precision rectifier improves the accuracy of the ac electronic voltmeter.

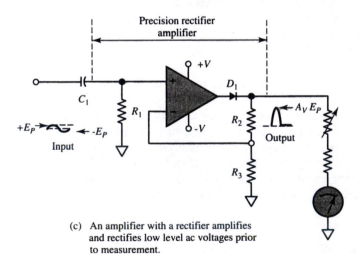

(c) An amplifier with a rectifier amplifies and rectifies low level ac voltages prior to measurement.

Figure 4-12 Operational amplifier ac voltmeter circuits using dc voltmeters with half-wave rectification. The meter series resistance must be calculated to give a deflection proportional to the rms value of the input sine wave.

There is *no* rectifier voltage drop from input to output. The circuit is known as a *precision rectifier.*

The procedure for calculation of series resistance R_S for this circuit is just as explained in Section 3-5, with one exception; no diode voltage drop is involved. Note that capacitors C_2, C_3, and C_4 are connected across the attenuator resistors in Figure 4-12(b). These are normally employed with the attenuators on ac electronic voltmeters in order to compensate for the input capacitance of the amplifier. This input capacitance problem also occurs with oscilloscopes. Compensation capacitors are discussed further in Section 9-9.

Low-level ac voltages should be accurately amplified before being rectified and applied to a meter circuit. Amplification is combined with half-wave rectification in the circuit shown in Figure 4-12(c). With the diode omitted, the op-amp circuit is a noninverting amplifier as described in Section 4-2. Inclusion of D_1 causes the positive half-cycles of the input to be amplified by a factor $A_v = (R_2 + R_3)/R_3$. The amplification is precise and here again there is no rectifier voltage drop involved.

The procedure for calculating the meter series resistance in the circuit in Figure 4-12(c) is the same as described in Section 3-5. However, in this case, the peak voltage applied to the meter circuit is $A_v E_p$, and again the rectifier voltage drop does not enter into the calculations.

The circuit in Figure 4-13(a) is a *voltage-to-current converter* with half-wave rectification. The circuit functions exactly as explained in Section 4-2, with the exception that only the positive half-cycles of the ac input are effective in passing current through the meter. During the negative half-cycle, the diode is reverse biased and no current flows through the meter or through resistor R_3. The meter peak current is $I_p = E_p/R_3$, and the average meter current is $I_{av} = 0.5(0.637\ I_p)$.

A full-wave bridge rectifier is employed in the circuit of Figure 4-13(b). When the input voltage is positive, the operational amplifier output is positive. Diodes D_1 and D_4 are forward biased so that current flows through the meter from top $(+)$ to bottom $(-)$. When the input is negative, D_2 and D_3 are forward biased. Once again current passes through the meter from the $+$ to the $-$ terminal. Whether the input is positive or negative, the meter peak current is again limited to $I_p = E_p/R_3$. The average meter current in the full-wave rectifier circuit is $I_{av} = 0.637\ I_p$.

Instead of a full-wave bridge rectifier, some electronic instruments use the half-bridge full-wave circuit of Figure 3-19. As explained in Section 3-5, this arrangement helps to correct for differences in the diodes. However, the circuit in Figure 4-13(b) tends to be unaffected by diode differences, because the meter current is stabilized by the amplifier feedback circuitry. So the half-bridge circuit would not improve the performance of an IC op-amp voltage-to-current converter type of voltmeter. A half-bridge rectifier could be useful in an electronic voltmeter that uses a transistor amplifier.

Figure 4-13(c) shows a half-bridge rectifier circuit connected to the output stage of a transistor amplifier. Output transistor Q_3 passes direct current through resistor R_5, producing a dc voltage at the collector of the transistor which might be above or below the zero voltage level. Capacitors C_1 and C_2 block the direct current from the meter circuit and pass the alternating current when an ac voltage appears at the amplifier output. The circuit then functions exactly as described in Section 3-5.

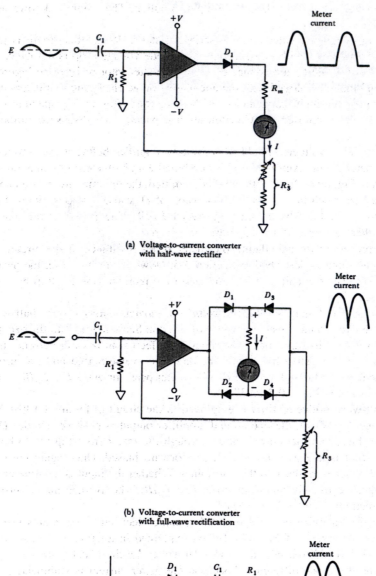

(a) Voltage-to-current converter
with half-wave rectifier

(b) Voltage-to-current converter
with full-wave rectification

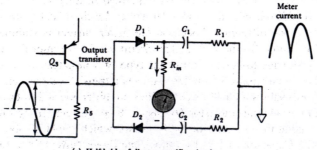

(c) Half-bridge full-wave rectifier circuit
with dc blocking capacitors

Figure 4-13 Ac electronic voltmeters can be constructed using rectification with a voltage-to-current converter, full-wave bridge rectification, or half-bridge full-wave rectification.

Analog Electronic Volt-Ohm-Millimeters   Chap. 4

Example 4-7

The half-wave rectifier electronic voltmeter circuit in Figure 4-13(a) uses a meter with a FSD current of 1 mA. The meter coil resistance is 1.2 kΩ. Calculate the value of R_3 that will give meter full-scale pointer deflection when the ac input voltage is 100 mV (rms). Also determine the meter deflection when the input is 50 mV.

Solution   *For* FSD, *the average meter current is*

$$I_{av} = 1 \text{ mA}$$

With half-wave rectifiers,

$$I_{av} = \frac{0.637}{2} I_p$$

or

$$I_p = \frac{2}{0.637} I_{av} = \frac{2}{0.637} \times 1 \text{ mA}$$

$$= 3.14 \text{ mA}$$

Peak value of $E_{R3} = $ input peak voltage:

$$E_p = \frac{E}{0.707} = \frac{100 \text{ mV}}{0.707}$$

$$= 141.4 \text{ mV}$$

$$R_3 = \frac{E_P}{I_p} = \frac{141.4 \text{ mV}}{3.14 \text{ mA}}$$

$$= 45 \text{ }\Omega$$

When E = 50 mV:

$$E_p = \frac{50 \text{ mV}}{0.707}$$

$$= 70.7 \text{ mV}$$

$$I_p = \frac{70.7 \text{ mV}}{45 \text{ }\Omega}$$

$$= 1.57 \text{ mA}$$

$$I_{av} = \frac{0.637}{2} \times 1.57 \text{ mA}$$

$$= 0.5 \text{ mA (half scale)}$$

4-5 CURRENT MEASUREMENT WITH ELECTRONIC INSTRUMENTS

Recall that the two reasons for introducing electronic devices into voltmeters are (1) to produce a very high input resistance, and (2) to amplify very small voltages to measurable levels. Item 1 does not apply in the case of current measurement; on the contrary, ammeters should normally have the lowest possible resistance. Item 2 can apply in the case of very low current levels.

The basic circuit of an analog electronic ammeter for measurement of very low currents is shown in Figure 4-14. The small voltage drop across shunt resistor R_s is amplified before being applied to the deflection instrument. This approach is just as applicable to the measurement of low-level alternating currents as it is to direct-current measurements. For alternating currents, an ac electronic voltmeter using rectifiers (see Section 4-4) is used instead of a dc instrument.

For medium or high current measurements there is no need to use electronic amplifiers. In fact, any electronic voltmeter could have several current measurement ranges, if appropriate shunts and rectifiers are included, as explained in Sections 3-3 and 3-6. The deflection instrument would be involved, but the electronic circuitry could be switched out.

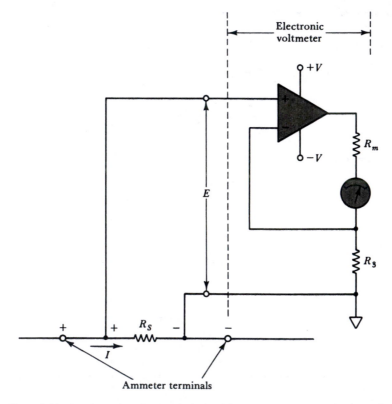

Figure 4-14 An electronic voltmeter can be used for current measurement by measuring the voltage drop across a shunt (R_s). The instrument scale is calibrated to indicate current.

Many electronic multirange instruments do not have any current-measuring facilities. Those that do measure current generally have very low-level current ranges, and some have relatively high resistances when operating as ammeters. For example, the meter resistance on one instrument is specified as 9 kΩ when operating on a 1.5 µA range. This must be taken into account when the instrument is connected in series with a circuit in which the current is to be measured. The instrument terminal voltage drop when used as an ammeter is termed the *burden voltage*. For a 9 kΩ resistance on a 1.5 µA range, the burden voltage is

$$V_B = 9 \text{ k}\Omega \times 1.5 \text{ }\mu\text{A} = 13.5 \text{ mV}$$

Other typical burden voltage specifications are *250 mV max, 2 V on a 10 A range,* and *6 mV/mA.* These voltages drops may or may not be important, depending on the circuit under test.

4-6 ANALOG ELECTRONIC MULTIMETERS

Laboratory-Type Electronic Multimeter

The Hewlett-Packard (HP) model 427A electronic voltmeter is representative of laboratory-type analog electronic instruments. The front panel illustrated in Figure 4-15 shows that it can measure dc voltage (*DCV* ±), ac voltage (*ACV*), and resistance (*OHMS*). A knife-edge pointer and mirror are provided for precise reading on two voltage scales: 0 to 1 and 0 to 3. The ohmmeter scale has its 1 position at the scale center, as in Figure 4-9(b). A decibel scale (*DBM*) is also provided. Voltage measurements are made using the

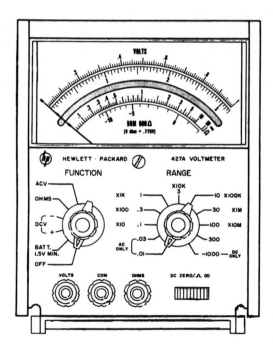

Figure 4-15 Front panel of the Hewlett-Packard model 427A electronic voltmeter. This instrument can measure dc voltage, ac voltage, and resistance. The range selection as a multiple of 10 or a multiple of 3 determines which of the two voltage scales [(0 to 1) or (0 to 3)] is used. (Courtesy of Hewlett-Packard.)

VOLTS and COM (common) terminals. For resistance measurements, the COM and OHMS terminals are employed. An electrical zero control (DC ZERO/Ω ∞) is included on the front panel, as well as a mechanical zero control for the pointer.

The HP427A has nine dc voltage ranges, from 0.1 V (full scale) to 1000 V. The ±DCV positions on the FUNCTION SWITCH permit either positive or negative polarity voltages to be measured. The measurement accuracy is ±2% of full scale, and the input resistance is 10 MΩ on all ranges. There are 10 ac voltage ranges, the lowest being 10 mV, and the highest 300 V. The frequency range for ac voltage measurements is 10 Hz to 1 MHz. This can be extended by the use of a high-frequency probe that has a peak detector circuit. The available probe is designed for use with voltages ranging from 0.25 V to 30 V, and it extends the frequency range of the HP427A to 500 MHz. The measurement accuracy for ac voltages is ±2% of full scale. However, the input impedance is stated as *10 MΩ shunted by a 40 pF capacitance* on ranges up to and including 1 V, and *10 MΩ shunted by 20 pF* on 3 V and greater ranges.

There are seven resistance-measuring ranges on the HP427A, starting at 10 Ω (center scale) and going to a maximum of 10 MΩ. The accuracy of resistance measurements is ±5% of the reading at midscale.

Procedure for dc Voltage Measurements

1. With the instrument switched off, check the pointer zero position. Adjust the mechanical zero as required.
2. If the instrument is battery operated, set the FUNCTION switch to BATT, and check that the battery voltage is a minimum of 1.5 V. For instruments with an internal power supply and line cord, this step is not necessary.
3. Set the FUNCTION switch to DCV + or DCV – as required.
4. Set the RANGE switch to 0.1, and short-circuit the VOLTS and COM terminals. Adjust the DC ZERO/Ω ∞ control to set the pointer precisely to zero on the scale, then remove the short-circuit connection.
5. Select a voltage range greater than the voltage to be measured. Where the approximate value of the voltage is not known, rotate the RANGE switch to the highest range.
6. Connect the input voltage to the VOLTS and COM terminals, and adjust the range switch to give the greatest on-scale pointer deflection. (Where there is a grounded point in a circuit, the COM terminal should be connected to that point. Where there is more than one electronic instrument involved, all of the COM terminals should be connected to a single point. The reasons for this are discussed in Section 4-1.)

The procedure for ac voltage measurements is exactly as for dc voltage, but with the FUNCTION switch set to ACV and no need for DC ZERO adjustment.

Resistance Measurement Procedure

1. With the instrument switched off, check the pointer zero and adjust the mechanical zero control as necessary.
2. Check the battery voltage as explained for dc voltage measurements.

3. Set the FUNCTION switch to OHMS, and with the instrument terminals *open-circuited,* adjust the DC ZERO/Ω ∞ control until the pointer indicates infinity (∞) on the resistance scale.

4. Select a resistance range to suit the approximate value of the resistance to be measured.

5. Connect the resistance to the COM and OHMS terminals, and adjust the RANGE control to give a resistance reading as close as possible to *center scale.*

Decibel (dB) and Decibel-Milliwatt (dBm) Measurements. Decibel measurement is essentially the same as ac voltage measurements. On the HP427A the decibel scale is read directly when the instrument is set to the 1 V (ACV) range. Since each range position above 1 V is 10 dB above 1 V, 10 dB must be added to the scale reading. For each range position below 1 V, 10 dB must be subtracted from the dB measurements. For example, if the pointer is indicating − 1 dB on the 10 V range, the dB measurement is −1 dB + 20 dB = 19 dB. A −5 dB scale reading on the 0.3 V range represents (−5 dB − 10 dB) = −15 dB.

The decibel scale on the HP427A is based on 0 dB as 1 mW dissipated in a 600 Ω load resistance. Where the load resistance is other than 600 Ω, the scale readings must be corrected to obtain the absolute dB measurements. However, scale changes can be read directly as changes in dB levels.

The *decibel-milliwatt* (dBm) term is sometimes employed because the absolute dB measurements are measurements of *changes* in power level from a starting point of 1 mW power dissipation.

VTVM and FET-Input Instruments

A *vacuum-tube voltmeter (VTVM)* and a *FET-input* multimeter are illustrated in Figure 4-16. Both have large scales for easy reading. The VTVM has dc and ac voltage ranges from 0.5 V to 1500 V, with an input resistance of 11 MΩ. It can also measure peak-to-peak voltages up to 400 V, and its frequency response is 40 H to 4 MHz. Resistance can be measured on ranges of 10 Ω to 10 MΩ (midscale). The FET-input voltmeter illustrated in Figure 4-16(b) has dc and ac voltage ranges extending from 300 mV to 1000 V, and ac peak-to-peak voltage ranges from 900 mV to 2.8 kV. Its input impedance is 10 MΩ, and its frequency response goes up to 100 kHz. Resistance can be measured from 10 Ω to 10 MΩ, and dc current can be measured from 100 μA to 1 A. Decibel measurement can also be performed.

4-7 MULTIMETER PROBES

There are many probes and adapters available for use with multimeters (and for use with analog and digital electronic instruments) that can extend the ranges of measurement, or adapt the instrument for measurement of temperature or other quantities. Some of these are illustrated in Figure 4-17.

(a) BK model 177 vacuum-tube voltmeter
(courtesy of BK Precision Test instruments)

(b) Simpson model 313-3 FET-input VOM (Courtesy of bach-simpson limited)

Figure 4-16 Vacuum-tube voltmeter (VTVM) (a) and FET-input electronic multimeter (b). Both instruments can measure dc voltage, ac voltage, ac peak-to-peak voltage, and resistance. The FET input meter also has dc current and decibel ranges.

High-Voltage Probe

The *high-voltage probe* (also known as a *voltage multiplier*) shown in Figure 4-17(a) is essentially a potential divider, well insulated for safety. The voltage to be measured is usually divided by a factor of 1000, so that the instrument scales are effectively multiplied by 1000; a 50 V scale becomes a 50 kV scale.

High-Current Probes

High levels of alternating current can be reduced by the use of a current transformer, and this principle is used in the *ac current probe* illustrated in Figure 4-17(b). The transformer core opens, as shown, to close around a conductor carrying the current to be measured. The conductor can be treated as a single-turn primary on the transformer, and the secondary winding then determines the measurable current level. Typically, the 1 mA ac scale on the multimeter is converted into a 1 A scale by the use of the current probe. The *Hall-effect probe* shown in Figure 4-17(c) has a similar appearance to the transformer-type ac current probe. However, it produces an output voltage for both dc and ac currents in a conductor. The Hall-effect transducer contained in the probe operates on the principle that a small voltage is produced at the edges of a flat current-carrying conductor in the presence of a magnetic field. The output is connected to the meter voltage terminals, typically resulting in the millivolt scale being read as a current scale with 1 mV representing 1 A.

(a) A high-voltage probe uses a
    resistive divider to reduce
    the voltage to a measurable
    level.

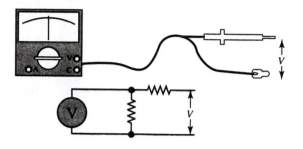

(b) A high-current probe uses a
    current transformer to reduce
    an ac current to a measurable
    level.

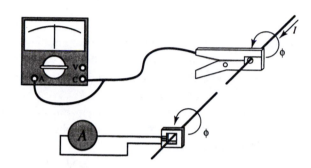

(c) A Hall effect high-current probe
    uses a magnetic flux (ac or dc)
    to produce a measurable voltage
    at the edges of a thin current-
    carrying conductor.

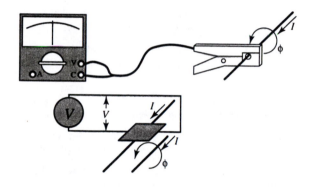

(d) An RF wave is rectified and
    smoothed in an RF probe to
    produce a measurable dc
    voltage.

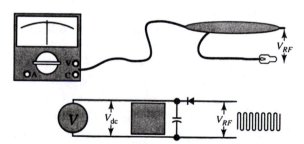

Figure 4-17 Various probes are available to extend multimeter measurement ranges.
High-voltage and high-current probes can multiply voltage and current ranges by 1000.
An RF probe allows the meter to measure the voltage level of a waveform with a
frequency well beyond its upper cutoff frequency.

RF Probe

The level of a radio-frequency (RF) voltage cannot be measured directly on a multimeter because the upper-frequency limit of the instrument is not high enough. In an *RF probe* [Figure 4-17(d)], the RF waveform is rectified and converted to a dc voltage equal to its peak level. With appropriate attenuation, the voltage can be measured on the meter and read as an rms quantity. Peak detector probes are discussed further in Section 15-2.

REVIEW QUESTIONS

4-1 Sketch the circuit of a simple transistor emitter-follower voltmeter. Explain the circuit operation, and compare it to a nonelectronic voltmeter.

4-2 Sketch the circuit of a practical (two-transistor) emitter-follower voltmeter. Carefully explain the circuit operation.

4-3 Using illustrations, discuss the problems that can occur with ground terminals when using an electronic voltmeter. Explain the need for a floating power supply in an electronic instrument.

4-4 Sketch a typical input attenuator as used with an electronic voltmeter. Show typical resistor values and range terminals, and explain attenuator operation.

4-5 Sketch the complete circuit of an emitter-follower voltmeter using a FET input stage. Carefully explain the circuit operation and the effect of the FET.

4-6 Draw a complete diagram of a transistor difference-amplifier type of voltmeter. Explain the circuit operation and discuss its advantage over emitter-follower voltmeters.

4-7 Sketch the complete circuit of an op-amp voltage follower voltmeter. Explain the circuit operation, and compare it to the simple emitter-follower voltmeter.

4-8 Draw the circuit diagram of an op-amp noninverting amplifier voltmeter. Explain its operation, and compare it to the voltage-follower voltmeter.

4-9 Draw the circuit diagram of a voltmeter using an op-amp voltage-to-current converter. Explain the circuit operation.

4-10 Sketch a series ohmmeter circuit for use with an electronic voltmeter. Explain the circuit operation, and discuss the zero adjustment procedure. Sketch and explain a typical scale for this instrument.

4-11 Sketch a shunt ohmmeter circuit as used with an electronic voltmeter. Explain the circuit operation, and compare it to a series ohmmeter.

4-12 Sketch a linear ohmmeter circuit as used with an electronic voltmeter. Explain the circuit operation, and compare it to the shunt and series ohmmeters.

4-13 Sketch half-wave rectifier ac electronic voltmeters using **(a)** a voltage follower and **(b)** a precision rectifier. Explain the operation of each circuit, and compare their performance.

4-14 Sketch the circuit of an ac electronic voltmeter using a half-wave precision rectifier with amplification. Explain the circuit operation.

4-15 Sketch the circuit of an ac electronic voltmeter using a voltage-to-current converter with half-wave rectification. Explain the circuit operation.

4-16 Sketch the circuit of an ac electronic voltmeter using a voltage-to-current converter with full-wave rectification. Explain the operation of the circuit.

4-17 Sketch a half-bridge full-wave rectifier as used with an ac electronic voltmeter. Explain the circuit operation.

4-18 Draw a circuit diagram to show how current can be measured using an electronic voltmeter. Explain.

4-19 List the procedure for dc voltage measurement using the electronic voltmeter shown in Figure 4-15.

4-20 List the procedure for using the electronic voltmeter shown in Figure 4-15 for ac voltage measurement.

4-21 List the resistance measurement procedure for the electronic voltmeter shown in Figure 4-15.

4-22 Discuss the procedure for decibel measurement with the electronic voltmeter in Figure 4-15.

4-23 Draw illustrations to show how the ranges of a multimeter can be extended by the use of a high-voltage probe and an ac high-current probe. Explain the operation of each type of probe.

4-24 Draw illustrations to show how the ranges of a multimeter can be extended by the use of a Hall-effect high-current probe, and an RF probe. Explain the operation of each type of probe.

PROBLEMS

4-1 A simple emitter-follower voltmeter circuit as in Figure 4-1 has $V_{CC} = 12$ V, $R_m = 1$ kΩ, a 2 mA meter, and a transistor with $h_{FE} = 80$. Calculate a suitable resistance for R_s to give full scale deflection when $E = 5$ V. Also, determine the voltmeter input resistance.

4-2 An emitter-follower voltmeter circuit, as in Figure 4-2, has the following components: $R_1 = 12$ kΩ, $R_2 = R_3 = 2.7$ kΩ, $R_4 = R_6 = 3.3$ kΩ, $R_5 = 500$ Ω, and $R_s + R_m = 10$ kΩ. A 100 μA meter is used, the supply voltage is ± 9 V, and the transistors have $h_{FE} = 75$. Determine V_p, I_{B1}, I_{B2}, I_2, I_3, and I_4 when $E = 0$. Also, calculate the range of adjustment for V_p.

4-3 Calculate the meter deflections for the circuit in Problem 4-2 when the input voltage levels are 0.6 V, 0.75 V, and 1 V.

4-4 A 3.5 V input (E) is applied to the input attenuator shown in Figure 4-4. Calculate the voltage E_G on each range selection.

4-5 The FET input voltmeter circuit in Figure 4-4 has the following components: $R_1 = 6.8$ kΩ, $R_2 = R_3 = 4.7$ kΩ, $R_4 = 1.5$ kΩ, $R_5 = 500$ Ω, $R_6 = 3.3$ kΩ, $R_s + R_m = 20$ kΩ. The meter full-scale current is 50 μA, the supply voltage is ± 10 V, the transistors have $h_{FE} = 80$, and the FET gate–source voltage is $V_{GS} = -3$ V. Determine V_p, I_s, I_2, I_3, and I_4 when $E = 0$. Also, calculate the range of adjustment for V_p.

4-6 Calculate the meter deflections for the circuit in Problem 4-5 when the attenuator is set to its 5 V range, and the input voltage levels are 1 V, 3 V, and 4 V.

4-7 The difference amplifier voltmeter in Figure 4-5(a) has the following components: $R_1 = R_2 = 15$ kΩ, $R_{L1} = R_{L2} = 3.9$ kΩ, $R_E = 3.3$ kΩ, $R_s = 33$ kΩ, and $R_m = 750$ Ω. The meter full-scale current is 50 μA, and the supply voltage is ± 12 V. Calculate the transistor voltage levels when $E = 0$.

4-8 The circuit in Problem 4-7 has transistors with $h_{FE} = 100$ and $h_{ie} = 1.2$ kΩ. Determine the input voltage (E) that will give full-scale deflection on the meter.

4-9 An op-amp voltage-follower voltmeter, as in Figure 4-6, has $R_a = 800$ kΩ, $R_b = 100$ kΩ, $R_c = 60$ kΩ, and $R_d = 40$ kΩ. A 50 μA meter is used with a resistance of $R_m = 750$ Ω. Determine the required resistance for R_s to give full-scale deflection when $E = 10$ V and the range switch is as illustrated.

4-10 The noninverting amplifier voltmeter circuit in Figure 4-7 uses an op-amp with $I_B = 300$ nA, and a 50 μA meter with $R_m = 100$ kΩ. Determine suitable resistances for R_3 and R_4 to give full-scale deflection when the input is 300 mV.

4-11 The voltage-to-current converter circuit in Figure 4-8 uses a 37.5 μA (FSD) deflection meter with $R_m = 900$ Ω. If $R_3 = 80$ kΩ, determine the required input voltage levels to give FSD and 0.5 FSD.

4-12 Determine the new resistance for R_3 for the circuit in Problem 4-11 to give FSD when $E = 1$ V. Also, calculate the op-amp output voltage.

4-13 Calculate the resistance scale markings at 25% and 75% of full scale for the series ohmmeter circuit in Figure 4-9.

4-14 Determine the percentage meter deflection in the circuit of Figure 4-9 when the 100 kΩ standard resistor is switched into the circuit and $R_x = 166$ kΩ.

4-15 Calculate the meter deflection for the shunt ohmmeter circuit in Figure 4-10 when $R_x = 2$ kΩ and when $R_x = 300$ Ω.

4-16 A 16.67 kΩ resistor is substituted for R_E in the linear ohmmeter circuit in Figure 4-11. Calculate the measured resistance when the meter indicates 3.9 V.

4-17 The half-wave rectifier electronic voltmeter in Figure 4-12(b) uses a 500 μA deflection meter with a 460 Ω coil resistance. If $R_s = 450$ Ω, calculate the rms input voltage required to give full-scale deflection.

4-18 The components used in Problem 4-17 are reconnected as in Figure 4-13(a) with $R_3 = R_s$. Determine the new rms input voltage required to give full-scale deflection. Also determine the meter deflections when the input is 100 mV and 200 mV.

4-19 The ac electronic voltmeter circuit in Figure 4-12(c) uses the following components: $R_1 = 22$ kΩ, $R_2 = 2.25$ kΩ, $R_3 = 6.8$ kΩ, $R_s + R_m = 1$ kΩ, and a 300 μA meter. Calculate the rms input voltages for meter full-scale deflection and for 0.5 FSD.

4-20 The full-wave rectifier voltmeter circuit in Figure 4-13(b) uses a 500 μA meter with $R_m = 460$ Ω together with $R_3 = 450$ Ω (as for Problems 4-17 and 4-18). Determine the rms input voltage for FSD on the meter.

Digital Instrument Basics

5

Objectives

You will be able to:

1. Sketch the circuits of various logic gates. Explain the operation of each gate, and sketch its logic symbols.
2. Sketch a basic transistor flip-flop circuit, and explain its operation. Draw the logic symbols for T and RST flip-flops, and explain the function of each.
3. Using illustrations, explain the operation of a light-emitting diode (LED) and a liquid crystal display (LCD). Discuss how they are used in seven-segment numerical displays.
4. Draw the logic diagrams for scale-of-16, decade, and scale-of-2000 counters. Explain their operation, and prepare tables showing the counter states after each input pulse.
5. Show how decade counters may be used for frequency division.
6. Draw circuit diagrams for analog-to-digital converters (ADC) and digital-to-analog converters (DAC). Show the system waveforms, explain how conversion is achieved, and discuss conversion accuracy and resolution.

Introduction

An understanding of basic digital circuits is required before the operation, applications, and performance of digital instruments can be investigated. Digital circuits involve the use of logic gates and flip-flops. A logic gate is a circuit that produces a desired (*high* or *low*) output voltage when its several input terminals are all at the prescribed (*high* or *low*) input voltage level. A flip-flop is basically a two-transistor circuit that has two states: Q_1 *on* and Q_2 *off*, or Q_1 *off* and Q_2 *on*. Four flip-flops may be connected as a decade counter, or scale-of-10. Three decade counters and one additional flip-flop can be used to count to

2000. Seven-segment displays, in light-emitting diode (LED) and liquid crystal display (LCD) format, are available for reading out a digital count. Analog voltages are converted to digital representations in the process of digital measurement. This involves the use of analog-to-digital converters (ADCs). Digital-to-analog converters (DACs) are also important.

5-1 BASIC LOGIC GATES

The circuit of a diode *AND* gate with three input terminals is shown in Figure 5-1(a). If one or more of the input terminals (i.e., diode cathodes) is grounded, the diodes are forward biased. Consequently, current I_1 flows from V_{CC}, and the output voltage V_o is equal to the diode forward voltage drop V_D. Suppose that the supply is $V_{CC} = 5$ V, and an input of 5 V is applied to terminal *A*, while terminals *B* and *C* are grounded. Diode D_1 is reverse biased while D_2 and D_3 remain forward biased, and V_o remains equal to V_D. If levels of 5 V are applied to all three inputs, no current flows through R_1, and $V_o = V_{CC} = 5$ V. Thus, a *high* output voltage is obtained from the *AND* gate only when high input voltages are present at input *A*, *AND* at input *B*, *AND* at input *C*. Hence the name *AND* gate.

An *AND* gate may have as few as two, or a great many input terminals. In all cases an output is obtained only when the correct input voltage level is provided simultaneously at every input terminal. Figure 5-1(b) shows the graphic symbol employed to represent an AND gate in logic system diagrams.

A three-input diode *OR* gate and its logic symbol are shown in Figure 5-2. It is obvious from the gate circuit that the output is zero when all three inputs are at ground level. If a 5 V input is applied to terminal *A*, diode D_1 is forward biased, and V_o becomes $(5 \text{ V} - V_D)$. If terminals *B* and *C* are grounded at this time, diodes D_2 and D_3 are reverse biased. Instead of terminal *A*, the positive input might be applied to terminal *B* or *C* to obtain a positive output voltage. A *high* output voltage is obtained from an OR gate, when a

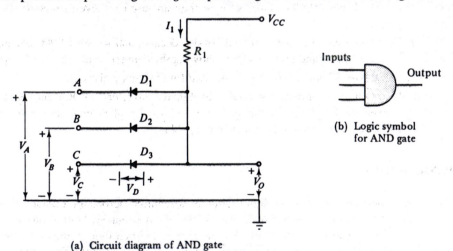

(a) Circuit diagram of AND gate

Figure 5-1 Circuit diagram and logic symbol for a three-input diode *AND* gate. A *high* output is produced only when *high* levels are present at input *A*, *and* at input *B*, *and* at input *C*.

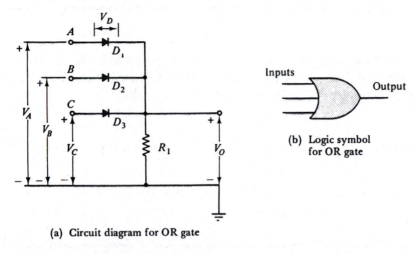

(a) Circuit diagram for OR gate

(b) Logic symbol for OR gate

Figure 5-2 Circuit diagram and logic symbol for a three-input diode *OR* gate. A *high* output is produced when *high* levels are present at input A, *or* at input B, *or* at input C.

high input is applied to terminal A, *OR* to terminal B, *OR* to terminal C; hence the name *OR* gate. As in the case of the AND gate, an OR gate may have only two or a great many input terminals.

As already explained, a diode AND gate has a low voltage output when one or more of its inputs are low, and a high output when all inputs are high. If a transistor is connected to invert the output of the AND gate, the transistor output is *high* when one or more of the AND gate inputs are low, and *low* when all AND gate inputs are high. Used in this fashion, the inverting stage is termed a *NOT* gate. The combination of the NOT gate and the AND gate is then referred to as a *NOT-AND* gate, or *NAND* gate. The logic symbol employed for a NAND gate is shown in Figure 5-3(a). The symbol is simply that of an AND gate with a small circle at the output to indicate that the output voltage is inverted.

A transistor inverter (or NOT gate) connected at the output of an OR gate produces a zero output when any one of the inputs is high. The circuit is termed a *NOT-OR* gate or *NOR* gate. A NOR gate logic symbol is shown in Figure 5-3(b). As in the case of the NAND gate, a small circle is employed to denote the polarity inversion at the output.

(a) Logic symbol for NAND gate

(b) Logic symbol for NOR gate

Figure 5-3 Logic symbols for *NAND* and *NOR* gates. A *NAND (NOT-AND)* gate is an *AND* gate with its output voltage inverted. A *NOR (NOT-OR)* gate is an *OR* gate with its output voltage inverted.

Example 5-1

The AND gate in Figure 5-1 has $V_{CC} = 5$ V, $R_1 = 1$ kΩ, and silicon diodes with $V_D = 0.7$ V. The *high* output current is 1 mA, and the *low* and *high* input voltages are 0 and 5 V, respectively. Determine the *low* and *high* output voltage levels.

Solution

High output voltage, $\quad\quad V_{OH} = V_{CC} - (I_0 \times R_1) = 5 \text{ V} - (1 \text{ mA} \times 1 \text{ k}\Omega)$

$$= 4 \text{ V}$$

Low output voltage, $\quad\quad V_{OL} = V_{i(\text{low})} + V_D = 0 \text{ V} + 0.7 \text{ V}$

$$= 0.7 \text{ V}$$

5-2 FLIP-FLOPS

The basic circuit of a *flip-flop*, or *bistable multivibrator*, in Figure 5-4(a) has two stable states: either Q_1 is *on* and Q_2 is *off*, or Q_2 is *on* and Q_1 is biased *off*. The circuit is completely symmetrical. Collector resistors R_{C1} and R_{C2} are equal, and bias resistors (R_1, R_2) and (R_1', R_2') constitute similar potential dividers at the transistor bases. Each transistor base is biased from the collector of the other device, so that when any one transistor is *on*, the other one is *off*. The outputs, identified as Q and $\overline{Q}$, are taken from the transistor collector terminals. When the output voltage at Q is *low*, that at $\overline{Q}$ is *high*, and vice versa. Triggering circuits connected to the flip-flop cause it to change state, or *toggle*. A square-wave signal, known as a *clock*, is normally used to toggle flip-flops from one state to the other.

Thelogic symbol for a *triggered*, or *clocked*, *flip-flop* is shown in Figure 5-4(b). A clocked flip-flop changes state each time an appropriate voltage is applied at the flip-flop input. The clock *(CLK)* terminal has a small circle to indicate that negative-going input voltages are required. The absence of the circle would show that positive-going toggling voltages must be used. Most clocked flip-flops also have a *control* input, as shown dashed in Figure 5-4(b). This can be used to ensure that the outputs remain in a particular state (Q *high* and $\overline{Q}$ *low*, or Q *low* and $\overline{Q}$ *high*) after toggling.

Figure 5-4(c) shows the logic symbol for a flip-flop that has *set (S)* and *clear (C)* terminals as well as the toggling input terminal. An input pulse to the S terminal sets the flip-flop into a desired state, while an input to the C terminal *clears* the flip-flop back to its other state. This is known as a *clocked-SC flip-flop*.

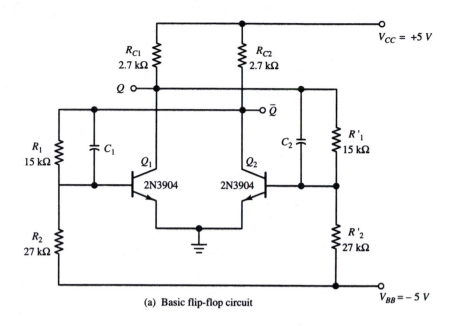

(a) Basic flip-flop circuit

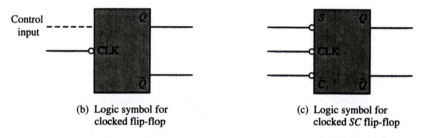

(b) Logic symbol for
clocked flip-flop

(c) Logic symbol for
clocked SC flip-flop

Figure 5-4   Basic transistor flip-flop circuit and logic symbols. When transistor Q_1 is *on*, Q_2 is biased *off*, and vice versa. Thus, if output Q is *low*, output $\bar{Q}$ is high, and vice versa. A *clocked* flip-flop changes state each time an input (clock) pulse is applied. A *clocked* SC flip-flop has *set (S)* and *clear (C)* inputs.

Example 5-2

Using the component values and supply voltage shown, calculate the collector and base voltages for the *off* transistor in the flip-flop in Figure 5-4(a). Assume that V_{BE} is 0.7 V and $V_{CE(SAT)}$ is 0.2 V for the *on* transistor.

Solution

$$\text{With } Q_2 \text{ on,} \quad V_{C2} = V_{CE(sat)} = 0.2 \text{ V}$$

$$\text{and} \quad V_{R1R2} = V_{C2} - V_{BB} = 0.2 \text{ V} - (-5 \text{ V})$$

$$= 5.2 \text{ V}$$

$$V_{R1} = \frac{R_1}{R_1 + R_2} \times V_{R1R2}$$

$$= \frac{15 \text{ k}\Omega}{15 \text{ k}\Omega + 27 \text{ k}\Omega} \times 5.2 \text{ V}$$

$$= 2 \text{ V}$$

$$V_{B1} = V_{C2} - V_{R1} = 0.2 \text{ V} - 2 \text{ V}$$

$$= -1.8 \text{ V}$$

With Q_1 off, $\quad V_{RC1} = \dfrac{R_{C1}}{R_{C1} + R_1' + R_2'} \times (V_{CC} - V_{BB})$

$$= \frac{2.7 \text{ k}\Omega}{2.7 \text{ k}\Omega + 15 \text{ k}\Omega + 27 \text{ k}\Omega} \times [5 \text{ V} - (-5 \text{ V})]$$

$$= 0.6 \text{ V}$$

$$V_{C1} = V_{CC} - V_{RC1} = 5 \text{ V} - 0.6 \text{ V}$$

$$= 4.4 \text{ V}$$

5-3 DIGITAL DISPLAYS

Light-Emitting-Diode Displays

Charge carrier recombination occurs at a forward-biased *pn*-junction as electrons cross from the *n*-side and recombine with *holes* on the *p*-side. When recombination takes place, the charge carriers give up energy in the form of heat and light. If the semiconductor material is translucent, the light is emitted and the junction is a light source, that is, a *light-emitting diode* (LED). When forward biased, the device is *on* and glowing; when reverse biased, it is *off.*

Figure 5-5(a) shows a cross-section view of a typical LED. Charge carrier recombinations take place in the *p*-type material; therefore, the *p*-region becomes the surface of the device. For maximum light emission, a metal film anode is deposited around the edge of the *p*-type material. The cathode connection for the device is a metal film at the bottom of the *n*-type region. Various types of semiconductor material are used to give red, yellow, green, or blue light emission. Infrared LEDs are also available.

Figure 5-5(b) illustrates the arrangement of a *seven-segment* LED numerical display. Passing a current through the appropriate segments allows any numeral from 0 to 9 to be displayed. The LEDs in a seven-segment display either have all of the anodes connected together, *common anode* [see Figure 5-5(c)], or all of the cathodes connected, *common cathode*. The typical voltage drop across a forward-biased LED is 1.2 V, and typical forward current for reasonable brightness is about 20 mA. This relatively large current requirement is a major disadvantage of LED displays. Some advantages of LEDs

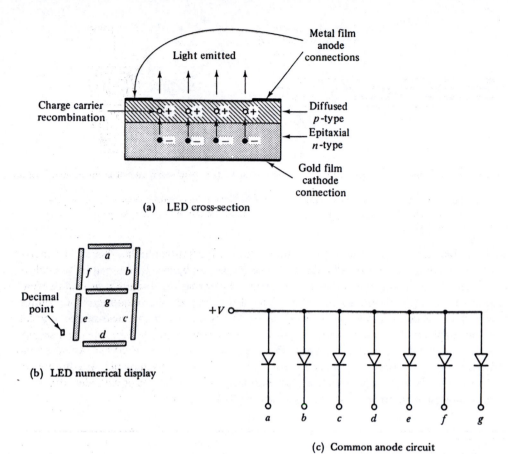

(a) LED cross-section

Metal film anode connections

Light emitted

Charge carrier recombination

Diffused p-type

Epitaxial n-type

Gold film cathode connection

(b) LED numerical display

Decimal point

(c) Common anode circuit

+V

a b c d e f g

Figure 5-5 A *light emitting diode (LED)* produces a bright glow when a suitable level of forward current flows. Selection of *on*-biased segments in a *seven-segment* numerical display allows any numeral from 0 to 9 to be displayed.

over other types of displays are: the ability to operate from a low-voltage dc supply, ruggedness, rapid switching ability, and small physical size.

Liquid Crystal Displays

Liquid crystal cell displays (LCD) are usually arranged in the same seven-segment numerical format as the LED display. The cross section of a *field-effect* liquid crystal cell is illustrated in Figure 5-6(a). The liquid crystal material may be one of several organic compounds that exhibit the optical properties of a crystal. Liquid crystal material is layered between glass sheets with transparent electrodes deposited on the inside faces. Two thin polarizing optical filters are placed at the surface of each glass sheet. The liquid crystal material actually twists the light passing through the cell when the cell is not energized. This allows light to pass through the optical filters, and the cell disappears into the background. When the cell is energized, no twisting of the light occurs and the energized cells in a seven-segment display stand out against their background.

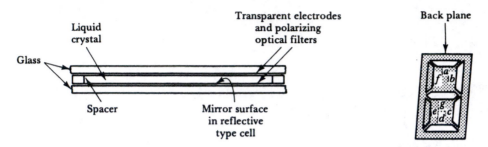

(a) Construction of liquid crystal cell	(b) Liquid crystal cell **seven-segment display**

Figure 5-6 A seven-segment *liquid crystal display (LCD)* usually reflects light from each of the energized cells (or segments) to display a desired numeral. LCD displays use much less energy than LED displays.

Since liquid crystal cells are light reflectors or transmitters rather than light generators, they consume very small amounts of energy. The only energy required by the cell is that needed to activate the liquid crystal. The total current flow through four small seven-segment displays is typically about 300 μA. However, the LCD requires an ac voltage supply, either in the form of a sine wave or a square wave. This is because a direct current produces plating of the cell electrodes, which could damage the device. A typical supply for a LCD is an 8 V peak-to-peak square wave with a frequency of 60 Hz. As in seven-segment LED displays, one terminal of each cell in a liquid crystal display is commoned. In the LCD display the cell terminals cannot be identified as anodes or cathodes; the common terminal is referred to as the *back plane* [see Figure 5-6(b)].

Example 5-3

(a) A $3\frac{1}{2}$-digit seven-segment LED display uses diodes that require a 20 mA forward current. Calculate the total supply current required.

(b) Determine the supply current required for a similar LCD display that uses 300 μA per segment.

Solution

    (a) For the LED display:

    I for each 7-segment display, $I_7 = 7I_F = 7 \times 20 \text{ mA}$
                                                                     $= 140 \text{ mA}$

    I for the $\frac{1}{2}$ (2-segment) display, $I_{1/2} = 2I_F = 2 \times 20 \text{ mA}$
                                                                      $= 40 \text{ mA}$

    Total current for the $3\frac{1}{2}$ digits, $I_T = 3I_7 + I_{1/2} = (3 \times 140 \text{ mA}) + 40 \text{ mA}$
                                                                      $= 460 \text{ mA}$

    (b)   For the LCD display:

    I for each 7-segment display, $I_7 = 7I_F = 7 \times 300 \text{ μA}$

$$= 2.1 \text{ mA}$$

I for the $\frac{1}{2}$ *(2-segment) display,* $I_{1/2} = 2I_F = 2 \times 300 \text{ µA}$

$$= 600 \text{ µA}$$

Total current for the $3\frac{1}{2}$ *digits,* $I_T = 3I_7 + I_{1/2} = (3 \times 2.1 \text{ mA}) + 600 \text{ µA}$

$$= 6.9 \text{ mA}$$

5-4 DIGITAL COUNTING

Scale-of-16 Counter

The logic diagram in Figure 5-7(a) shows four flip-flops (*FF*) connected in cascade. FF_A is toggled by negative-going input pulses, and each time the Q output of FF_A switches *low,* it applies a negative-going voltage to the clock input of FF_B. Thus, FF_A toggles FF_B, FF_B toggles FF_C, and FF_C toggles FF_D. The various combinations of flip-flop states that can exist in the four-stage circuit are shown by the table in Figure 5-7(b).

The state of the four flip-flops is indicated by the binary number system, in which *0* represents a *low* voltage (close to ground level), and *1* represents a *high* level (near the supply voltage). If all Q outputs are *low* before any input pulses are applied, the binary count is *0000.* See the first horizontal column in the table.

The first input pulse causes Q_A to switch to *high,* or logic level *1.* The positive-going output at Q_A has no effect on FF_B, so the count at this time is *0001.* The second input pulse changes Q_A to *low,* (logic level *0*). This produces a negative-going voltage (from Q_A) which toggles FF_B to give a *high* level at Q_B, and a count of *0010.* The third input pulse changes the state of FF_A once again but does not affect FF_B. The count is now *0011.* The fourth pulse changes Q_A to *low,* toggling FF_B to Q_B *low,* which toggles FF_C, giving a *high* level at Q_C and a count of *0100.*

The toggling process continues in this way with each input pulse until the maximum count of *1111* is reached when 15 input pulses have been applied. The 16th pulse switches Q_A to *low* once more, toggling Q_B to *low:* Q_B output toggles Q_C to *low;* and Q_C output toggles Q_D to *low.* Thus, the four flip-flops are returned to their original states, and the binary count returns to *0000.* It is seen that the cascade of four flip-flops has 16 different states. Therefore, the circuit is termed a *scale-of-16* counter, or a *modulus-16* counter.

Decade Counter

A *scale-of-10 counter,* or *decade counter,* is produced by modifying the scale-of-16 circuit. Six of the 16 states must be eliminated, to leave only 10 possible states. In the method shown in Figure 5-8(a), clocked SC flip-flops are used together with a NAND gate. The NAND gate produces a *low* output when Q_B and Q_D outputs are *high.* As reference to the table in Figure 5-8(b) reveals, this occurs only on the 10th input pulse. The NAND gate output is normally high, so it does not affect the toggling of the flip-flops until the 10th input pulse arrives. The *low* output from the NAND gate *clears* all four flip-flops to the starting state of *0000* once again. Thus, the last six states of the scale-of-16 are eliminated to produce the decade counter.

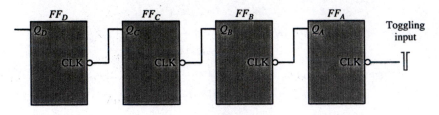

(a) Scale-of-16 counter

Flip-flop outputs

Q_D	Q_C	Q_B	Q_A	Input pulses
0	0	0	0	0
0	0	0	1	1
0	0	1	0	2
0	0	1	1	3
0	1	0	0	4
0	1	0	1	5
0	1	1	0	6
0	1	1	1	7
1	0	0	0	8
1	0	0	1	9
1	0	1	0	10
1	0	1	1	11
1	1	0	0	12
1	1	0	1	13
1	1	1	0	14
1	1	1	1	15
0	0	0	0	16

(b) States of the flip-flop Q outputs

Figure 5-7 *Scale-of-16 counter* constructed of four flip-flops connected in cascade. FF_A is toggled by the input pulses. Each of the other flip-flops changes state only when the Q output of the previous flip-flop goes *low*.

Scale-of-2000 Counter

The output condition of the flip-flops in a decade counter is in a form known as *binary-coded decimal (BDC)*. This must be converted into a form that will drive a seven-segment display. Arrangements of logic gates for this purpose, known as *BCD-to-seven-segment drivers,* are available in integrated circuit form.

One decade counter together with a digital display and the necessary conversion circuitry can be employed to count from 0 to 9. Each time the tenth input pulse is applied, the display goes from 9 to 0 again. When this occurs, the output of the final flip-flop in the decade counter goes from 1 to 0 [see Figure 5-8(b)]. This is the only time that the decade counter produces a negative-going output, and this output can be used to toggle another decade counter.

Consider the block diagram of the *scale-of-2000* counter shown in Figure 5-9. The system consists of three complete decade counters, BCD-to-seven-segment drivers and

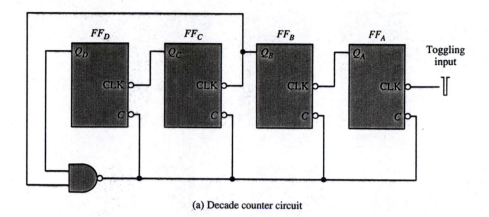

(a) Decade counter circuit

Flip-flop outputs				Input pulses
Q_D	Q_C	Q_B	Q_A	
0	0	0	0	0
0	0	0	1	1
0	0	1	0	2
0	0	1	1	3
0	1	0	0	4
0	1	0	1	5
0	1	1	0	6
0	1	1	1	7
1	0	0	0	8
1	0	0	1	9
1	0	1	0	10
0	0	0	0	0

(b) States of the flip-flop Q outputs

Figure 5-8 A *decade counter* is a modified scale-of-16 counter. When Q_B and Q_D each have a *high* output, the NAND gate produces a *low* to clear all Q outputs to 0. Thus, the last six states of the scale-of-16 are eliminated to give only 10 possible states.

displays, and one flip-flop controlling a display which indicates only numeral 1 when *on*. Starting from *0*, all three counters are set at their normal starting conditions, and the numeral 1 indicator is *off*. This gives an indication of 000. The first nine input pulses register only on the first (right-hand-side) seven-segment display. On the 10th input pulse, the first display goes to 0, and a negative-going pulse output from the first decade counter triggers the second decade counter. The display of the second counter now registers 1, so that the complete display reads 010. The counter has counted to 10, and has also registered 10 on the display system.

The next nine input pulses cause the first counter to go from 0 to 9 again, so that the display reads 019 on the 19th pulse. The 20th pulse causes the first display to go to 0 once again. At this time, the first decade counter puts out another negative pulse, which again triggers the second decade counter. The total display now reads 020, which indicates the fact that 20 pulses have been applied to the input of the first decade

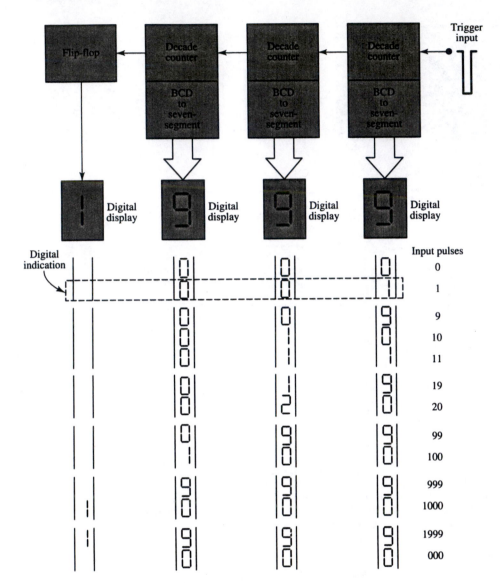

Figure 5-9 *Scale-of-2000* counter using three decade counters, one flip-flop, BCD-to-seven-segment drivers, and a $3\frac{1}{2}$-digit numerical display. The first (right-hand) stage is toggled by the input pulses, and each successive stage is toggled from the previous stage. Maximum count is 1999.

counter. It is seen that the second decade counter and display is counting *tens* of input pulses.

Counting continues as described until the display indicates 099 after the 99th input pulse. The 100th input pulse causes the first two displays (from the right) to go to 0. The second decade counter emits a negative pulse at this time, which triggers the third decade counter. Therefore, the count reads 100. On the 1000th pulse, the first three decade counters go from 999 to 000, and the negative pulse emitted from the third decade counter triggers the flip-flop and turns on the 1 display. The display now reads 1000. It is seen that the system shown in Figure 5-9 can count to a maximum of 1999. One more pulse causes the display to return to its initial 000 condition. The three components of the display that indicate up to 999 are referred to as a *three-digit display*. With the additional 1 component included, the complete numerical display is termed a $3\frac{1}{2}$-*digit display*.

Digital Frequency Division

Many digital measuring techniques require an accurate time period, or *time base*. Figure 5-10 shows the block diagram and waveforms for a digitally produced time base. The 1 MHz crystal-controlled oscillator (or *clock*) provides an extremely accurate output frequency (with a time period of 1 μs). The output frequency from the final flip-flop of a decade counter is exact one-tenth of the input (toggling) frequency, which means that the output time period is 10 times the time period of the input waveform. Thus, the output time period from decade counter 1 in Figure 5-10 is precisely 10 μs. Similarly, the time periods of the output waveforms generated by decade counters 2 and 3 are 100 μs and 1 ms, respectively. Additional decade counters may be included to further multiply the time period. If scale-of-16 or scale-of-12 counters are used instead of decade counters, the

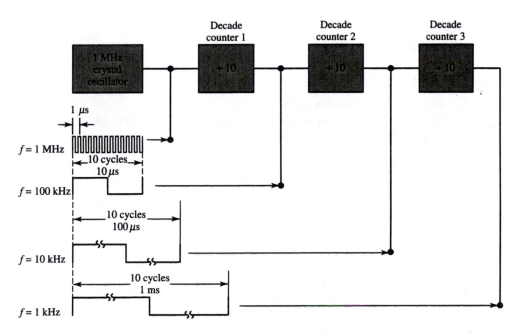

Figure 5-10 Digital frequency divider. The output frequency from each decade counter is exactly one-tenth of the input frequency, and its time period is 10 times the input time period.

oscillator time period is multiplied by 16 or 12 or whatever modulus number is involved. The equation for the output time period is

$$T = T_o \times N^n \qquad (5\text{-}1)$$

where T_o is the oscillator time period, N is the modulus number of the counters, and n is the number of counters.

Example 5-4

Determine the final output frequency from the frequency-divider system in Figure 5-10 if the three decade counters are replaced by scale-of-16 counters.

Solution

Equation 5-1, $\qquad T = T_o \times N^n = 1\ \mu s \times 16^3$

$$= 4.1\ ms$$

$$f = \frac{1}{T} = \frac{1}{4\ ms}$$

$$= 244\ Hz$$

5-5 ANALOG-TO-DIGITAL CONVERTER

Ramp-Type Analog-to-Digital Converter

The *analog-to-digital converter* (ADC) circuit in Figure 5-11(a) is known as a *ramp-type ADC*, because it uses a ramp generator to produce a time period that is proportional to the analog input voltage. The ramp voltage starts from ground level and increases at a constant rate, as illustrated in Figure 5-11(b). The ramp is fed to one input of a *voltage comparator*, and the analog input voltage (V_i) is applied to the other comparator input terminal. During the time that the ramp voltage (V_r) is below the level of V_i, the comparator output is *high;* and this allows pulses from the clock generator to pass through the AND gate to the counting circuits (*register*). When V_r becomes exactly equal to V_i, the comparator output switches to a *low* level, thus stopping further clock pulses from toggling the counting circuits. The time period (t_1) of the comparator *high* output is directly proportional to input voltage V_i. Thus, because the counting circuits are toggled only during t_1, the count is the digital equivalent of the analog input.

The negative-going voltage step at the end of the ramp resets the register to its zero condition before the cycle of ramp generation and counting recommences. If the counting circuits are a simple cascade of (four or more) flip-flops, as in Figure 5-7, the digital output is in binary form. If a cascade of decade counters is employed, as discussed for the counter in Figure 5-9, the output is in binary-coded decimal (BCD).

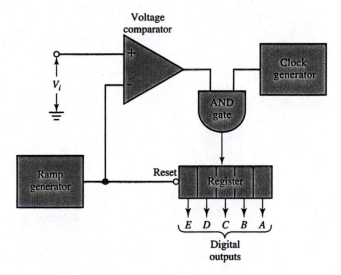

(a) *ADC* system

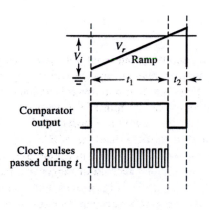

(b) System waveforms

Figure 5-11 Ramp-type *analog-to-digital converter (ADC)*. The voltage comparator detects equality between the ramp and the input voltage, thus producing a time period (t_1) that is proportional to V_i. The clock generator toggles the counting circuits for time t_1 to produce the digital equivalent of V_i.

Example 5-5

The ADC system in Figure 5-11 uses a 1 MHz clock generator and a ramp voltage that increases from 0 V to 1.25 V in a time of 125 ms. Determine the number of clock pulses counted into the register when $V_i = 0.9$ V, and when it is 0.75 V.

Solution

$$\text{For } V_r = 1.25 \text{ V,} \qquad t_r = 125 \text{ ms}$$

So for $V_i = 0.9$ V,	$t_1 = \dfrac{t_r}{V_r} \times V_i = \dfrac{125 \text{ ms}}{1.25 \text{ V}} \times 0.9 \text{ V}$
	$= 90 \text{ ms}$
For the clock pulses,	$T = \dfrac{1}{f} = \dfrac{1}{1 \text{ MHz}}$
	$= 1 \text{ }\mu\text{s}$
Pulses counted,	$N = \dfrac{t_1}{T} = \dfrac{90 \text{ ms}}{1 \text{ }\mu\text{s}}$
	$= 900$
For $V_i = 0.75$ V,	$t_1 = \dfrac{t_r}{V_r} \times V_i = \dfrac{125 \text{ ms}}{1.25 \text{ V}} \times 0.75 \text{ V}$
	$= 75 \text{ ms}$
Pulses counted,	$N = \dfrac{t_1}{T} = \dfrac{75 \text{ ms}}{1 \text{ }\mu\text{s}}$
	$= 750$

Quantizing Error

In the process of conversion from analog to digital, the input voltage is *quantized* into a number of discrete levels. For example, suppose that an input voltage which ranges from zero up to a maximum of 10 V is converted into a digital output which is an integer from 0 to 10. Clearly, the output cannot accurately represent an input of 9.1 V, or 9.9 V. Both are likely to produce a 9 output, giving an error that can be a maximum of 1 in 10, or 10%. This is known as the conversion error, or *quantizing error* of the ADC.

Now refer to the table of outputs for the scale-of-16 counter in Figure 5-7, and recall that the counter has a *4-bit* output. As illustrated, the 16th count represents the zero condition; therefore, when used as the register in the ADC in Figure 5-11, the 4-bit output represents only 15 levels of input. For any ADC, the number of quantized levels of input is given by

$$\boxed{N = 2^n - 1} \tag{5-2}$$

where n is the number of bits.

For an ADC with a 4-bit output, the quantizing error is 1 in 15, or 6.7%. If the output is a *5-bit* number, the number of input levels represented by the output is 31. In this case, the quantizing error is 1 in 31, or approximately 3%. Obviously, the greatest number of bits in the digital output gives the smallest quantizing error. The quantizing error is not the only error in an ADC; component errors can also be important, especially as the bit number (and number of quantized levels) increases.

Digital Instrument Basics   Chap. 5

Example 5-6

Determine the number of output bits required for an ADC to give a quantizing error less than 1%.

Solution   *For 1% quantizing error, count ≥ 100.*

        Equation 5-2, $N = 2^n - 1$

        For $n = 6$, $N = 2^6 - 1 = 63$

        For $n = 7$, $N = 2^7 - 1 = 127$

        For less than 1% error, use $n = 7$.

5-6 DIGITAL-TO-ANALOG CONVERTER

R/2R Network

The simplest kind of *digital-to-analog converter* (DAC) consists of an *R/2R* network, as illustrated in Figure 5-12(a). Since there are four digital input terminals, the circuit is known as a *4-bit* DAC. Resistors R_3, R_5, and R_7 each have a resistance R, as shown. (The resistance R might be anywhere in the range from 1 kΩ to 100 kΩ, depending on the load current to be supplied at the circuit output.) Resistors R_1, R_2, R_4, R_6, and R_8 each have a resistance of *2R*. Each digital input voltage is either zero (ground level) or V_i, which might typically be 5 V. The analog output voltage depends on the presence or absence of each digital input bit.

To understand how the *R/2R* circuit converts digital inputs into an analog voltage, consider the situation when only input A is present, and all other inputs are at zero. This is illustrated in Figure 5-12(b). Input voltage V_i is potentially divided by resistors R_1 and R_2 to give an *open-circuit* voltage of $V_i/2$ at the junction of R_1 and R_2 (i.e., for R_3 disconnected from R_1 and R_2). Looking back into the junction of R_1 and R_2, it is seen that the output resistance at this point, is $R_1 \| R_2 = R$. Since $(R_3 + R) = 2R$, $V_i/2$ is further divided by R_4 and $(R_3 + R)$ to give an open-circuit voltage of $V_i/4$ at the junction of R_3 and R_4 [Figure 5-12(c)]. Now note that the source resistance at the junction of R_3 and R_4 is $R_4 \| (R_3 + R) = R$. Consequently, $V_i/4$ is further divided by $(R_5 + R)$ and R_6 to produce an *open-circuit* voltage of $V_i/8$ at the R_5/R_6 junction [Figure 5-12(d)]. Once again, the potential divider source resistance is R, at the junction of R_5 and R_6, so $(R_7 + R)$ combines with R_8 to divide $V_i/8$ down to $V_i/16$ at the output [Figure 5-12(e)].

By similar reasoning it can be shown that an output voltage of $V_i/8$ is produced when only input B is present. Also, $V_o = V_i/4$ when only input C is present, and $V_o = V_i/2$ when there is an input only at terminal D.

The table in Figure 5-12(f) shows the analog output voltage levels for all combinations of digital inputs. The output voltage for any given input combination can easily be

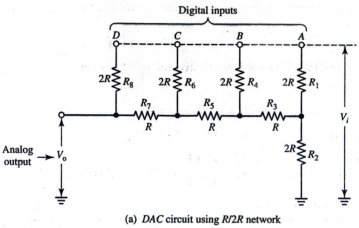

Digital inputs

(a) DAC circuit using R/2R network

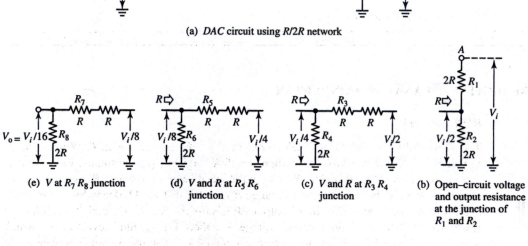

(e) V at $R_7 R_8$ junction

(d) V and R at $R_5 R_6$ junction

(c) V and R at $R_3 R_4$ junction

(b) Open–circuit voltage and output resistance at the junction of R_1 and R_2

Digital inputs (V_i at each input terminal)				Analog output
D	C	B	A	
0	0	0	0	0
0	0	0	1	$V_i/16$
0	0	1	0	$2(V_i/16)$
0	0	1	1	$3(V_i/16)$
0	1	0	0	$4(V_i/16)$
0	1	0	1	$5(V_i/16)$
0	1	1	0	$6(V_i/16)$
0	1	1	1	$7(V_i/16)$
1	0	0	0	$8(V_i/16)$
1	0	0	1	$9(V_i/16)$
1	0	1	0	$10(V_i/16)$
1	0	1	1	$11(V_i/16)$
1	1	0	0	$12(V_i/16)$
1	1	0	1	$13(V_i/16)$
1	1	1	0	$14(V_i/16)$
1	1	1	1	$15(V_i/16)$

(f) Digital inputs and analog outputs

Figure 5-12 *Digital-to-analog converter (DAC)* using an *R/2R* network. With an input only at terminal A, the output is $V_i/16$. An input only at B gives an output of $2V_i/16$. A and B inputs together with no others present give $3V_i/16$.

determined by assigning the appropriate (analog equivalent) numeral to each input according to its level of importance. Thus, *bit A* [known as the *least significant bit (LSB)*] is assigned numeral 1, while *bit B,* which is twice as important as *A* [see Figure 5-12(f)], is represented by numeral 2. Similarly, *bit C* (twice as important as *B*) is assigned 4, and *bit D* [the *most significant bit (MSB)*] is given 8. The output voltage is

$$V_o = (D + C + B + A)\frac{V_i}{16}$$ (5-3)

where either zero or the appropriate numerals are substituted according to the digital input.

Example 5-7

Calculate the output voltage from the DAC in Figure 5-12 when the digital input is 1010 and $V_i = 10$ V.

Solution

Equation 5-3, $\qquad V_o = (D + C + B + A)\dfrac{V_i}{16}$

$$= (8 + 0 + 2 + 0)\frac{V_i}{16}$$

$$= 10\frac{V_i}{16} \qquad \text{[see Figure 5-12(f)]}$$

$$= 10 \times \frac{10\text{ V}}{16}$$

$$= 6.25\text{ V}$$

Resolution

Consideration of the DAC in Figure 5-12 shows that the maximum output is $15(V_i/16)$, when all input bits are present. Also, the smallest output voltage change $(V_i/16)$ occurs when *bit A* changes between *0* and *1*. When the input is gradually increased from zero to maximum, the output increases in steps of $V_i/16$. Thus, the number of output steps is 15, and the *resolution* of the output is 1 in 15. This is the same as the quantizing effect when converting from analog to digital. The resolution can be determined from the number of quantized levels given by Equation 5-2.

REVIEW QUESTIONS

5-1 Sketch the circuit and logic symbol for a four-input diode AND gate. Explain the circuit operation.

5-2 Sketch the circuit and logic symbol for a two-input diode OR gate. Explain the circuit operation.

5-3 Sketch the logic symbol for three-input NAND and NOR gates. Explain the NAND and NOR functions.

5-4 Draw the circuit diagram of a basic two-transistor flip-flop. Explain the conditions that keep one transistor *on* and the other one *off*.

5-5 Draw the logic symbols for T and RST flip-flops. Explain the function of each type of flip-flop.

5-6 Sketch the cross section of a light-emitting diode, and explain its operation. Sketch an LED seven-segment display. Explain common-cathode and common-anode LED displays.

5-7 Sketch the cross section of a liquid crystal cell, and explain its operation. Sketch an LCD seven-segment display. Compare the supply current requirements for LED and LCD displays.

5-8 Draw the logic diagram for a scale-of-16 counter, and explain its operation. Prepare a table showing the counter output states for each input pulse.

5-9 Draw the logic diagram for a decade counter, and explain its operation. Prepare a table showing the counter output states for each input pulse.

5-10 Draw the complete block diagram for a scale-of-2000 counter, and explain how it operates. Prepare a partial table showing the counter output states for various input pulses.

5-11 Draw a block diagram to show how decade counters may be used for frequency division. Show the system waveforms, and explain its operation.

5-12 Draw the logic diagram for a ramp ADC. Show the system waveforms, and explain how conversion is achieved.

5-13 Explain the quantizing error that occurs with an ADC, and show how it affects the accuracy of the output.

5-14 Sketch the circuit of a 4-bit DAC using an $R/2R$ network. Explain how the circuit operates, and discuss the resolution of DAC outputs.

5-15 Sketch the circuit of an 8-bit DAC using an $R/2R$ network. Explain briefly.

PROBLEMS

5-1 The AND gate in Figure 5-1 has $V_{CC} = 9$ V, $R_1 = 1.5$ kΩ, and diodes with $V_D = 0.2$ V. The *low* and *high* input voltages are 0 and 9 V, respectively, and an 8.2 kΩ resistor is connected at the output terminals. Determine the *low* and *high* output voltage levels.

5-2 A four-input diode AND gate with a 5 V supply has a 6.8 kΩ resistor connected across its output terminals. Determine a suitable resistance for R_1 (as in Figure 5-1) if the *high* output voltage is to be minimum of 3.5 V. Also, calculate the individual diode forward currents when all input are *low*.

5-3 Determine the *low* and *high* output voltage levels for the OR gate in Figure 5-2 if it uses the components and input voltage levels in Example 5-1.

5-4 Each diode in the *OR* gate in Figure 5-2 is to have a minimum forward current of 1 mA. Determine a suitable resistance for R_1 if the *high* input voltage level ranges from 4 V to 5V.

5-5 If the supply voltage for the flip-flop in Figure 5-4(a) is changed to ±9 V, calculate the collector and base voltages for the *off* transistors. Take V_{BE} as 0.7 V and $V_{CE(SAT)}$ as 0.2 V.

5-6 Determine the collector and base voltages for a flip-flop as in Figure 5-4(a) if $R_{C1} = R_{C2} = 4.7$ kΩ, $R_1 = R_1' = 27$ Ω, $R_2 = R_2' = 40$ kΩ, and $V_{CC} = ±9$ V.

5-7 A $4\frac{1}{2}$-digit seven-segment LED display draws a maximum supply current of 450 mA. Calculate the current taken by each segment. Also, determine the maximum supply current required for a similar LCD display that uses 350 μA per segment.

5-8 A frequency-divider circuit is made up of a 2 MHz oscillator, a divide-by-16 counter, and two decade counters. Determine the time period of the outputs from each counter.

5-9 A time period of 5.12 ms is to be produced at the output of a digital frequency divider that uses one scale-of-16 counter, one scale-of-12 counter, and one decade counter. Determine the oscillator frequency required. Also calculate the other available time periods.

5-10 The ramp ADC in Figure 5-11 has a 1.5 MHz clock frequency and a ramp voltage which rises linearly from 0 to 2 V in 1.333 ms. Determine the number of clock pulses counted into the register when the input voltage is 1.2 V.

5-11 A ramp ADC with a 2 MHz clock generator is to have 1000 clock pulses representing a 1 V input. Specify the required ramp voltage.

5-12 Determine the minimum number of output bits required for an ADC to give a quantizing error less than 0.5%.

5-13 Determine the quantizing error for an ADC with a 16-bit output.

5-14 Calculate the analog output voltages from the DAC in Figure 5-12 for digital inputs of 0111 and 1110 when $V_i = 5$ V.

5-15 Determine the digital inputs for the DAC in Figure 5-12 to give analog outputs of 7.5 V and 1.25 V if $V_i = 10$ V.

5-16 Determine the resolution of the output from a DAC that has a 12-bit input.

Digital Voltmeters and Frequency Meters

6

Objectives

You will be able to:

1. Sketch the block diagram of a digital voltmeter (DVM) using an analog-to-digital converter. Show the system waveforms, and explain how a DVM operates.

2. Sketch the block diagram and system waveforms for a DVM using a dual-slope integrator. Explain its operation, and discuss the advantages of the dual-slope system.

3. Discuss the terminals and controls on the front panels of typical digital multimeters (DMMs). Discuss the applications and performance of DMMs.

4. Draw the basic block diagram of a digital frequency meter, sketch the system waveforms, and explain the system operation.

5. Discuss the errors that occur in digital frequency meters, and explain the method used for specifying measurement accuracy.

6. Explain how time period, pulse width, and frequency ratio may be measured on a digital frequency meter.

Introduction

Digital voltmeters are essentially analog-to-digital converters with digital displays to indicate the measured voltage. Digital multimeters are electronic volt-ohm-milliameters with digital displays.

If a pulse waveform is used to toggle a digital counter for a time period of exactly 1 second, the counter registers the number of pulses that occur per second; that is, it registers the frequency of the input waveform. If the count is 1000, the frequency is 1000 pulses per second. A digital frequency meter is a digital counter combined with an accurate timing system. Conversely, an accurate frequency source combined with a digital counter can be used for time measurements. If a 1 MHz frequency (time period of 1 μs) is used to toggle the counting circuits for the time to be measured, the counter registers the time directly in microseconds.

6-1 DIGITAL VOLTMETER SYSTEMS

Ramp-Type Digital Voltmeters

A digital voltmeter (DVM) essentially consists of an analog-to-digital converter, a set of seven-segment numerical displays, and the necessary BCD-to-seven-segment drivers. Figure 6-1 illustrates the use of a ramp-type ADC (see Section 5-5) as a digital voltmeter.

The counting circuitry, BCD-to-seven-segment decoder/drivers, and digital readouts in Figure 6-1 constitute a scale-of-2000 counter as in Figure 5-9, with the exception that a *latch* is included in the system. If the latch were not present, the digital readouts would be reset to zero at the commencement of the counting time (t_1), change rapidly as the count progresses throughout t_1, and remain constant for t_2. Thus, the display would be virtually unreadable. The latch circuits isolate the display from the counting circuits during the time that counting is in progress. The positive-going edge of the comparator output waveform at the end of the time t_1 briefly triggers the latch to connect the decoder/drivers to the counting circuits when counting has ceased. This corrects (or updates) the display, if necessary, and otherwise allows it to remain constant and readable. Latch circuits are basically special kinds of flip-flops.

A *display/enable* control, which open-circuits the supply voltage to the display devices, is sometimes used instead of a latch. The display is simply switched *off* during the counting time, and *on* when counting ceases. When the display time and counting time are brief enough, the *on/off* frequency of the display is so high that the eye sees only a constant display.

Example 6-1

Calculate the maximum time t_1 for the digital voltmeter in Figure 6-1 if the clock generator frequency is 1.5 MHz. Also, suggest a suitable frequency for the ramp generator.

Solution

$$\text{Maximum pulses counted,} \quad N = 1999$$

$$\text{clock time period} = \frac{1}{f}$$

$$t_1 = N \times \frac{1}{f} = 1999 \times \frac{1}{1.5 \text{ MHz}}$$

$$= 1.33 \text{ ms}$$

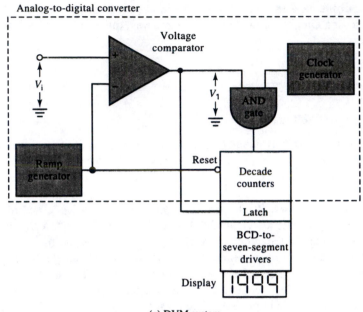

(a) DVM system

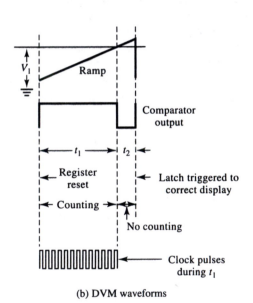

(b) DVM waveforms

Figure 6-1 A ramp-type digital voltmeter (DVM) uses a (ramp-type) ADC with decade counters. BCD-to-seven-segment drivers together with a numerical display provides the measured voltage indication.

Select

$$t_2 \approx 0.25 t_1 = 0.25 \times 1.33 \text{ ms}$$

$$\approx 0.33 \text{ ms}$$

$$t_1 + t_2 = 1.33 \text{ ms} + 0.33 \text{ ms}$$

$$= 1.66 \text{ ms}$$

Ramp generator frequency,

$$f_R = \frac{1}{t_1 + t_2} = \frac{1}{1.66 \text{ ms}}$$

$$\approx 600 \text{ Hz}$$

Dual-Slope-Integrator DVM

For accurate voltage measurements, ramp-type DVMs require precise ramp voltages and precise time periods, both of which can be difficult to maintain. The *dual-slope-integrator DVM* virtually eliminates these requirements by using a special type of ramp generator circuit (or integrator). The integrator capacitor is first charged from the analog input voltage, and then discharged at a constant rate to give a time period that is measured digitally. The charge and discharge result in two (voltage versus time) slopes, which gives the circuit its name *dual-slope*.

Consider the block diagram and waveforms for the dual-slope-integrator DVM illustrated in Figure 6-2. The *control* waveform for the integrator is derived from the clock generator by use of a frequency divider (see Section 5-4). During time t_1, the integrator capacitor is charged negatively from V_i, giving a negative-going ramp, as illustrated. This produces a voltage (V_o) that is directly proportional to V_i. The constant-current source is then switched into the circuit to discharge the capacitor, thus producing a positive-going ramp voltage.

The *zero-crossing detector* is a voltage comparator that gives a *high* (positive) output when the integrator output waveform is negative, and a low output at the end of the positive-going ramp (when the ramp voltage crosses the zero level). The AND gate has *high* inputs from both the zero-crossing detector and the control waveform only during the positive-going ramp time, that is, during time t_2. Pulses from the clock generator pass through the AND gate to the counting circuits during this time. The counting circuits are reset to zero by the positive-going edge of the control wave (see dashed line) at the commencement of t_2, so the output of the counting circuits is a digital measurement of time t_2. Since t_2 is directly proportion to V_o, and V_o is directly proportional to V_i, the output is a digital measurement of the analog input.

It is obvious that the dual-slope-integrator DVM must have a reliable constant-current source. However, the voltmeter is quite easily calibrated by applying an accurately known input voltage, and adjusting the constant current source to give an output reading that equals the applied voltage. If the clock generator frequency drifts, it increases or decreases times t_1 and t_2 in proportion, and does not affect the accuracy of measurement.

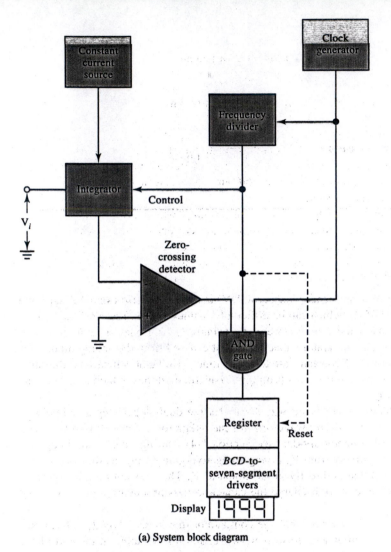

(a) System block diagram

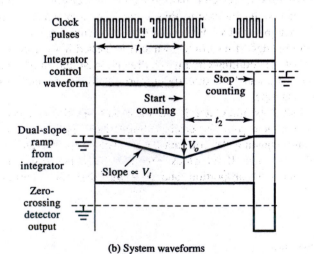

(b) System waveforms

Figure 6-2 A dual-slope integrator DVM charges a capacitor to a voltage (V_o) proportional to V_i, then discharges it at a constant rate. The discharge time (t_2) is measured digitally. The system is largely independent of variations in the clock frequency, and is easily calibrated.

Range Changing

Figure 6-3 shows a switching arrangement for DVM range selection. The attenuator functions exactly as explained for an analog electronic voltmeter (Section 4-1). An input of 1.999 V or less is applied directly to the comparator of the DVM circuit (Figure 6-1). At this time the *ganged* rotary switch selects the decimal point after the first numeral in the display. The decimal point may be a LED or LCD that has one terminal energized, so that it switches *on* when the other terminal is grounded, as illustrated. When the voltage range is switched to 19.99 V, the decimal point after the second numeral is selected. When the range is 199.9 V, the third decimal point is energized.

Many DVMs have automatic ranging circuits *(auto-ranging)* which evaluate the input voltage, select the appropriate range, and illuminate the correct decimal point. When the range is manually selected, greatest accuracy is achieved by using the range that gives the largest possible numerical display. This is the same as adjusting the range of an analog instrument for the greatest on-scale pointer deflection. For example, a measurement of 1.533 V is obviously more accurate than a reading of 01.5 V. When the selected range is too low, the DVM display usually *blanks* or flashes *on* and *off* continuously.

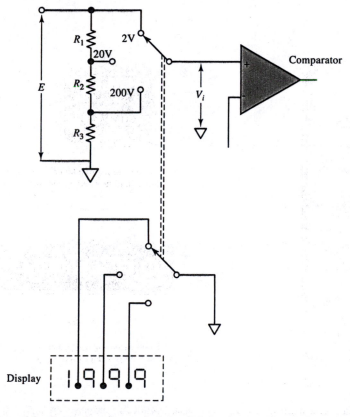

Figure 6-3 Range-changing method for digital voltmeter. The decimal point is switched at the same time as the voltage range.

6-2 DIGITAL MULTIMETERS

Basic Hand-Held Digital Multimeter

The hand-held (or pocket-type) digital multimeter (DMM) illustrated in Figure 6-4(a) has a single function switch with the following selections: OFF, V *(ac volts)*, V *(dc volts)*, Ω *(resistance)*, A *(ac milliamperes)*, A *(dc milliamperes)*. Three connecting terminals are provided, identified as *COM (common)*, *V/Ω (volts/ohms)*, and *A (current)*. The V/Ω and *COM* terminals are employed for both voltage and resistance measurements (the appropriate function selection—*V ac, V dc, or Ω*—determines the terminal use), and the *A* and *COM* terminals are used when measuring current.

Current and resistance measurement methods used in digital multimeters are similar to those employed in analog electronic instruments (see Sections 4-3 and 4-5). In each case, the quantity is converted into a voltage, and the voltage is measured by the digital voltmeter circuitry.

Digital multimeters normally have auto-ranging, so that range selection is not required. The appropriate *annunciator* appears alongside the 3½-digit display, as shown, to indicate the units measured: *VAC, VDC*, and so on. A minus sign also appears in front

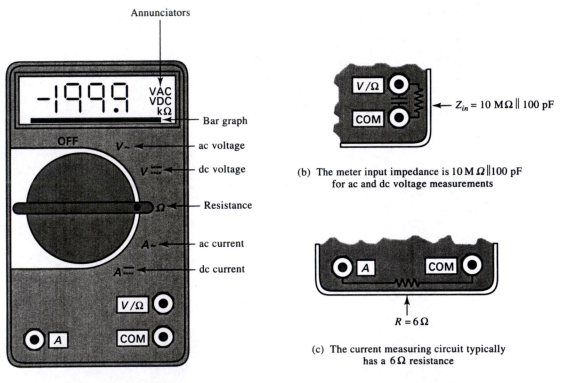

(a) Basic digital multimeter

Figure 6-4 Basic hand-held, or pocket-type digital multimeter (DMM). This instrument has voltage, current, and resistance ranges. Annunciators alongside the 3½-digit display indicate the type of measurement being made. Typical frequency range for ac measurements is 40 Hz to 500 Hz.

Digital Voltmeters and Frequency Meters   Chap. 6

of the numerical display, when necessary, to indicate a negative measured quantity (e.g., when the V/Ω terminal is connected to a battery negative, and the *COM* terminal is connected to the positive).

As already discussed for other multifunction instruments, it is very important to use the correct terminals and to set the function switch correctly for the meter application desired. Incorrect connection and function selection may result in damage to the circuit being tested and damage to the instrument.

The following is a typical specification for the basic DMM illustrated in Figure 6-4:

Range of measurement

V ac	0 to 1.999/19.99/199.9/750 V
V dc	0 to 1.999/19.99/199.9/1000 V
Ω	0 to 199.9 Ω, 0 to 1.999/19.99/199.9 kΩ, 0 to 1.999/19.99 MΩ
A ac	0 to 1.999/19.99/199.9/1999 mA
A dc	0 to 1.999/19.99/199.9/1999 mA

Frequency range for ac measurements	40 Hz to 500 Hz
Burden voltage for current measurements	6 mV/mA
Input impedance	10 MΩ‖100 pF

It should be noted that 40 Hz to 500 Hz is the typical useful frequency range for ac measurements on low-cost DMMs. The instrument should not be used for electronic circuitry measurements involving frequencies outside the specified range. The meter input impedance is typically 10 MΩ in parallel with 100 pF, as illustrated in Figure 6-4(b). These quantities may be important when making high-impedance circuit measurements.

The *burden voltage* is a method of stating the instrument terminal voltage drop when used as an ammeter [Figure 6-4(c)]. A burden voltage of 6 mV/mA represents an ammeter resistance of 6 Ω. For a 6 mV/mA burden voltage, the terminal voltage drop produced by a 200 mA measured current is

$$V_T = 200 \text{ mA} \times 6 \text{ mV/mA}$$

$$= 1.2 \text{ V}$$

Since this voltage drop is inserted into the circuit under test, its effect on the circuit must be considered. Most DMMs have the current-measuring circuits fused for protection against excessive current flow.

Accuracy

The accuracy of a DMM depends on the type of measurement being made; however, basic dc accuracy is usually ±0.7% of the reading, or better. Digital instrument accuracy is usually stated as ±(0.5% rdg + 1d), or simply as (0.5 + 1). This means ±(0.5% of the reading + 1 digit), where the 1 digit refers to the extreme right (or least significant) numeral of the display. For an accuracy of (0.5 + 1), the maximum error in a 1.800 V reading would be

$$\text{error} = \pm[(0.5\% \text{ of } 1.8 \text{ V}) + 0.001 \text{ V}]$$
$$= \pm 0.01 \text{ V}$$

or
$$\text{error} \approx \pm 0.56\% \text{ of the reading}$$

High-Performance Hand-Held DMMs

Many hand-held DMMs have more functions and considerably better performance than the basic multimeter discussed above. One such instrument, the Fluke 25, is shown in Figure 6-5. The Fluke 25 has a 4-digit display with dc and ac voltage ranges of *320.0 mV, 3.200 V, 32.00 V, 320.0 V, and 1000 V.* Current ranges (dc and ac) are *320.0 μA, 3.200 mA, 32.00 mA, 320 mA, and 10 A.* The resolution is the least significant digit in each range: 100 μV for the lowest voltage range, 1 V for the highest. The accuracy is specified as (0.1% + 1) for all of the dc voltage ranges. Alternating voltages may be measured over a frequency range of 40 Hz to 30 KHz, with a measurement accuracy of (0.5% + 3) at low frequencies, and (4% + 10) at the high frequency end. The burden voltage is 0.5 V/μA when measuring up to 3.2 mA, 5.6 mV/mA for the 32 and 320 mA ranges, and 50 mV/A on the 10 A range. A *bar graph* is included as part of the display to give an analog indication of the quantity being measured.

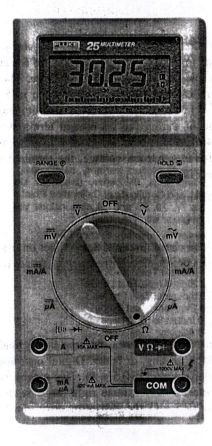

Figure 6-5 The Fluke 25 is a high-performance hand-held DMM. Its accuracy is specified as ±(0.1% + 1) for all dc voltages ranges. It has a frequency response of 40 Hz to 30 kHz for ac measurements. Other features include a bar graph to give an analog indication of the measured quantity, manual range selection (as well as auto-ranging), and a hold button to hold the indicated measurement. (© 1991, John Fluke Mfg. Co., Inc. All rights reserved. Reproduced with permission.)

Although the meter in Figure 6-5 has auto-ranging, there is a *RANGE* button for locking into any particular range. The *HOLD* button can be used to hold a measured quantity, so that the probes need only be touched to the circuit momentarily when measuring voltage, for example. In both cases (*RANGE* and *HOLD*) an annunciator appears on the display to indicate the condition of the meter.

As well as voltage, current, and resistance ranges, the Fluke 25 has a selection that provides for an audible *continuity test* and a *diode test*. For continuity testing, the meter beeps when a short circuit is detected between the $V\Omega$ and *COM* terminals. This might typically be used for testing fuses and switch contacts, and for identifying the ends of a single conductor contained in a bundle of conductors. When testing diodes, the diode forward voltage drop (typically 0.6 V to 0.7 V for a silicon diode) should show when forward biased, and an open-circuit should be indicated for reverse bias.

Bench-Type DMM

Bench-type (or laboratory-type) digital multimeters are large enough to have displays with many digits ($6\frac{1}{2}$ digits is not unusual); consequently, they can have better resolution and greater accuracy than hand-held instruments. The $4\frac{1}{2}$ digit display on the Simpson 460-6 multimeter illustrated in Figure 6-6 provides a resolution of 1 in 20,000, and the measurement accuracy for dc voltages is $\pm(0.07\% + 1)$. $V\Omega$, *COM, mA,* and *10 A* terminals are provided, and function selection is by pushbutton switches. The frequency response for ac measurements is 20 Hz to 100 kHz. For current measurements, the burden

Figure 6-6 Simpson 460-6 bench/portable-type digital multimeter. This instrument has an accuracy of $\pm(0.07\% + 1)$ for dc voltage measurements, and the frequency range for ac measurements is 20 Hz to 100 kHz. (Courtesy of bach-simpson limited.)

voltage ranges from 250 mV to 750 mV; these are fixed voltage drops, not mV/mA quantities.

Additional Features

Some additional features found on individual digital multimeters (both hand-held and bench type) include measurement of decibels, frequency, duty cycle, capacitance, conductance, true rms values. As discussed earlier for analog instruments, digital meter ranges can be extended by the use of high-frequency probes, high-voltage probes, and high-current probes (see Figure 6-7).

Comparison of Digital and Analog Multimeters

Digital multimeters are superior to analog instruments in at least two important categories: accuracy and durability. The accuracy of good-quality analog instruments is typically ±2% *of full scale,* which means ±4% on a half-scale reading, and worse farther down the scale. The least expensive digital meters can have an accuracy of better than ±0.6% *of the measured quantity.* Many analog instruments that use taut-band suspension can survive drops to a floor from bench-top levels, but they are likely to be damaged by greater drops. The mechanism of jeweled-bearing instruments will almost certainly suffer damage when the instrument is dropped. Digital multimeters can handle much tougher treatment and still give good service, and many are designed to be waterproof. Analog instruments may also suffer damage if connected with the wrong polarity or if the measured voltage or current exceeds the selected range (see Figure 6-8). A digital instrument will simply indicate a negative quantity when connected in reverse, and will switch automatically to an appropriate range or indicate overload when the measured quantity is excessive. Monitoring a changing condition of a measured quantity is one application for which many people prefer analog instruments. The pointer of the analog instrument seems to respond more quickly than the digital display. However, digital instruments have greater resolution than analog meters, so a very small change in a measured quantity will be most clearly indicated by a digital meter.

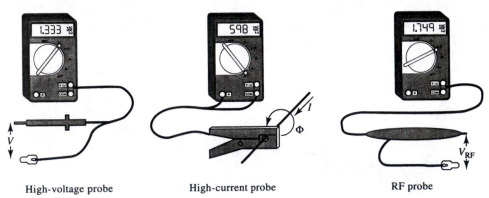

High-voltage probe   High-current probe   RF probe

Figure 6-7 The voltage range of multimeter can be extended by a high-voltage probe. High current levels can be measured by the use of a high-current probe. A high-frequency probe allows radio-frequency measurements to be made.

 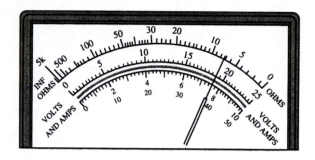

Digital Meter

Leaves no doubt about the
measured quantity.

Superior resolution and
accuracy.

Indicates a negative quantity
when the terminal polarity
is reversed.

Not usually damaged by rough
treatment.

Analog Meter

Wrong scale might be used,
or might be read incorrectly.

Inferior resolution and
accuracy.

Pointer attempts to deflect
to the left of zero when the
polarity is reversed.

Can be irreparably damaged
when dropped from bench levels.

Figure 6-8 Digital instruments can survive rough treatment that would destroy analog meters. Digital meters are also more accurate than analog.

Example 6-2

A 20 V dc voltage is measured by analog and digital multimeters. The analog instrument is on its 25 V range, and its specified accuracy is ±2%. The digital meter has a $3\frac{1}{2}$-digit display and an accuracy of ±(0.6 + 1). Determine the measurement accuracy in each case.

Solution

Analog instrument:

$$voltage\ error = \pm2\%\ of\ 25\ V$$

$$= \pm0.5\ V$$

$$error = \frac{\pm0.5\ V}{20\ V} \times 100\%$$

$$= \pm2.5\%$$

Digital instrument:

For 20 V displayed on a $3\frac{1}{2}$ digit display,

$$1 \ digit = 0.1 \ V$$

$$voltage \ error = \pm(0.6\% \ of \ the \ reading + 1 \ d)$$

$$= \pm(0.6\% \ of \ 20 \ V + 0.1 \ V)$$

$$= \pm 0.12 \ V + 0.1 \ V$$

$$= \pm 0.22 \ V$$

$$error = \frac{\pm 0.22 \ V}{20 \ V} \times 100\%$$

$$= \pm 1.1 \ \%$$

6-3 DIGITAL FREQUENCY METER SYSTEM

Basic Frequency Meter

The digital frequency meter illustrated in Figure 6-9(a) consists of an accurate timing source (or *time base*), digital counting circuits, circuitry for *shaping* the input waveform, and a circuit for *gating* the shaped waveform to the counter. The input is first amplified or attenuated, as necessary, and then fed to the *wave-shaping* circuit, which converts it into a square or pulse waveform with the same frequency as the input [Figure 6-9(b)]. The presence of this wave-shaping circuit means that the input can be sinusoidal, square, triangular, or can have any other repetitive-type waveform. The shaped waveform is fed to one input terminal of a two-input AND gate, and the other AND gate input is controlled by the Q output from a flip-flop. Consequently, the pulses to be counted pass through the AND gate only when the flip-flop Q terminal is *high*.

The flip-flop is controlled by the timing circuit, changing state each instant that the timer output waveform goes in a negative direction (a *negative-going edge*). When the timing circuit output frequency is 1 Hz, as illustrated, the flip-flop Q output terminal is alternately *high* for a period of 1 s and *low* for 1 s. In this case, the counting circuits are toggled (by the pulses from the wave-shaping circuit) for a period (termed the *gate time*) of 1 s, and the total count indicates the frequency directly in hertz. The counting circuits are reset to the zero-count condition by the negative-going edge of the $\overline{Q}$ output from the flip-flop, so that the count always starts from zero.

As in the case of the digital voltmeters discussed in Section 6-1, latch or display enable circuits are employed to make the digital display readable. For the system illustrated in Figure 6-9, the latch circuits are briefly triggered at the end of the counting time by the positive-going edge of the flip-flop $\overline{Q}$ output. The display is corrected at this instant (if necessary), and then remains constant until the next latch trigger input.

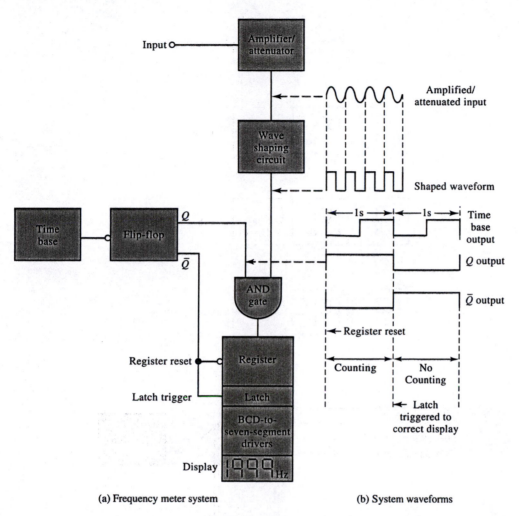

(a) Frequency meter system

(b) System waveforms

Figure 6-9 Basic block diagram and waveforms for a digital frequency meter. Cycles of the frequency to be measured are counted over a known time period. One thousand cycles counted over a period of 1 s gives a 1 kHz measured frequency.

Range Changing

As explained in Section 5-4, accurate time periods of 10 μs, 100 μs, 1 ms, and so on, can be produced by the use of a crystal oscillator and several decade counters. Such a time base can be used with a digital frequency meter to give several ranges of frequency measurement. When a 1 s time period is used for counting the input pulses, the $3\frac{1}{2}$-digit display in Figure 6-9 might have a *Hz* unit identification alongside it, as illustrated. Alternatively, as shown in Figure 6-10(a), the frequency units could be identified as *kHz* if a decimal point is included after the first numeral.

Now consider the effect of using a 100 ms counting time instead of a 1 s time period. A display of 1999 indicates 1999 cycles of input waveform per 100 ms, or 19.99 kHz

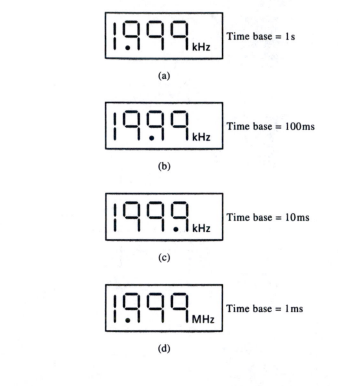

(a) Time base = 1s

(b) Time base = 100ms

(c) Time base = 10ms

(d) Time base = 1ms

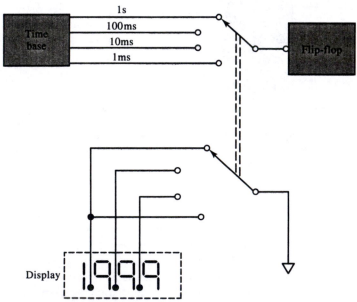

(e) Selection of time period and decimal point

Figure 6-10 Time period and decimal point selection for a digital frequency meter. When the time period (or gate time) is 1 s, a count of 1999 pulses is displayed as 1.999 kHz. The decimal point and units display are selected according to the time period.

[Figure 6-10(b)]. Thus, when the time base is switched to 100 ms, the decimal point must also be switched. Similarly, if the time base is switched to 10 ms, the decimal point is moved right once again, so that the maximum measurable frequency is 199.9 kHz [Figure 6-10(c)]. A further switch of the time base to a period of 1 ms gives a maximum pulse count of 1999 pulses per 1 ms, or 1.999 MHz [Figure 6-10(d)]. Figure 6-10(e) shows a switching arrangement for the selection of time period and decimal point.

6-4 FREQUENCY METER ACCURACY

Range Selection Error

Just as in the case of other instruments, the lowest possible (frequency) range should be used for the greatest measurement accuracy with a digital frequency meter. This is demonstrated in Example 6-3.

Example 6-3

A digital frequency meter has a time base derived from a 1 MHz clock generator frequency-divided by decade counters. Determine the measured frequency when a 1.512 kHz sine wave is applied and the time base uses (a) six decade counters and (b) four decade counters.

Solution

(a) *Using six decade counters:*

Time base frequency, $\qquad f_1 = \dfrac{1 \text{ MHz}}{10^6} = 1 \text{ Hz}$

Time period, $\qquad t_1 = \dfrac{1}{1 \text{ Hz}} = 1 \text{ s}$

Cycles counted, $\qquad n_1 = (\text{input frequency}) \times t_1$

$\qquad\qquad = 1.512 \text{ kHz} \times 1 \text{ s}$

$\qquad\qquad = 1512 \text{ cycles}$

Measured frequency, $\qquad f = 1.512 \text{ kHz}$

(b) *Using four decade counters:*

Time base frequency, $\qquad f_2 = \dfrac{1 \text{ MHz}}{10^4} = 100 \text{ Hz}$

Time period, $\qquad t_2 = \dfrac{1}{100 \text{ Hz}} = 10 \text{ ms}$

$$\text{Cycles counted,} \qquad n_2 = (\text{input frequency}) \times t_2$$

$$= 1.512 \text{ kHz} \times 10 \text{ ms}$$

$$= 15 \text{ cycles}$$

$$\text{Measured frequency} \qquad f = 01.5 \text{ kHz}$$

In part (b) of Example 6-3, note that the first digit (1) of the $3\frac{1}{2}$-digit display is *off* when the count is less than 1000 cycles of input, while the second digit indicates 0 when the count is less than 100 cycles. Clearly, the greatest measurement occurs when the greatest number of cycles are counted, that is, when the lowest-frequency range (longest counting time period) is used.

Accuracy Specification

In the system described in Section 6-3, the time base could switch the AND gate *on* or *off* while an input pulse (from the wave-shaping circuit) is being applied. The partial pulses that get through the AND gate may or may not succeed in triggering the counting circuits. So there is always a possible *gating* error of ±1 cycle in the count during the timing period. This (one count) is defined as the *least significant digit (LSD)*. Thus, the accuracy of a digital frequency meter is usually stated as

$$\pm 1 \text{ LSD} \pm \text{ time base error,}$$

Errors in the time base generated by a crystal controlled oscillator are normally the result of variations in temperature, supply voltage changes, and aging of crystals. With reasonable precautions, the total time base error might typically be $< 1 \times 10^{-6}$, or *less than 1 part in 10^6 parts*. (Higher-quality time bases have smaller errors.) The total measurement error depends on the actual frequency measured. This is demonstrated by Example 6-4.

Example 6-4

A frequency counter with an accuracy of ± 1 LSD $\pm(1 \times 10^{-6})$ is employed to measure frequencies of 100 Hz, 1 MHz, and 100 MHz. Calculate the percentage measurement error in each case.

Solution

At $f = 100$ Hz,

$$error = \pm(1 \text{ count} + 100 \text{ Hz} \times 10^{-6})$$

$$= \pm(1 \text{ count} + 1 \times 10^{-4} \text{ counts})$$

$$\simeq \pm 1 \text{ count}$$

$$\% \ error = \pm\left(\frac{1}{100 \ Hz} \times 100\%\right)$$

$$= \pm 1\%$$

At f = 1 MHz,

$$error = \pm(1 \ count + 1 \ MHz \times 10^{-6})$$

$$= \pm(1 \ count + 1 \ count)$$

$$\simeq \pm 2 \ counts$$

$$\% \ error = \frac{\pm 2}{1 \ MHz} \times 100\%$$

$$= \pm 2 \times 10^{-4}\%$$

At f = 100 MHz,

$$error = \pm(1 \ count + 100 \ MHz \times 10^{-6})$$

$$= \pm(1 \ count + 100 \ counts)$$

$$= \pm 101 \ counts$$

$$\% \ error = \pm\left(\frac{101}{100 \ MHz} \times 100\%\right)$$

$$= \pm 1.01 \times 10^{-4}\%$$

Example 6-4 shows that at a frequency of 100 Hz the error due to ±1 count is ±1%, while that due to the time base is insignificant. At 1 MHz, the error due to one count is equal to that due to the time base. At 100 MHz, the time base is responsible for an error of ±100 counts, although the total error is still a very small percentage of the measured frequency. Therefore, at high frequencies the time base error is larger than the ±1 count error, while at low frequencies the ±1 count error is the larger of the two (see Figure 6-11). At frequencies lower than 100 Hz, the percentage error due to ±1 count is greater than 1%, so the greatest measurement error occurs at low frequencies. The low-frequency error can be greatly reduced by the *reciprocal counting technique*.

Reciprocal Counting

In the rearranged frequency meter system shown in Figure 6-12, the 1 MHz oscillator frequency from the time base (see Figure 5-10) is applied directly to the AND gate in place of the output from the wave-shaping circuit. The reshaped input wave (which is to have its frequency measured) is employed to toggle the flip-flop. As illustrated by the waveforms, this arrangement results in the AND gate passing pulses from the 1 MHz oscillator to the counting circuits during the time period (*T*) of the input wave.

When a 100 Hz frequency is to be measured, the AND gate passes pulses to the counting circuits for a time of 1/100 Hz = 10 ms. The time period of each cycle from the 1 MHz oscillator is 1 μs. Therefore, the number of pulses counted during *T* is

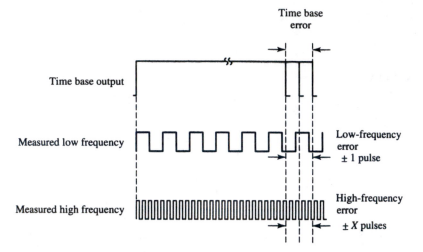

Figure 6-11 The counting error (as a number of cycles) that results from the time base error is largest when measuring a high frequency.

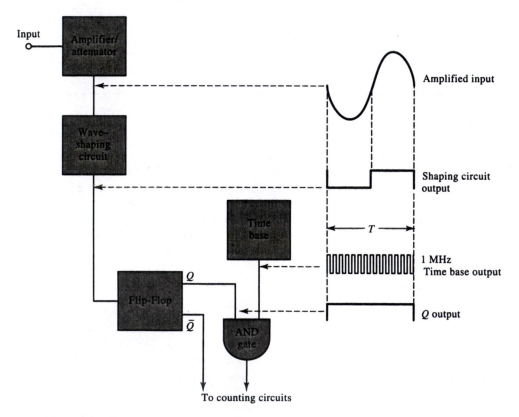

Figure 6-12 Digital frequency meter system rearranged for reciprocal counting. The time period of low-frequency inputs can be accurately measured by counting the clock pulses during the period.

$$n = \frac{10 \text{ ms}}{1 \ \mu\text{s}} = 10 \ 000$$

This shows on the display as 10 000 μs, and is inverted to determine the input frequency. (A $4\frac{1}{2}$-digit display should be used.) The accuracy of measurement of the 100 Hz frequency is now ± 1 count in 10 000, or

$$\text{error} = \frac{\pm 1}{10 \ 000} \times 100\%$$
$$= \pm 0.01\%$$

A 0.01% error is a big improvement over the 1% error that occurs with the straight counting technique (see Example 6-4). The accuracy improvement with the reciprocal counting method is even better at frequencies lower than 100 Hz. For high-frequency measurements, the straight counting method gives the most accurate result.

6-5 TIME AND RATIO MEASUREMENTS

Pulse Time Measurement

The time period of any input waveform can be measured by means of the reciprocal counting technique described in Section 6-4, [see Figure 6-13(a)]. If the flip-flop in Figure 6-12 is made to toggle on positive-going inputs as well as on negative-going signals, the width (or duration) of an input pulse can be measured [Figure 6-13(b)]. Most digital counters have a *start input* and a *stop input,* so that the *time between events* can be measured.

Frequency Ratio Measurement

The ratio of two frequencies can be measured by a counter. The lowest of the two frequencies is applied as an input to the flip-flop in Figure 6-12, and the higher frequency is applied (via appropriate shaping circuitry) to the AND gate input. The instrument now counts the number of high-frequency cycles that occur during the time period of the low frequency. The waveforms are illustrated in Figure 6-13(c). If the high frequency is 100 times the low input frequency, 100 cycles of high frequency are counted. Thus, the displayed number is the ratio of the two input frequencies. The waveform illustrated in Figure 6-13(c) are for the case when one frequency is very much greater than the other. When this is *not* the case (e.g., when $f_1/f_2 = 1.11$), the cycles of f_1 are counted over perhaps 100 cycles of f_2 [Figure 6-13(d)]. With the decimal point selected correctly, the displayed ratio is 1.110.

6-6 COUNTER INPUT STAGE

The input stage to a counter is an amplifier/attenuator feeding into a wave-shaping circuit which usually employs a Schmitt trigger. The Schmitt trigger circuit has an *upper trigger point* (UTP), or upper level of input voltage at which it triggers. It also has a *lower trigger point* (LTP), or lower voltage level at which its output switches back to the original state. The difference between the UTP and LTP is termed *hysteresis*. The effect of the hysteresis

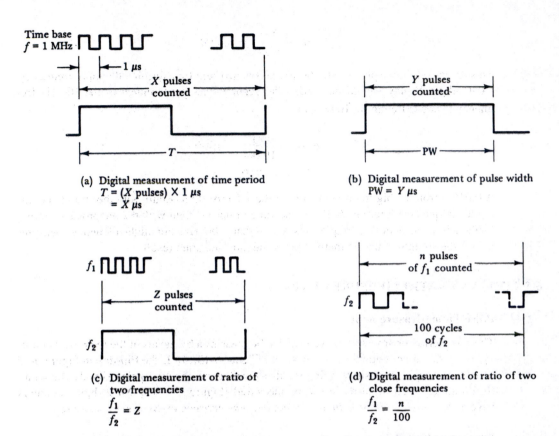

Figure 6-13 As well as frequency measurement, a digital frequency meter can be used to measure waveform time period, pulse width, and the ratio of two frequencies.

is illustrated in Figure 6-14(a). Assuming that the amplifier/attenuator stage is set for a gain of 1, the Schmitt output goes positive when the input sine wave passes the UTP. When the sine wave goes below the LTP, the Schmitt output returns to its previous level.

Now look at the effect of the noisy input signal shown in Figure 6-14(b). The noise spikes cause the signal to cross the hysteresis band more than twice in each cycle. Thus, unwanted additional output pulses are generated which introduce errors in the frequency measurement. This difficulty could be overcome by expanding the hysteresis band of the Schmitt circuit. Alternatively, as illustrated in Figure 6-14(c), the input signal can be attenuated. Most frequency counters have a continuously variable input attenuator. The attenuator should be adjusted to set the signal at the lowest level which will satisfactorily trigger the counting circuits. Some counters have *low-pass filters* that may be switched into the input stage to attenuate high-frequency noise, and thus further improve the noise immunity of the counter.

Figure 6-15(a) illustrates a problem that occurs when the counter is used as a timer to measure pulse width (PW). The average width of the pulse is the PW illustrated in Figure 6-15(a). However, because of the hysteresis of the Schmitt trigger circuit there is an error in the measured pulse width (see illustration). This error can be minimized either by reducing the hysteresis, or by amplifying the pulse as shown in Figure 6-15(b).

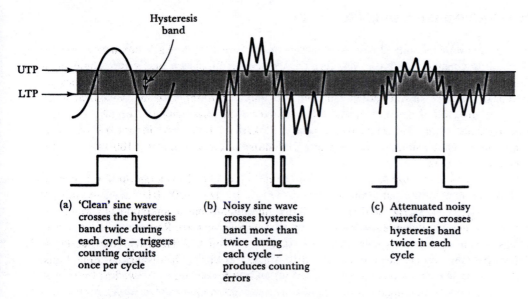

(a) 'Clean' sine wave crosses the hysteresis band twice during each cycle — triggers counting circuits once per cycle

(b) Noisy sine wave crosses hysteresis band more than twice during each cycle — produces counting errors

(c) Attenuated noisy waveform crosses hysteresis band twice in each cycle

Figure 6-14 Noisy input signals can produce counting errors on a digital frequency meter. Signal attenuation adjustment usually eliminates the problem.

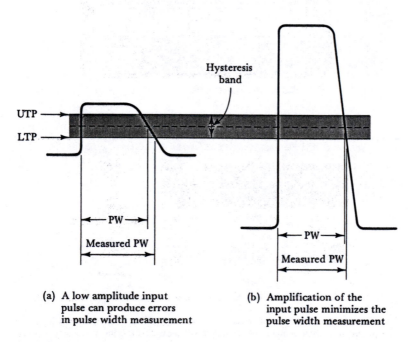

(a) A low amplitude input pulse can produce errors in pulse width measurement

(b) Amplification of the input pulse minimizes the pulse width measurement

Figure 6-15 Long rise or fall times can cause errors in the measurement of pulse width measurement on a digital frequency meter. Signal amplification usually minimizes the error.

6-7 COUNTER/TIMER/FREQUENCY METERS

Two digital frequency meters are illustrated in Figure 6-16: a basic frequency counter and a multifunction counter. The basic instrument in Figure 6-16(a) has a single input (coaxial) terminal, an ON/OFF switch, and an 8-digit display to measure frequency up to 100 MHz. Using this counter involves no more than applying the waveform to be measured. The gate time is fixed at 1 s, and the time base error is typically ±10 ppm (parts per million), or 10×10^{-6}, which is ±1 kHz in 100 MHz. Input *sensitivity* (the minimum input amplitude to trigger the counting circuits) might be 100 mV rms for sine-wave inputs.

The *universal frequency counter* shown in Figure 6-16(b) is a high-performance instrument. It has two input terminals: input *A* for the 10 Hz to 120 MHz range, and input *B* for 70 MHz to 1.3 GHz. Sensitivity is 10 mV, and input *A* has a continuously variable attenuator and a switchable low-pass noise filter (see Section 6-6). Resolution is stated as at least seven digits in a 1 s measuring time; this is ±1 Hz in 10 MHz. Time base stability is 2×10^{-7}, which translates into an error of ± 2 Hz in 10 MHz. The display has nine digits, some of which can be blanked out when not required for accuracy. As well as frequency, measurements can be made of *time period, pulse width, frequency ratio, frequency difference, rpm,* and *totalizing* (event counting between start and stop).

(a) Basic digital frequency counter.

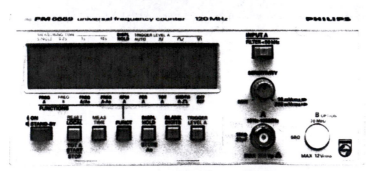

(b) PM 6669 universal frequency counter (reproduced with permission).

Figure 6-16 Basic digital frequency meter that only measures frequency (a), and a high-performance multifunction counter (b). The PM6669 measures frequency (to 1.3 GHz), time period, pulse width, frequency ratio, and frequency difference. (© 1991, John Fluke Mfg. Co., Inc. All rights reserved. Reproduced with permission.)

REVIEW QUESTIONS

6-1 Sketch the block diagram and system waveforms for a DVM using an analog-to-digital converter. Show the system waveforms, and explain its operation.

6-2 Define dual-slope integrator and zero-crossing detector. Sketch the block diagram and system waveforms for a digital voltmeter that uses a dual-slope integrator. Explain how it operates, and discuss the advantages of the dual-slope system.

6-3 Sketch a range-changing circuit for a DVM, and explain how it operates.

6-4 Draw the front panels of two typical digital multimeters, showing the terminals and controls. Explain terminal connections, function selection, range selection, and meter readings.

6-5 State typical digital instrument accuracy specifications. Compare the accuracy of digital and analog multimeters.

6-6 Draw the basic block diagram of a digital frequency meter, sketch the system waveforms, and carefully explain its operation.

6-7 Sketch a switching arrangement for changing the measured frequency range and displayed measurement for a digital frequency meter. Explain.

6-8 Discuss the errors that occur in digital frequency meters, and explain the method of specifying measurement accuracy.

6-9 Define reciprocal counting. Draw the basic block diagram of the digital frequency meter rearranged for reciprocal counting. Explain its operation, and show why reciprocal counting is sometimes used in preference to the straight counting method.

6-10 Draw waveforms to show how time period, pulse width, and frequency ratio may be measured on a digital frequency meter. Explain each case.

6-11 Discuss the need for input attenuation and amplification with a digital frequency meter. Draw waveforms to illustrate the errors that can be produced by noisy waveforms, and the method of dealing with them.

6-12 Discuss the frequency range, accuracy, and sensitivity of typical basic and high-performance frequency counters. Also, list the various measurements that can be made on a high-performance instrument.

PROBLEMS

6-1 Determine the ramp time required for the digital voltmeter in Figure 6-1 to register 1999 V if the clock generator frequency is 1 MHz. Also determine a suitable frequency for the ramp generator.

6-2 Recalculate the measured voltage for the DVM in Problem 6-1 if the clock frequency drifts by −5%.

6-3 The DVM in Figure 6-2 has a 200 kHz clock, and the integrator control waveform frequency is 45 Hz. Calculate the number of clock pulses that occur during t_1, and determine a suitable time duration for t_2 when the input is 1 V. Recalculate t_1, t_2, and the measured voltage if the clock frequency drifts by −5%.

6-4 Calculate the maximum measurement error for a digital voltmeter with an accuracy of ±(0.1% rdg + 1d),when indicating 1.490V.

6-5 Determine the possible maximum and minimum measured voltage when the instrument in Problem 6-4 indicates 1.255 V.

6-6 A digital frequency meter uses a time base consisting of a 1 MHz clock generator frequency-divided by six decade counters. Determine the meter indication (a) when the input frequency is 5 kHz and the time base output is selected at the sixth decade counter and (b) when the input frequency is 2.9 kHz and the time base output is at the fifth decade counters.

6-7 A frequency meter with an accuracy of ±1 LSD ±(1×10^{-5}) is used to measure frequencies of 30 Hz, 30 MHz, and 300 MHz. Calculate the percentage error for each measurement.

6-8 The frequency meter in Problem 6-6 is rearranged for reciprocal counting. Determine the error that can occur when a 30 Hz frequency is measured on this system.

6-9 A frequency meter with a 1 MHz clock source is used for measuring the time period of an input wave.

(a) Determine the measured time period when 1560 pulses are registered on the display.

(b) Determine the new display reading for the same input wave if the clock generator is replaced with a 1.5 MHz source.

6-10 A frequency meter measuring the ratio of two frequencies displays 1133 when the pulses of the unknown frequency (f_2) are counted over 1000 cycles of the known frequency (f_1). If f_1 is 33 kHz, determine f_2.

6-11 Determine the accuracy of measurement for each of the two instruments discussed in Section 6-7 when measuring 750 kHz.

Low, High, and Precise Resistance Measurements

7

Objectives

You will be able to:

1. Explain ammeter and voltmeter methods and the substitution method of resistance measurement.

2. Sketch the circuit diagram of a Wheatstone bridge, explain its operation, and derive its balance equation.

3. Determine the Wheatstone bridge range, measurement accuracy, and sensitivity.

4. Explain four-terminal resistors. Draw the circuit of a Kelvin bridge for low resistance measurement, and derive its balance equation.

5. Explain very high resistance measurement methods. Show how volume resistance and surface leakage resistance may be separated when measuring insulation resistance.

6. Sketch the circuit diagram of a megohmmeter, and explain its operation.

7. Discuss electronic meters for measurement of very high and very low resistances.

Introduction

Although the ohmmeter is convenient for measuring resistance, it is not very accurate. Resistance can be measured by ammeter and voltmeter methods, but there are errors that must be taken into account. Precise measurements of resistance in the medium range are best made using the Wheatstone bridge. For measurement of the very low resistance of ammeter shunts, a modification of the Wheatstone bridge, known as the Kelvin bridge, is used. Measurement of very high resistances, such as insulation resistance, requires high voltages, low-current ammeters, and special techniques to separate surface leakage resistance from volume resistance. Electronic instruments are available for direct measurement of very high and very low resistances.

7-1 VOLTMETER AND AMMETER METHODS

One way of determining the value of a resistance is simply to apply a voltage across its terminals and use an ammeter and voltmeter to measure the current and voltage. The measured quantities may then be substituted into Ohm's law to calculate the resistance $R = E/I$. Although very accurate measuring instruments may be used, there can be serious error in the resistance value arrived at in this way.

Consider the circuit in Figure 7-1(a). The voltmeter measures the voltage E across the terminals of the resistor, but the ammeter indicates the resistor current I *plus* the voltmeter current I_v. The calculated resistance now comes out as

$$R = \frac{E}{I + I_v} \qquad (7\text{-}1)$$

Since it is obvious that R actually equals E/I, the presence of I_v constitutes an error. Of course, if I_v is very much smaller than I, the error may be negligible.

Now look at Figure 7-1(b), in which the ammeter is directly in series with resistor R. In this case the ammeter measures only the resistor current I; the voltmeter current I_v is not involved. However, the voltmeter now indicates the resistor voltage E *plus* the ammeter voltage drop E_A. Consequently, the calculated resistance is

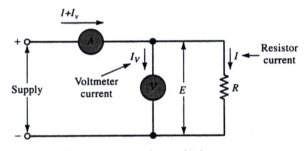

(a) Voltmeter connected across load

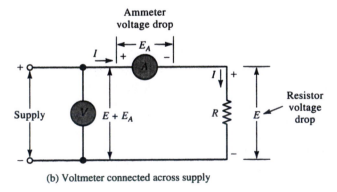

(b) Voltmeter connected across supply

Figure 7-1 Ammeter and voltmeter method of measuring resistance. When the voltmeter is connected across the load (a), the ammeter measures the voltmeter current and the resistor current. When connected across the supply (b), the voltmeter measures the ammeter voltage drop as well as the resistor voltage.

$$R = \frac{E + E_A}{I} \qquad (7\text{-}2)$$

Once again this is not the same as $R = E/I$, and an error results because of the presence of E_A. Clearly, if E_A is very much smaller than E, the error may be small enough to neglect.

The circuit in Figure 7-1(a) is accurate when resistor R has a value very much smaller than the voltmeter resistance. When this is the case, I_v is very much smaller than I, and

$$R = \frac{E}{I + I_v} \approx \frac{E}{I}$$

The arrangement in Figure 7-1(b) is most suitable where the value of R is very much larger than the ammeter resistance. This makes E_A very small compared to E, and gives

$$R = \frac{E + E_A}{I} \approx \frac{E}{I}$$

To select the best circuit [(a) or (b)] for measuring any given resistance, first connect up the instruments as shown in Figure 7-1(a). Carefully note the ammeter indication, then disconnect one terminal of the voltmeter. If the ammeter indication does not appear to change when the voltmeter is disconnected, I_v is very much less than I, and Figure 7-1(a) is the correct connection for accurately measuring R. If the indicated current drops noticeably when the voltmeter is disconnected, I_v is *not* very much smaller than I. In this case, the circuit in Figure 7-1(b) affords the most accurate determination of R.

Example 7-1

A resistance is measured by the ammeter and voltmeter circuit illustrated in Figure 7-1(b). The measured current is 0.5 A, and the voltmeter indication is 500 V. The ammeter has a resistance of $R_a = 10\ \Omega$, and the voltmeter on a 1000 V range has a sensitivity of 10 kΩ/V. Calculate the value of R.

Solution

$$E + E_A = 500 \text{ V}$$

$$I = 0.5 \text{ A}$$

$$R_a + R = \frac{E + E_A}{I}$$

$$= \frac{500 \text{ V}}{0.5 \text{ A}} = 1000\ \Omega$$

$$R = 1000\ \Omega - R_a$$

$$= 1000\ \Omega - 10\ \Omega$$

$$= 990\ \Omega$$

Example 7-2

If the ammeter, voltmeter, and resistance R in Example 7-1 are reconnected in the form of Figure 7-1(a), determine the ammeter and voltmeter indications.

Solution

$$R_v = 1000 \text{ V} \times 10 \text{ k}\Omega/\text{V}$$
$$= 10 \text{ M}\Omega$$

$$R_v \| R = 10 \text{ M}\Omega \| 990 \ \Omega$$
$$= 989.9 \ \Omega$$

$$supply \ voltage = 500 \text{ V}$$

$$voltmeter \ reading = E = \frac{500 \text{ V} \times R_v \| R}{R_a + R_v \| R}$$

$$= \frac{500 \text{ V} \times 989.9 \ \Omega}{10 \ \Omega + 989.9 \ \Omega}$$

$$= 495 \text{ V}$$

$$ammeter \ reading = I + I_v = \frac{E}{R_v \| R}$$

$$= \frac{495 \text{ V}}{989.9 \ \Omega}$$

$$\approx 0.5 \text{ A}$$

Example 7-3

Referring to Example 7-1 and 7-2, determine which of the two circuits, Figure 7-1(a) or Figure 7-1(b), gives the most accurate measurement of R when R is calculated without taking R_a and R_V into account.

Solution

For Figure 7-1(a) (Example 7-2):

$$voltmeter \ reading = 495 \text{ V}$$

$$ammeter \ reading = 0.5 \text{ A}$$

$$R = \frac{495 \text{ V}}{0.5 \text{ A}} = 990 \ \Omega$$

For Figure 7-1(b) (Example 7-1):

$$voltmeter \ reading = 500 \text{ V}$$

$$ammeter\ reading = 0.5\ A$$

$$R = \frac{500\ V}{0.5\ A} = 1000\ \Omega$$

In this case, the circuit of Figure 7-1(a) gives the most accurate determination of R.

7-2 SUBSTITUTION METHOD

A very simple method of measuring resistance is illustrated in Figure 7-2. In Figure 7-2(a) the resistor R which is to be measured is connected in series with an ammeter and a dc supply. The ammeter is adjusted to a suitable range, and its pointer position on the scale is carefully noted. The resistor is now disconnected and a *decade resistance box* (a precision variable resistor) is substituted in its place. The decade resistance box should initially be set to its highest possible value. Once in the circuit, the decade box is adjusted until the ammeter indicates precisely the same level of current as when R was in the circuit. The resistance of the decade box (indicated by the switch positions) is now exactly equal to R. The accuracy of the substitution method depends upon the accuracy of the decade box and the sensitivity of the ammeter.

7-3 WHEATSTONE BRIDGE

Circuit

Very accurate resistance measurements can be made using a *Wheatstone bridge*. As shown in Figure 7-3(a), the bridge circuit consists of an unknown resistance (to be measured) R, two precision resistors P and Q, an adjustable resistor S, and a galvanometer G. Supply voltage E produces current flow through the resistors. Actually, the resistance to be measured could be in any one of the four positions.

To determine the resistance of R, the variable resistance is adjusted until the galvanometer indicates null, or zero voltage. Initially, the galvanometer should be shunted to

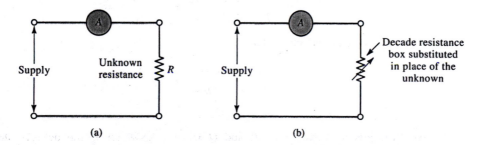

(a)   (b)

Figure 7-2 Substitution method of resistance measurement. The decade resistance box is adjusted until the ammeter indicates the same current level as when the unknown resistance was connected. Then the unknown equals the decade box resistance.

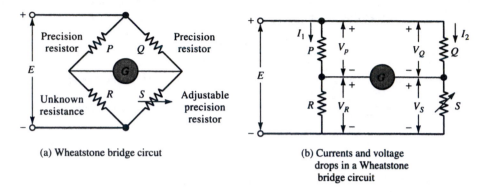

(a) Wheatstone bridge circut

(b) Currents and voltage
drops in a Wheatstone
bridge circuit

Figure 7-3   Wheatstone bridge circuit for accurate resistance measurement. *P*, *Q*, and *S* are precision resistors, and *R* is the unknown. When the galvanometer indicates zero voltage, $V_P = V_Q$ and $V_R = V_S$. This gives $R = SP/Q$.

protect it from excessive current levels (as in Figure 3-8). As null is approached, the shunting resistance is gradually made larger until the galvanometer indicates zero with the resistor open-circuited. The bridge is now said to be *balanced*. Before galvanometer null is achieved, the bridge is said to be in an *unbalanced* state. The circuit is redrawn in Figure 7-3(b) to show the voltages and currents throughout the bridge when it is balanced.

When the galvanometer indicates zero, the voltages at each of its terminals must be equal. Thus, in Figure 7-3(b),

$$V_P = V_Q \qquad \text{and} \qquad V_R = V_S$$

Also, because there is no current flow through the galvanometer, I_1 flows through both *P* and *R*, and I_2 flows through *Q* and *S*. Therefore,

$$I_1 R = I_2 S \tag{1}$$

and

$$I_1 P = I_2 Q \tag{2}$$

Dividing Equation 1 by Equation 2 gives

$$\frac{I_1 R}{I_1 P} = \frac{I_2 S}{I_2 Q} \qquad \text{or} \qquad \frac{R}{P} = \frac{S}{Q}$$

which gives the unknown resistance,

$$\boxed{R = \frac{SP}{Q}} \tag{7-3}$$

With the precise values of *S*, *P*, and *Q* known, resistance *R* can be accurately determined.

Note that the supply voltage *E* is not involved in the calculation. As will be explained, the supply voltage does affect the bridge *sensitivity*.

Example 7-4

A Wheatstone bridge as in Figure 7-3 has $P = 3.5$ kΩ, $Q = 7$ kΩ, and galvanometer null is obtained when $S = 5.51$ kΩ.

    (a)   Calculate the value of R.

    (b)   Determine the resistance measurement range for the bridge if S is adjustable from 1 kΩ to 8 kΩ.

Solution

    (a)   *Equation 7-3,*

$$R = \frac{SP}{Q}$$

$$= \frac{5.51 \text{ k}\Omega \times 3.5 \text{ k}\Omega}{7 \text{ k}\Omega}$$

$$= 2.755 \text{ k}\Omega$$

    (b)   *When $S = 1$ kΩ:*

$$R = \frac{1 \text{ k}\Omega \times 3.5 \text{ k}\Omega}{7 \text{ k}\Omega}$$

$$= 500 \ \Omega$$

When $S = 8$ kΩ:

$$R = \frac{8 \text{ k}\Omega \times 3.5 \text{ k}\Omega}{7 \text{ k}\Omega}$$

$$= 4 \text{ k}\Omega$$

The measurement range is 500 Ω to 4 kΩ.

Accuracy

In Chapter 2 it is explained that the maximum possible error in the product and quotient of quantities is the sum of the errors in all components involved. Thus, in the case of the Wheatstone bridge, where $R = SP/Q$, the percentage error in R is the sum of the percentage errors in each of S, P, and Q.

Example 7-5

In Example 7-4, P and Q have accuracies of $\pm 0.05\%$, and S has an accuracy of $\pm 0.1\%$. Calculate the accuracy of the measured resistance of R and its upper and lower limits for the P, Q, and S values stated.

Solution

$$\text{Error in } R = (P \text{ error}) + (Q \text{ error}) + (S \text{ error})$$

$$= \pm (0.05\% + 0.05\% + 0.1\%)$$

$$= \pm 0.2\%$$
$$R = 2.755 \text{ k}\Omega \pm 0.2\%$$
$$\approx 2.755 \text{ k}\Omega \pm 5.5 \ \Omega$$
$$= 2.7495 \text{ k}\Omega \text{ to } 2.7605 \text{ k}\Omega$$

Sensitivity

Refer to Figure 7-3 and to Example 7-4 once again. If S is a decade resistance box, its lowest adjustment decade might be 0 to 10 Ω in 1 Ω steps. A more precise decade box could have a decade of 0 to 1 Ω in 0.1 Ω steps.

If the galvanometer pointer is clearly deflected from zero when S is adjusted by $\pm 0.1 \ \Omega$, the bridge can be described as *sensitive to* $\pm 0.1 \ \Omega$ in S. The deflection from zero may be as small as 1mm, as long as the pointer clearly moves off zero when S is adjusted by 0.1 Ω, and moves back again to zero when S is readjusted.

The bridge sensitivity to the measured resistance R may be calculated as

$$\Delta R = \frac{\Delta S \, P}{Q} = \frac{0.1 \ \Omega \times 3.5 \text{ k}\Omega}{7 \text{ k}\Omega} = 0.05 \ \Omega$$

Now it can be stated that the sensitivity of this particular bridge to measured resistances is $\pm 0.05 \ \Omega$. As explained in Section 2-3, this *sensitivity* is also a definition of the *precision* with which the bridge makes resistance measurements. The accuracy of measurements is still dependent on the accuracies of the individual components, as demonstrated in Example 7-5. However, the error due to bridge sensitivity (in this case $\pm 0.05 \ \Omega$) should always be much less than that due to the component accuracies.

The sensitivity of a Wheatstone bridge depends on the current sensitivity of the galvanometer (see Section 3-2), the galvanometer resistance, and the bridge supply voltage. To calculate the bridge sensitivity it is necessary to "look into" the bridge circuit from the galvanometer terminals. The bridge circuit is replaced with its *Thévenin equivalent circuit*, that is, by its open-circuit output voltage in series with its internal resistance. The effect of minimum detectable galvanometer current can now be calculated.

Figure 7-4(a) shows the bridge with the galvanometer removed. The open-circuit output voltage across the terminals at which the galvanometer is connected is

$$V_R - V_S = \frac{E_B R}{R + P} - \frac{E_B S}{Q + S}$$

or

$$V_R - V_S = E_B \left(\frac{R}{R + P} - \frac{S}{Q + S} \right) \tag{7-4}$$

The internal resistance of the bridge (i.e., "seen" from the galvanometer terminals) is determined by first assuming that the voltage supply E_B has an internal resistance which is very much smaller than the resistance of the bridge components. This assump-

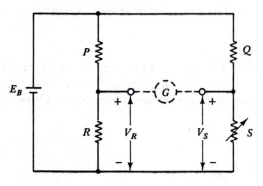

(a) Open-circuit voltage at the galvanometer terminals is $(V_R - V_S)$

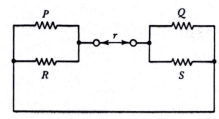

(b) Internal resistance is $r = P \| R + Q \| S$

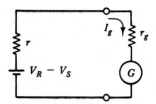

(c) Thévenin equivalent circuit

Figure 7-4 Development of the Thévenin equivalent circuit for the Wheatstone bridge. The sensitivity of the bridge to changes in the unknown resistance can be determined using the equivalent circuit, component values, and galvanometer specification.

tion gives the circuit of Figure 7-4(b) when E_b is replaced with a short circuit. Thus, it is seen that the internal resistance of the bridge is

$$r = P\|R + Q\|S \qquad (7\text{-}5)$$

The complete equivalent circuit of the bridge and galvanometer is drawn in Figure 7-4(c). The galvanometer current is now

$$I_g = \frac{V_R - V_S}{r + r_g} \qquad (7\text{-}6)$$

Example 7-6 demonstrates the procedure for calculation of the bridge sensitivity.

Example 7-6

A Wheatstone bridge has $P = 3.5$ kΩ, $Q = 7$ kΩ, and $S = 4$ kΩ when $R = 2$ kΩ. The supply is $E_B = 10$ V, and the galvanometer has a current sensitivity of 1 μA/mm and a resistance $r_g = 2.5$ kΩ. Calculate the minimum change in R which is detectable by the bridge.

Solution

Equation 7-5, $\qquad r = P\|R + Q\|S$

$$= 3.5 \text{ k}\Omega \| 2 \text{ k}\Omega + 7 \text{ k}\Omega \| 4 \text{ k}\Omega$$

$$= 3.82 \text{ k}\Omega$$

From Equation 7-6,

$$V_R - V_S = I_g(r + r_g)$$

$$= 1 \ \mu\text{A} \ (3.82 \text{ k}\Omega + 2.5 \text{ k}\Omega)$$

$$= 6.32 \text{ mV}$$

When the bridge is balanced,

$$V_R = \frac{E_B \times R}{P + R} = \frac{10 \text{ V} \times 2 \text{ k}\Omega}{3.5 \text{ k}\Omega + 2 \text{ k}\Omega}$$

$$= 3.636 \ 36 \text{ V}$$

When the open-circuit galvanometer voltage is 6.32 mV,

$$V_R + \Delta V_R = 3.636 \ 36 \text{ V} + 6.32 \text{ mV}$$

$$= 3.642 \ 68 \text{ V}$$

and $\qquad V_P = E - (V_R + \Delta V_R) = 10 \text{ V} - 3.642 \ 68 \text{ V}$

$$= 6.357 \ 32 \text{ V}$$

$$I_P = I_R = \frac{V_P}{P} = \frac{6.357 \ 32 \text{ V}}{3.5 \text{ k}\Omega}$$

$$= 1.816 \text{ mA}$$

so $\qquad R + \Delta R = \frac{V_R + \Delta V_R}{I_R} = \frac{3.642 \ 68 \text{ V}}{1.816 \text{ mA}}$

$$= 2.0059 \text{ k}\Omega$$

$$\Delta R = (R + \Delta R) - R = 2.0059 \text{ k}\Omega - 2 \text{ k}\Omega$$

$$= 5.9 \ \Omega$$

Referring to Equation 7-6, it is seen that if the difference between V_R and V_s is increased, the galvanometer current I_g is increased. $V_R - V_S$ can be increased by using a larger supply voltage E_B. Also from Equation 7-6, if the lowest detectable change in I_g is made smaller by using a more sensitive galvanometer, $V_R - V_S$ can be reduced. Thus, the bridge sensitivity can be improved either by increasing E_B or by using a more sensitive galvanometer.

Range of Measurement

For accurate resistance measurements by the Wheatstone bridge, the resistances to be measured must always be much greater than contact and connecting-lead resistances. When these quantities are an appreciable fraction of the resistance to be measured, the lower limit of the bridge has been reached. Figure 7-5 illustrates the effect of connecting-lead resistance. The galvanometer could be connected to terminal a or b. If the connecting lead resistance is $Y\,\Omega$, from Equation 7-3 the measured resistance is either

$$R = \frac{(S+Y)P}{Q} \qquad \text{or} \qquad R = \frac{SP}{Q+Y}$$

The practical lower limit for accurate resistance measurement by the Wheatstone bridge is found to be approximately 5 Ω. Very low resistance measurement requires special techniques and a different kind of resistance bridge. This is discussed in Section 7-4. The Wheatstone bridge may be used for measurement of very high resistances, although again special techniques are required (see Section 7-5). The upper limit of measurement for the Wheatstone bridge is about $10^{12}\,\Omega$.

Using a Wheatstone Bridge

Commercially available Wheatstone bridges normally have terminals for connecting an external supply. Care must be taken to ensure that the supply voltage used does not exceed the specified maximum for the instrument. The supply voltage and the galvanometer are usually connected to the bridge circuit by means of pushbutton switches that must be held down to maintain contact. With the unknown resistor connected, the supply button is pressed and held, and the galvanometer button is tapped to check the direction and amplitude of deflection. The resistive components of the bridge are adjusted with the supply and galvanometer switched *off*, then the buttons are again pressed and the galvanometer deflection is observed. The procedure is continued until the galvanometer indicates null, and further continued as the galvanometer sensitivity is increased. This method protects the galvanometer from excessive current levels, and reduces the possibility of component resistance variations due to the heating effects of continuous current.

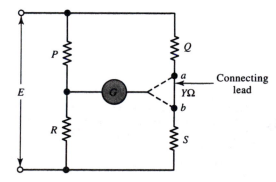

Figure 7-5 The connecting leads in a Wheatstone bridge circuit can introduce errors when measuring very low resistances. The $Y\,\Omega$ resistance could be taken as part of Q, or part of S.

7-4 LOW RESISTANCE MEASUREMENT

Four-Terminal Resistors

Low-value resistors such as ammeter shunts must have the resistor terminals accurately defined. This is to avoid errors introduced by contacts that are carrying heavy currents. Two sets of terminals are provided: *current terminals* and *potential terminals* (see Figure 7-6). The current terminals are the largest and are at the outer extremes of the resistor. Leads that carry large currents are connected to the current terminals. The potential terminals are situated between the two current terminals, and these normally handle currents in the microampere or milliampere range. Thus, there is no significant contact voltage drop at the potential terminals. The resistance of the component is defined as that existing between the potential terminals.

Kelvin Bridge

Referring to Figure 7-5, it was explained that the voltage drop across the connecting lead between a and b introduces serious error when measuring low-value resistances. The Kelvin bridge in Figure 7-7 shows essentially the same circuit as Figure 7-5, with additional resistors p and r. If the ratio p/r is exactly the same as P/R, the error due to the voltage drop across Y is eliminated. The balance equation for this bridge is a little more difficult to derive than that for the Wheatstone bridge.

When the galvanometer indicates null, there is zero current through the galvanometer, and the galvanometer terminal voltage is $V_g = 0$, as illustrated. With the bridge in this condition, a current i_1 flows through P, and the same current passes through R. Also, a current I flows through Q. This splits up at terminal a, so that i_2 flows through p and r, and $I - i_2$ flows through Y. The current flowing in S is again I.

Since there is no potential difference between the galvanometer terminals when the bridge is balanced, the voltage across R is equal to the sum of the voltage drops across r and S:

$$i_1 R = i_2 r + IS$$

from which

$$IS = i_1 R - i_2 r$$

or

$$IS = R\left(i_1 - \frac{i_2 r}{R}\right) \tag{1}$$

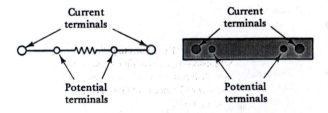

Current terminals

Potential terminals

Current terminals

Potential terminals

Figure 7-6 Four-terminal resistors have current terminals and potential terminals. The resistance is defined as that between the potential terminals, so that contact voltage drops at the current terminals do not introduce errors.

Low, High, and Precise Resistance Measurements   Chap. 7

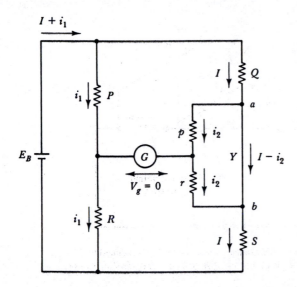

Figure 7-7 The Kelvin bridge is essentially a Wheatstone bridge with two additional resistors (*p* and *r*) included to eliminate the voltage drop across *Y* from the balance equation.

Also, the voltage drop across *P* equals the sum of the voltage drops across *p* and *Q*:

$$i_1 P = i_2 p + IQ$$

which gives

$$IQ = i_1 P - i_2 p$$

or

$$IQ = P\left(i_1 - \frac{i_2 p}{P}\right) \qquad (2)$$

Dividing Equation 2 by Equation 1 gives

$$\frac{IQ}{IS} = \frac{P(i_1 - i_2 p/P)}{R(i_1 - i_2 r/R)}$$

and *p/r* = *P/R* or *p/P* = *r/R*. Therefore, *Q/S* = *P/R*,

giving

$$\boxed{Q = \frac{SP}{R}} \qquad (7\text{-}7)$$

In this case, *Q* is the unknown resistance, *S* is a standard low-value resistor, and *P*, *R*, *p*, and *r* are precision adjustable resistors. The bridge may be balanced by adjustment of *P*, *R*, *p*, and *r*, always maintaining the ratio *p/r* = *P/R*. Then, the unknown low-value resistance can be determined by substituting into Equation 7-7.

Figure 7-8 shows the usual way that the Kelvin bridge circuit is shown. *S* and *Q* are seen to be four terminal resistors, and *P*, *p*, *R*, and *r* are connected at their potential terminals.

The range of measurement of a Kelvin bridge is typically from 10 $\mu\Omega$ (or 0.00001 Ω) to 1 Ω. Depending on the component accuracies, the bridge measurement accuracy can be ±0.2%. Even lower resistance values (down to 0.1 $\mu\Omega$) can be measured with reduced accuracy.

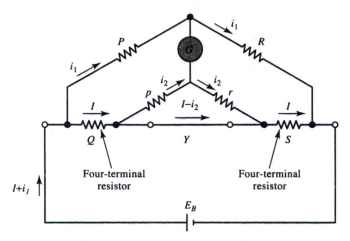

Figure 7-8 Kelvin bridge for very low resistance measurement. S is a standard four-terminal resistor, Q is the resistor to be measured, and $P, R, p,$ and r are precision resistors. When the bridge is balanced, $Q = SP/R$.

Like the Wheatstone bridge, the current sensitivity of the galvanometer is important in determining the measurement sensitivity (and precision) of the Kelvin bridge. The sensitivity of a Kelvin bridge is determined essentially by the same method as used for a Wheatstone bridge.

Example 7-7

A four-terminal resistor with an approximate value of $0.15 \ \Omega$ is to be measured by use of a Kelvin bridge. A standard resistor of $0.1 \ \Omega$ is available. Determine the required ratio of R/P and r/p.

Solution

$$S = 0.1 \ \Omega$$
$$Q \approx 0.15 \ \Omega$$

From Equation 7-7:
$$\frac{R}{P} = \frac{S}{Q} = \frac{0.1}{0.15}$$
$$= \frac{10}{15}$$

7-5 LOW-RESISTANCE MEASURING INSTRUMENTS

Low-Resistance Linear Ohmmeter

The linear ohmmeter circuit described in Section 4-3 can be applied to the measurement of low resistances. Figure 7-9(a) illustrates the method once again. A constant current is

Low, High, and Precise Resistance Measurements   Chap. 7

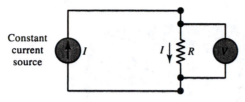

(a) Basic linear ohmmeter circuit

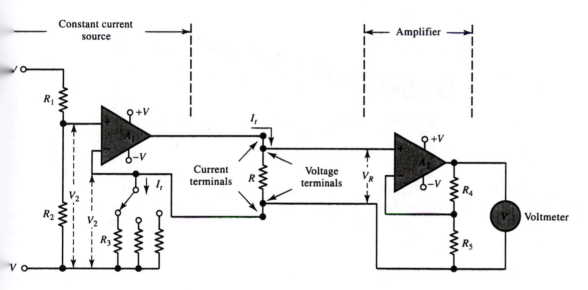

(b) Low-resistance ohmmeter circuit

Figure 7-9 A linear ohmmeter circuit consists of a constant-current source and a voltmeter. The measured voltage can be read as ohms when a suitable current level is employed. The small voltage drop across a very low resistance can be amplified for application to a voltmeter.

passed through the resistance to be measured, and the resistor voltage drop is monitored. For a 1 mA current, a 1 mV drop indicates a resistance of 1 Ω. A 0.1 mV drop indicates 0.1 Ω, and 0.5 mV gives 0.5 Ω. The voltages can be amplified to a measurable level and read as resistance values.

A possible circuit for a low-resistance meter is shown in Figure 7-9(b). Operational amplifier A_1 and its associated components constitute a constant-current circuit. The circuit is simply an op-amp noninverting amplifier. Because of the circuit feedback, voltage V_2 applied to the op-amp noninverting input terminal also appears across resistor R_3. Thus, with V_2 constant, the current through R_3 is a constant quantity (V_2/R_3), and this is the test current (I_t) which is passed through the resistance to be measured. The voltage drop across the resistance (V_R) is amplified by another noninverting amplifier (A_2), and passed to the voltmeter circuit, which might be either analog or digital.

Figure 7-10 Front panel of the Simpson model 444 micro-ohmmeter. Resistance measurement ranges are 20 mΩ to 20 Ω, and maximum sensitivity is 1 μΩ. (Courtesy of bach-simpson limited.)

Changing I_t up or down by a factor of 10 changes the measured resistances for a given voltage drop by a factor of 10. The current level is changed simply by selecting a new value of R_3. If the voltage drop is measured digitally, the decimal point position should be switched when the current level is altered. When measured by an analog meter, the new scale range must be noted.

Micro-ohmmeter

Figure 7-10 shows the front panel of a commercial digital micro-ohmmeter which has resistance ranges from 20 mΩ to 20 Ω. Because of the 4½-digit display, the instrument has a maximum sensitivity of 1 μΩ on the 20 mΩ range. Resistance measurement can be made in two modes: applying a dc voltage to the measured resistance, or applying a 40 Hz square-wave ac. An audible alarm tone is provided to indicate resistances below the selected range.

7-6 HIGH-RESISTANCE MEASUREMENT

Voltmeter and Ammeter Method

Very high resistances, such as insulation resistances, can be measured by means of a voltmeter and a microammeter connected as illustrated in Figure 7-11. A high-voltage supply must be used to produce a measurable current. The current is extremely low, so a microammeter is necessary for current measurement. The voltmeter must be connected across the supply. If the voltmeter is connected across the resistance, the microammeter will measure the voltmeter current, which is likely to be much greater than the resistor current. The value of the resistance is, of course, determined by substituting the measured current and voltage into Ohm's law.

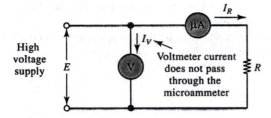

Guard Wire and Guard Ring

A problem occurs with the measurement of very high resistances because there are two resistive components: a *volume resistance* and a *surface leakage resistance*. Consider the metal-sheathed cable illustrated in Figure 7-12(a). When a voltage is applied, there are two components of current: a volume current I_v, which flows from the core through the insulation to the metal sheath, and a surface leakage current I_s, which flows across the surface of the insulation. Both currents flow through the microammeter. The consequence of this is that the resistance calculated from the instrument readings is the parallel combination of volume resistance and surface leakage resistance. For most practical purposes, this combined surface and volume resistance is the effective resistance of the insulation. However, in some circumstances, the two resistances must be separated.

To separate the two resistive components, a *guard wire* is employed, as illustrated in Figure 7-12(b). This is simply several turns of wire wrapped tightly around the insulation. The guard wire is connected to the supply, so I_s (from the guard ring to the sheath) no longer flows through the microammeter. There is no significant surface leakage current between the conductor and the guard wire, because the potential difference between conductor and guard wire is only the voltage drop across the microammeter. All of the

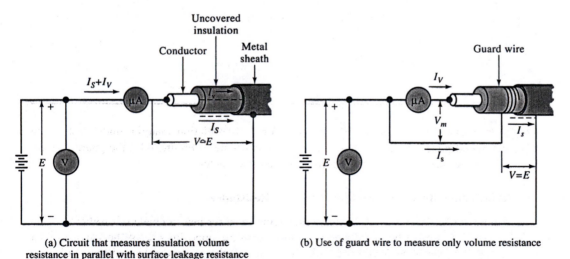

(a) Circuit that measures insulation volume resistance in parallel with surface leakage resistance

(b) Use of guard wire to measure only volume resistance

Figure 7-12 Guard-wire method of measuring the insulation resistance of a cable. The guard wire diverts the the surface leakage current so that it is not measured by the microammeter.

supply voltage appears across the insulation between the guard wire and sheath, so the surface leakage current here is still relatively large.

Since the microammeter measures only I_V in the arrangement shown in Figure 7-12(b), the volume resistance of the insulation is easily calculated:

$$r_V = \frac{E}{I_V} \qquad (7\text{-}8)$$

Example 7-8

The insulation of a metal-sheathed electric cable is tested using a 10 000 V supply and a microammeter. A current of 5 μA is measured when the components are connected as in Figure 7-12(a). When the circuit is arranged as in Figure 7-12(b), the current is 1.5 μA. Calculate (a) the volume resistance of the cable insulation and (b) the surface leakage resistance.

Solution

(a) *Volume resistance:*

$$I_V = 1.5 \ \mu A$$

$$r_V = \frac{E}{I_V} = \frac{10\ 000 \ V}{1.5 \ \mu A} = 6.7 \times 10^9 \ \Omega$$

(b) *Surface leakage resistance:*

$$I_V + I_S = 5 \ \mu A$$

$$I_S = 5 \ \mu A - I_V = 5 \ \mu A - 1.5 \ \mu A$$

$$= 3.5 \ \mu A$$

$$r_S = \frac{E}{I_S} = \frac{10\ 000 \ V}{3.5 \ \mu A} = 2.9 \times 10^9 \ \Omega$$

Figure 7-13 shows a disk-shaped sample of insulation material under test, using two metal plates and a *guard ring*. Like the guard wire in Figure 7-12, the guard ring removes the surface leakage current from the microammeter.

Wheatstone Bridge Measurement of High Resistance

Figure 7-14(a) shows a Wheatstone bridge employed for measurement of insulation resistance. In this case the guard ring is connected to the opposite side of the galvanometer from the upper plate. When the bridge is balanced, the galvanometer indicates null, and there is zero potential difference across its terminals. Thus, there is zero potential difference between the upper plate and the guard ring, and consequently, no surface leakage current flows from the upper plate to the guard ring.

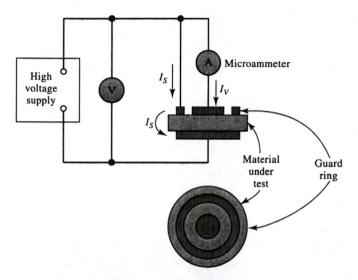

Figure 7-13 Guard-ring method of measuring the resistance of a sample of insulating material. The guard ring diverts the surface leakage current so that only the volume current passes through the microammeter.

In Figure 7-14(b) the insulation sample is replaced by its equivalent circuit. Resistance R represents the volume resistance of the material. Resistance b represents the surface resistance between the upper plate and the guard ring, while c is the surface resistance between the guard ring and lower plate. As already stated, when the bridge is in

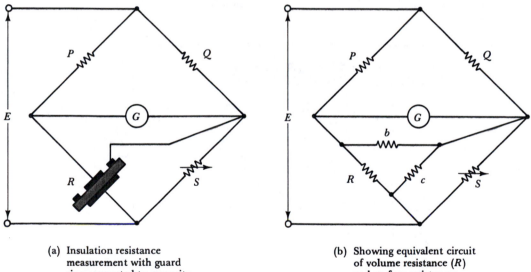

(a) Insulation resistance measurement with guard ring connected to opposite side of galvanometer from upper plate

(b) Showing equivalent circuit of volume resistance (R) and surface resistances (b and c)

Figure 7-14 Use of a Wheatstone bridge to measure very high resistances. The surface leakage resistance is eliminated when the bridge is balanced.

balance, there is zero voltage across *b*, so it can be ignored. Resistance *c* appears in parallel with the bridge resistor *S*. If *c* is very much larger than *S* (which is usually the case), *c* too can be ignored. The bridge is now reduced to its usual four components, with volume resistance *R* being the unknown resistance to be determined. Once again, Equation 7-3 is applicable.

When used for high-resistance measurements, a Wheatstone bridge must have a very high supply voltage. All components and terminals must be well insulated, and great care should be exercised to avoid electric shock.

7-7 HIGH-RESISTANCE MEASURING INSTRUMENTS

Hand-Cranked Megohmmeter

The *megohmmeter,* or *megger,* is a portable deflection instrument widely used to check the insulation resistance of electrical cables and equipment. The instrument has a constant high-voltage source, usually produced by a hand-cranked generator. The voltage may range anywhere between 100 V and 5000 V. The circuit diagram of a megger is shown in Figure 7-15(a), and a representative instrument is illustrated in Figure 7-15(b). The pointer on a hand-cranked megger is deflected by a PMMC system with two coils [see Figure 7-15(a)]. There is no mechanical controlling force; instead, the coils are connected to oppose each other. One coil is identified as a *control coil*, and the other as a *deflection coil.*

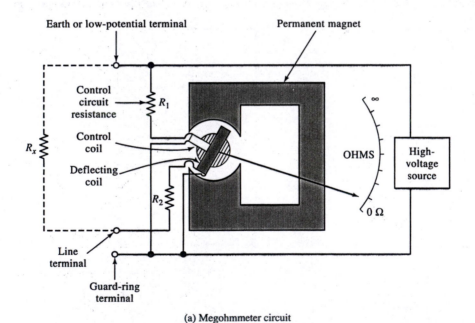

(a) Megohmmeter circuit

Figure 7-15 Megohmmeter that uses a hand-cranked generator to produce a high-voltage source. The forces from the control and deflecting coils partially balance each other to give a pointer deflection proportional to the measured resistance.

Voltage is applied to the control coil via standard resistor R_1, so that the controlling force is proportional to the generator voltage divided by R_1. The deflection coil is supplied via R_x, the resistance to be measured, and R_2, the internal deflection circuit resistance. The deflecting force is proportional to the generator voltage divided by $R_x + R_2$. Deflection is proportional to the difference between R_1 and $R_x + R_2$, and the instrument scale can be calibrated to directly indicate R_x.

When the megger is measuring an open circuit, no current flows in the deflecting coil. In this case, the force from the control coil causes the pointer to be deflected to one end of the scale. This end is marked *infinity* (∞). When measuring a short circuit, the deflecting coil force is very much greater than that from the control coil. Consequently, the pointer is deflected to the opposite end of the scale from infinity, and this end is marked 0 Ω. When the pointer is stationary at the center of the scale, the deflecting and control forces are equal, and $R_x + R_2 = R_1$, or $R_x = R_1 - R_2$. The scale is marked equal to $(R_1 - R_2)$ Ω at this point and is proportionately marked at other points. Range changing can be effected by switching to different values of R_2. The megohmmeter in Figure 7-15(b) has two ranges: (0–500 MΩ) with 5 MΩ at midscale, and (0–200 Ω) with 10 Ω at midscale. The higher resistance range uses 1000 V, and the lower range has a 6 V supply.

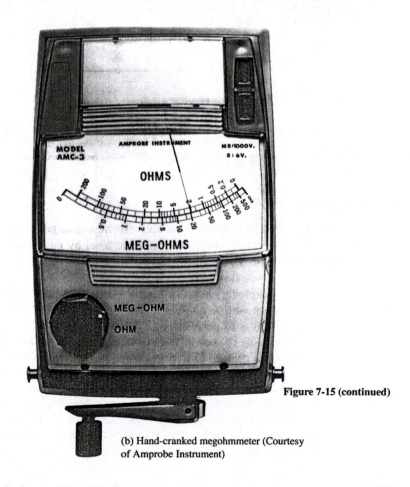

Figure 7-15 (continued)

(b) Hand-cranked megohmmeter (Courtesy of Amprobe Instrument)

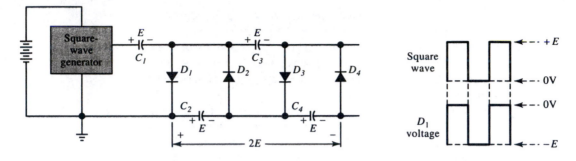

(a) Diode-capacitor voltage multiplying circuit and waveforms

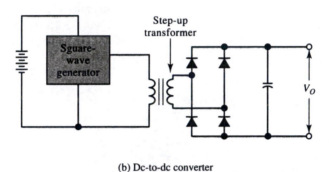

(b) Dc-to-dc converter

Figure 7-16 Circuits for producing a high voltage from a battery. In the voltage-multiplying circuit, the capacitor voltages add up to give the desired high voltage. The dc-to-dc converter circuit uses a transformer to step-up the square-wave voltage. This is rectified and smoothed to convert it back to a dc voltage.

Voltage-Multiplying Circuit

Instead of a hand-cranked generator to produce the high voltage required for insulation testing, a battery voltage can be raised to a suitable level by means of a diode/capacitor *voltage-multiplying circuit.* The circuit is shown in Figure 7-16 together with a possible alternative circuit, a *dc-to-dc converter.*

   In the voltage-multiplying circuit in Figure 7-16(a), a square-wave generator produces an output waveform with an amplitude that approximately equals the battery voltage (E). When the wave is at its positive peak, diode D_1 is forward biased, and capacitor C_1 is charged with the polarity shown. For convenience, assume that the diode voltage drop is zero and that C_1 is charged to E volts. When the square wave is at its zero level, the positive side of C_1 is at zero; consequently, the negative side is at $-E$. This causes diode D_2 to be forward biased and capacitor C_2 to be charged (from C_1) to approximately E volts with the polarity illustrated. Now, when the square-wave amplitude goes to E once again, D_1 is forward biased, D_2 is reverse biased, and capacitor C_3 is charged from C_2 via D_3. Similarly, when D_2 is forward biased and D_3 is reversed, C_4 is charged from C_3 via diode D_4. As shown, the total voltage across C_2 and C_4 is now 2E. Actually it is less than $2E$ because of the loss of voltage across the forward-biased diodes. However, the voltage is increased substantially above the batery voltage level. When a large number of diodes and transistors are involved, the battery voltage can be multiplied to a high enough level for insulation testing.

Direct voltage can also be increased to a high dc level by means of the dc-to-dc converter in Figure 7-16(b). The output of the square-wave generator in this circuit is applied to a step-up transformer to produce a high-voltage secondary. This is then rectified and smoothed to convert it to a high dc voltage.

Battery-Powered Megohmmeters

The two battery-powered megohmmeters shown in Figure 7-17 have pushbutton switches for closing the resistance measuring circuits. The circuit may be closed for only the brief time necessary to take a resistance reading, thus minimizing the current drain on the batteries. The digital instrument in Figure 7-17(a) has a 1000 V maximum test voltage, and resistance ranges from 0–200 Ω up to 0–2000 MΩ. The specified accuracy of measurement is $\pm$(1% of reading + 2 LSD) for the lowest resistance range, and $\pm$(3.5% of reading + 3 LSD) for the highest range. The analog instrument in Figure 7-17(b) has a nonlinear scale. Its maximum range is 0–100 MΩ with 2.5 MΩ at center scale, and the maximum dc test voltage is 500 V. Measurement accuracy is specified as $\pm$5% of the reading up to 10 MΩ, and $\pm$10% for resistances above 10 MΩ.

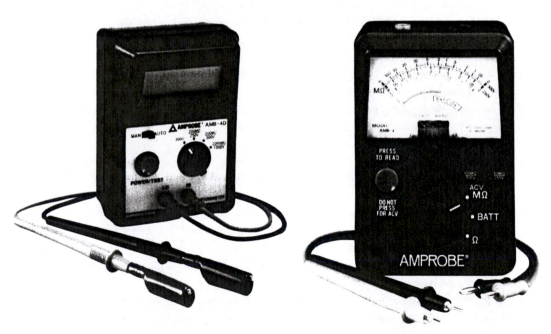

(a) Digital battery-powered megohmmeter   (b) Analog battery-powered megohmmeter

Figure 7-17 Digital and analog battery-powered megohmmeters. The digital instrument uses a 1000 V maximum test voltage, and measures resistance up to 2000 MΩ. The analog meter uses 500 V, and its maximum range goes to 100 MΩ. (Courtesy of Amprobe Instrument.)

REVIEW QUESTIONS

7-1 Show how an ammeter, a voltmeter, and a dc supply can be used to measure a resistance. Show the two possible connections, write the resistance equations for each, and discuss the errors.

7-2 Sketch circuit diagrams to show how a voltmeter and ammeter should be connected to measure **(a)** a very high resistance and **(b)** a very low resistance. Explain briefly.

7-3 Explain the substitution method of resistance measurement. Draw an appropriate diagram.

7-4 Sketch the circuit diagram of a Wheatstone bridge, explain its operation, and derive the equation for the unknown resistance.

7-5 Using the component tolerances, write an equation for the measurement error in a Wheatstone bridge.

7-6 Explain what is meant by the sensitivity of a Wheatstone bridge.

7-7 Draw the Thévenin equivalent circuit for a Wheatstone bridge, and derive an equation for the galvanometer current.

7-8 Explain the source of error in a Wheatstone bridge used to measure very low resistances. Discuss the range of resistance measurement with a Wheatstone bridge.

7-9 Draw the circuit symbol for a four-terminal resistor. Explain briefly.

7-10 Draw the circuit of a Kelvin bridge, explain its operation, and derive the equation for the unknown resistance.

7-11 Sketch a basic circuit for a low-resistance ohmmeter. Explain its operation.

7-12 Show how an ammeter and voltmeter should be used for the measurement of very high resistances. Explain briefly.

7-13 Draw a circuit diagrams to show how the insulation resistance of a cable should be measured. Show how the volume resistance and the surface leakage resistance can be separated. Explain.

7-14 Sketch a diagram to illustrate the guard ring method of measuring a sample of insulating material. Explain how the surface leakage and volume resistances are determined.

7-15 Sketch a circuit diagram to show how a Wheatstone bridge should be used for very high resistance measurements. Explain how the surface leakage resistance is eliminated from the equation for the unknown resistance.

7-16 Redraw the diagram for Question 7-15 to show how only the surface leakage resistance may be measured by the Wheatstone bridge.

7-17 Sketch the circuit diagram of a megohmmeter, and explain how it operates.

7-18 Draw the circuit diagram of a diode/capacitor voltage multiplying circuit. Explain how this circuit produces a high dc voltage from a battery.

7-19 Sketch the basic circuit of a dc-to-dc converter, and explain its operation.

7-20 Discuss typical measuring voltages and resistance ranges for **(a)** a hand-cranked megger and **(b)** a battery-powered megger.

PROBLEMS

7-1 An ammeter is connected in series with an unknown resistance (R_x) and a dc supply. A voltmeter is connected directly across the supply. The ammeter resistance is 10 Ω, and the voltmeter sensitivity is 10 kΩ/V. Determine R_x if the ammeter indicates 0.5 A and the voltmeter reading is 500 V.

7-2 In Problem 7-1, the ammeter is on a 1 A range, and the voltmeter is on a 1000 V range. Calculate the maximum and minimum resistances for R_x if both instruments are accurate to ±1%.

7-3 A voltmeter is connected directly in parallel with an unknown resistance (R_x), and an ammeter is directly in series with a dc supply connected to the resistance and voltmeter. The ammeter is on a 1 mA range and its resistance is 0.1 Ω. The voltmeter, which has a sensitivity of 10 kΩ/V, is on its 5 V range. Determine **(a)** the nominal resistance of R_x and **(b)** the maximum and minimum resistance of R_x if the voltmeter and ammeter accuracies are ±1%. The indicated quantities are 500 μA and 4 V.

7-4 A Wheatstone bridge, as in Figure 7-3, has $P = 100$ Ω, $Q = 150$ Ω, and $S = 119.25$ Ω when balanced. Determine the unknown resistance.

7-5 Calculate the maximum and minimum resistance that can be measured by a Wheatstone bridge (as in Figure 7-3) which can have $P = $ (1 kΩ, 5 kΩ, or 10 kΩ), $Q = $ (1 kΩ, 5 kΩ, or 10 kΩ), and S adjustable from 1 kΩ to 6 kΩ.

7-6 A Wheatstone bridge has a 10 V supply, all four resistors equal to 5 kΩ, and a galvanometer with a 25 Ω resistance and a sensitivity of 0.5 μA/mm. Determine the minimum detectable change in the measured resistance.

7-7 Wheatstone bridge, as in Figure 7-3, has $P = 1$ kΩ and $Q = $ (100 Ω, 1 kΩ, or 100 kΩ). S is adjustable from 1 kΩ to 5 kΩ. Determine the range of resistance that can be measured.

7-8 All four resistors in a Wheatstone bridge are 1 kΩ, the galvanometer has a 100 Ω resistance and 0.05 μA/mm sensitivity, and the supply is 20 V. Determine the minimum change that can be detected in the measured resistance.

7-9 Determine the accuracy of measurement of a Wheatstone bridge that uses precision resistors with an accuracy of ±0.025% and an adjustable precision resistor that has an accuracy of ±0.05%.

7-10 A four-terminal resistor with a nominal resistance of 0.025 Ω is to be measured on a Kelvin bridge (as in Figure 7-8). A 0.01 Ω standard resistor is available. Determine the required ratio of R/P and r/p.

7-11 Calculate the measured resistance for a Kelvin bridge that has $P = p = 12$ kΩ, $R = r = 3.673$ kΩ, and $S = 0.015$ Ω when balanced.

7-12 Determine the accuracy of measurement of the Kelvin bridge in Problem 7-11 if the precision resistors (P, p, R, r) have an accuracy of ±0.05% and the accuracy of S is ±0.1%.

7-13 The resistance of a disk of insulating material is measured as in Figure 7-13. With a 12 000 V supply and the guard ring not connected, the measured current is 1.2 μA.

The indicated current falls to 0.045 µA when the ring is connected directly to the supply. Calculate the volume and surface leakage resistances of the material.

7-14 The insulation resistance of a metal-sheathed electric cable is to be measured as in Figure 7-12. A 15 kV supply is used, and the measured current is 3.5 µA without the guard wire, and 2 µA with the guard wire. Calculate the volume resistance and the surface leakage resistance of the cable insulation.

Inductance and Capacitance Measurements

8

Objectives

You will be able to:

1. Sketch *RC* series and parallel equivalent circuits for a capacitor, and write equations relating the two circuits.

2. Sketch *RL* series and parallel equivalent circuits for an inductor, and write equations relating the two circuits.

3. Explain the *Q* factor of an inductor and the *D* factor of a capacitor, and write the equations for each factor.

4. Draw circuit diagrams for the following ac bridges: simple capacitance bridge, series-resistance capacitance bridge, parallel-resistance capacitance bridge, inductance comparison bridge, Maxwell bridge, and Hay inductance bridge.

5. Explain the operation of each of the bridges listed above, derive the equations for the quantities to be measured, and discuss the advantages and disadvantages of each bridge.

6. Sketch ac bridge circuit diagrams showing how a commercial multifunction impedance bridge uses a standard capacitor and three adjustable standard resistors to measure a wide range of capacitances and inductances. Explain.

7. Discuss the problems involved in measuring small *R, L,* and *C* quantities, explain suitable measuring techniques, and calculate measured quantities.

8. Sketch and explain the basic circuits for converting inductance and capacitance into voltages for digital measurements. Discuss the specification and performance of a digital *RLC* meter.

9. Draw the circuit diagram for a *Q* meter, explain its operation and controls, and determine the *Q* of a coil from the *Q* meter measurements.

Introduction

Inductance, capacitance, inductor Q factor, and capacitor D factor can all be measured precisely on ac bridges, which are adaptations of the Wheatstone bridge. An ac supply must be used, and the null detector must be an ac instrument. A wide range of ac bridge circuits are available for various specialized measurements. Some commercial ac bridges use only a standard capacitor and three adjustable standard resistors to construct several different types of inductance and capacitance bridge circuits. Special techniques must be employed for measuring very small inductance and capacitance quantities. For digital measurement, inductance, capacitance, and resistance are first applied to circuits that convert each quantity into a voltage. Capacitors and inductors that are required to operate at high frequencies are best measured on a Q meter.

8-1 *RC* AND *RL* EQUIVALENT CIRCUITS

Capacitor Equivalent Circuits

The equivalent circuit of a capacitor consists of a pure capacitance C_P and a parallel resistance R_P, as illustrated in Figure 8-1(a). C_P represents the actual capacitance value, and R_P represents the resistance of the dielectric or *leakage resistance*. Capacitors that have a high leakage current flowing through the dielectric have a relatively low value of R_P in their equivalent circuit. Very low leakage currents are represented by extremely large values of R_P. Examples of the two extremes are electrolytic capacitors that have high leakage currents (low parallel resistance), and plastic film capacitors which have very low leakage (high parallel resistance). An electrolytic capacitor might easily have several microamperes of leakage current, while a capacitor with a plastic film dielectric could typically have a resistance as high as 100 000 MΩ.

A parallel *RC* circuit has an equivalent series *RC* circuit [Figure 8-1(b)]. Either one of the two equivalent circuits (series or parallel) may be used to represent a capacitor in a circuit. It is found that capacitors with a high-resistance dielectric are best represented by the series *RC* circuit, while those with a low-resistance dielectric should be represented by the parallel equivalent circuit. However, when the capacitor is measured in terms of the series C and R quantities, it is usually desirable to resolve them into the parallel

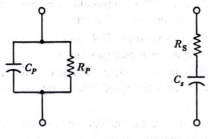

(a) Parallel equivalent circuit (b) Series equivalent circuit

Figure 8-1 A capacitor may be represented by either a parallel equivalent circuit or a series equivalent circuit. The parallel equivalent circuit best represents capacitors that have a low-resistance dielectric, while the series equivalent circuit is most suitable for capacitors with a high-resistance dielectric.

equivalent circuit quantities. This is because the (parallel) leakage resistance best represents the quality of the capacitor dielectric. Equations that relate the series and parallel equivalent circuits are derived below.

Referring to Figure 8-1, the series impedance is

$$Z_s = R_s - jX_s$$

and the parallel admittance is

$$Y_p = \frac{1}{R_p} + j\frac{1}{X_p} = G_p + jB_p$$

where G is conductance and B is susceptance. The impedances of each circuit must be equal.

Thus,
$$Z_s = \frac{1}{Y_p}$$

giving
$$R_s - jX_s = \frac{1}{G_p + jB_p}$$

or
$$G_p + jB_p = \frac{1}{R_s - jX_s}$$

$$= \frac{1}{R_s - jX_s}\left(\frac{R_s + jX_s}{R_s + jX_s}\right)$$

giving
$$G_p + jB_p = \frac{R_s + jX_s}{R_s^2 + X_s^2}$$

Equating the real terms,

$$G_p = \frac{R_s}{R_s^2 + X_s^2}$$

or
$$\boxed{R_p = \frac{R_s^2 + X_s^2}{R_s}}$$
(8-1)

Equating the imaginary terms,

$$B_p = \frac{X_s}{R_s^2 + X_s^2}$$

or
$$\boxed{X_p = \frac{R_s^2 + X_s^2}{X_s}}$$
(8-2)

The equations above can be shown to apply also to equivalent series and parallel *RL* circuits, as well as *RC* circuits.

Inductor Equivalent Circuits

Inductor equivalent circuits are illustrated in Figure 8-2. The series equivalent circuit in Figure 8-2(a) represents an inductor as a pure inductance L_s in series with the resistance of its coil. This series equivalent circuit is normally the best way to represent an inductor, because the actual winding resistance is involved and this is an important quantity. Ideally, the winding resistance should be as small as possible, but this depends on the thickness and length of the wire used to wind the coil. Physically small high-value inductors tend to have large resistance values, while large low-inductance components are likely to have low resistances.

The parallel RL equivalent circuit for an inductor [Figure 8-2(b)] can also be used. As in the case of the capacitor equivalent circuits, it is sometimes more convenient to use a parallel RL equivalent circuit rather than a series circuit. The equations relating the two are derived below.

Referring to Figure 8-2, the series circuit impedance is

$$Z_s = R_s + jX_s$$

and the parallel circuit admittance is

$$Y_p = \frac{1}{R_p} - j\frac{1}{X_p}$$

$$Y_p = G_p - jB_p$$

$$Z_s = Z_p$$

or

$$R_s + jX_s = \frac{1}{G_p - jB_p}$$

$$R_s + jX_s = \frac{1}{G_p - jB_p}\left(\frac{G_p + jB_p}{G_p + jB_p}\right)$$

giving

$$R_s + jX_s = \frac{G_p + jB_p}{G_p^2 + B_p^2}$$

R_S

L_S

(a) Series equivalent circuit

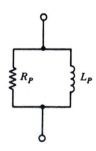

R_P L_P

(b) Parallel equivalent circuit

Figure 8-2 An inductor may be represented by either a parallel equivalent circuit or a series equivalent circuit. The series equivalent circuit is normally used, but it is sometimes convenient to employ the parallel equivalent circuit.

Inductance and Capacitance Measurements   Chap. 8

Equating the real terms,

$$R_s = \frac{G_p}{G_p^2 + B_p^2}$$

$$= \frac{1/R_p}{1/R_p^2 + 1/X_p^2} \left(\frac{R_p^2 X_p^2}{R_p^2 X_p^2} \right)$$

$$\boxed{R_s = \frac{R_p X_p^2}{X_p^2 + R_p^2}} \qquad (8\text{-}3)$$

Equating the imaginary terms,

$$X_s = \frac{B_p}{G_p^2 + B_p^2}$$

$$= \frac{1/X_p}{1/R_p^2 + 1/X_p^2} \left(\frac{R_p^2 X_p^2}{R_p^2 X_p^2} \right)$$

$$\boxed{X_s = \frac{R_p^2 X_p}{X_p^2 + R_p^2}} \qquad (8\text{-}4)$$

Like Equations 8-1 and 8-2, Equations 8-3 and 8-4 apply to both *RC* and *RL* circuits.

Q Factor of an Inductor

The quality of an inductor can be defined in terms of its power dissipation. An ideal inductor should have zero winding resistance, and therefore zero power dissipated in the winding. A *lossy* inductor has a relatively high winding resistance; consequently it does dissipate some power. The *quality factor*, or *Q factor*, of the inductor is the ratio of the inductive reactance and resistance at the operating frequency.

$$\boxed{Q = \frac{X_s}{R_s} = \frac{\omega L_s}{R_s}} \qquad (8\text{-}5)$$

where L_s and R_s refer to the components of an *RL* series equivalent circuit [Figure 8-2(a)]. Ideally, ωL_s should be very much larger than R_s, so that a very large *Q* factor is obtained. *Q* factors for typical inductors range from a low of less than 5 to as high as 1000 (depending on frequency).

As discussed earlier, an inductor may be represented by either a series equivalent circuit or a parallel equivalent circuit. When the parallel equivalent circuit is employed, the *Q* factor can be shown to be

$$Q = \frac{R_p}{X_p} = \frac{R_p}{\omega L_p} \qquad\qquad (8\text{-}6)$$

D Factor of a Capacitor

The quality of a capacitor can be expressed in terms of its power dissipation. A very pure capacitance has a high dielectric resistance (low leakage current) and virtually zero power dissipation. A *lossy* capacitor, which has a relatively low resistance (high leakage current), dissipates some power. The *dissipation factor D* defines the quality of the capacitor. Like the Q factor of a coil, D is simply the ratio of the component reactance (at a given frequency) to the resistance measurable at its terminals. In the case of the capacitor, the resistance involved in the D-factor calculation is that shown in the parallel equivalent circuit. (This differs from the inductor Q-factor calculation, where the resistance is that in the series equivalent circuit.) Using the *parallel equivalent circuit*:

$$D = \frac{X_p}{R_p} = \frac{1}{\omega C_p R_p} \qquad\qquad (8\text{-}7)$$

Ideally, R_p should be very much larger than $1/(\omega C_p)$, giving a very small dissipation factor. Typically, D might range from 0.1 for electrolytic capacitors to less than 10^{-4} for capacitors with a plastic film dielectric (again depending on frequency).

When a series equivalent circuit is used, the equation for dissipation factor can be shown to be

$$D = \frac{R_s}{X_s} = \omega C_s R_s \qquad\qquad (8\text{-}8)$$

Comparing Equation 8-7 to 8-6 and Equation 8-8 to 8-5, it is seen that in each case D is the inverse of Q.

Example 8-1

An unknown circuit behaves as a 0.005 μF capacitor in series with a 8 kΩ resistor when measured at a frequency of 1 kHz. The terminal resistance is measured by an ohmmeter as 134 kΩ. Determine the actual circuit components and the method of connection.

Solution

$$X_s = \frac{1}{2\pi f C} = \frac{1}{2\pi \times 1 \text{ kHz} \times 0.005 \text{ μF}}$$

$$= 31.8 \text{ k}\Omega$$

$$R_s = 8 \text{ k}\Omega$$

$$\text{Equation 8-1,} \qquad R_p = \frac{R_s^2 + X_s^2}{R_s} = \frac{(8 \text{ k}\Omega)^2 + (31.8 \text{ k}\Omega)^2}{8 \text{ k}\Omega}$$

$$= 134 \text{ k}\Omega$$

$$\text{Equation 8-2,} \qquad X_p = \frac{R_s^2 + X_s^2}{X_s} = \frac{(8 \text{ k}\Omega)^2 + (31.8 \text{ k}\Omega)^2}{31.8 \text{ k}\Omega}$$

$$= 33.8 \text{ k}\Omega$$

$$C_p = \frac{1}{2\pi f X_p} = \frac{1}{2\pi \times 1 \text{ kHz} \times 33.8 \text{ k}\Omega}$$

$$\approx 0.005 \text{ }\mu\text{F}$$

Since the measured terminal resistance is 134 kΩ, the circuit must consist of a 0.005 μF capacitor connected in parallel with a 134 kΩ resistor. For a series-connected circuit, the terminal resistance would be much higher than 134 kΩ.

8-2 AC BRIDGE THEORY

Circuit and Balance Equations

The basic circuit of an ac bridge is illustrated in Figure 8-3. This is exactly the same as the Wheatstone bridge circuit (Figure 7-3) except that impedances are shown instead of resistances, and an ac supply is used. The null detector must be an ac instrument such as an electronic galvanometer, headphones, or an oscilloscope.

When the null detector indicates zero in the circuit of Figure 8-3, the alternating voltage across points a and b is zero. This means (as in the Wheatstone bridge) that the voltage across Z_1 is exactly equal to that across Z_2, and the voltage across Z_3 equals the voltage drop across Z_4. Not only are the voltages equal in amplitude, they are also equal

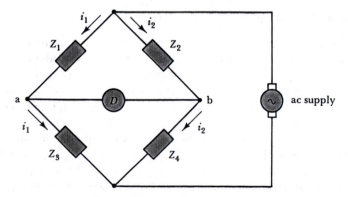

Figure 8-3 The basic ac bridge circuit is similar to the Wheatstone bridge except that impedances are involved instead of resistances. An ac supply must be employed, and the null detector must be an ac instrument.

in phase. If the voltages were equal in amplitude but not in phase, the ac null detector would not indicate zero.

$$V_{Z1} = V_{Z2}$$

so

$$i_1 Z_1 = i_2 Z_2 \qquad (1)$$

and

$$V_{Z3} = V_{Z4}$$

or

$$i_1 Z_3 = i_2 Z_4 \qquad (2)$$

Dividing Equation 1 by Equation 2,

$$\frac{i_1 Z_1}{i_1 Z_3} = \frac{i_2 Z_2}{i_2 Z_4}$$

giving

$$\boxed{\frac{Z_1}{Z_3} = \frac{Z_2}{Z_4}} \qquad (8\text{-}9)$$

As already stated, bridge balance is obtained only when the voltages at each terminal of the null detector are equal in phase as well as in magnitude. This results in Equation 8-9, which involves complex quantities. In such an equation, the real parts of the quantities on each side must be equal, and the imaginary parts of the quantities must also be equal. Therefore, when deriving the balance equations for a particular bridge, it is best to express the impedances in rectangular form rather than polar form. The real quantities can then be equated to obtain one balance equation, and the imaginary (or j quantities) can be equated to arrive at the other balance equation.

The need for two balance equations arises from the fact that capacitances and inductances are never pure; they must be defined as a combination of R and C or R and L (as discussed in Section 8-1). One balance equation permits calculation of L or C, and the other is used for determining the R quantity.

Balance Procedure

As already explained, two component adjustments are required to balance the bridge (or obtain a minimum indication on the null detector). These adjustments are *not* independent of each other; one tends to affect the relative amplitudes of the voltages at each terminal of the null detector, and the other adjustment has a marked effect on the relative phase difference of these voltages. For example, Z_4 in Figure 8-3 might consist of a variable capacitor in series with a variable resistor, as illustrated in Figure 8-4(a). Adjustment of C_4 may make V_{Z4} equal in amplitude to V_{Z3} without bringing it into phase with V_{Z3}. The result is, of course, that the null detector voltage $V_{Z3} - V_{Z4}$ is not zero [see Figure 8-4(b)]. Further adjustment of C_4 could alter the phase of V_{Z4} but will also alter its amplitude. If R_4 is now adjusted, $V_{Z3} - V_{Z4}$ might be further reduced by bringing the voltages closer together in phase. However, this cannot be achieved without altering the amplitude of V_{Z4}, which is the voltage drop across R_4 and C_4 [Figure 8-4(c)]. When the best null has been obtained by adjustment of R_4, C_4 is once again adjusted. This is likely to once more make

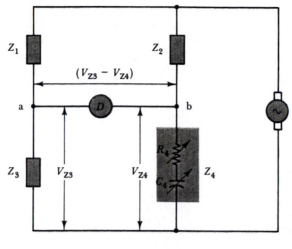

(a) Null detector voltage = $(V_{Z2} = V_{Z3})$

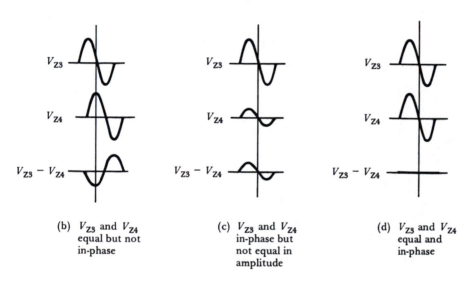

(b) V_{Z3} and V_{Z4}
equal but not
in-phase

(c) V_{Z3} and V_{Z4}
in-phase but
not equal in
amplitude

(d) V_{Z3} and V_{Z4}
equal and
in-phase

Figure 8-4   When an ac bridge is balanced, V_{Z3} must equal V_{Z4}, and the two voltages must be in phase. This requires alternately adjusting two quantities (R_4 and C_4 in this circuit) until the smallest possible null detector indication is achieved.

V_{Z4} close to V_{Z3} in amplitude, but again has an unavoidable effect on the phase relationship. The procedure of alternately adjusting R_4 and C_4 to minimize the null detector voltage is continued until the smallest possible indication is obtained. Then, V_{Z4} is equal to V_{Z3} both in magnitude and phase [Figure 8-4(d)].

AC Bridge Sensitivity

The same considerations that determined the sensitivity of a Wheatstone bridge apply to ac bridge circuits. The bridge sensitivity may be defined in terms of the smallest change in the measured quantity that causes the galvanometer to deflect from zero. Bridge sensitivity can be improved by using a more sensitive null detector and/or by increasing the level of supply voltage. The bridge sensitivity is analyzed by exactly the same method used for the Wheatstone bridge, except that impedances are involved instead of resistances. Accuracy of measurements is also determined in the same way as Wheatstone bridge accuracy.

8-3 CAPACITANCE BRIDGES

Simple Capacitance Bridge

The circuit of a *simple capacitance bridge* is illustrated in Figure 8-5(a). Z_1 is a standard capacitor C_1, and Z_2 is the unknown capacitance C_x. Z_3 and Z_4 are known variable resistors, such as decade resistance boxes. When the bridge is balanced, $Z_1/Z_3 = Z_2/Z_4$ (Equation 8-9) applies:

$$Z_1 = \frac{-j1}{\omega C_1} \qquad Z_2 = \frac{-j1}{\omega C_x}$$

$$Z_3 = R_3 \qquad \text{and} \qquad Z_4 = R_4$$

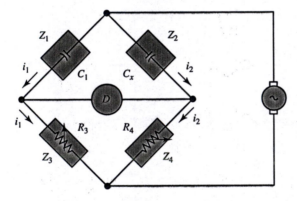

(a) Simple capacitance bridge

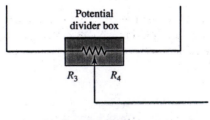

(b) Potential divider substituted for R_3 and R_4

Figure 8-5 The simple capacitance bridge measures the unknown capacitance C_x in terms of standard capacitor C_1 and adjustable precision resistors R_3 and R_4. At balance, $C_x = C_1 R_3 / R_4$. This circuit functions only with capacitors that have very high resistance dielectrics.

Inductance and Capacitance Measurements   Chap. 8

Therefore,
$$\frac{-j1/\omega C_1}{R_3} = \frac{-j1/\omega C_x}{R_4}$$

or
$$\frac{1}{C_1 R_3} = \frac{1}{C_x R_4}$$

giving
$$\boxed{C_x = \frac{C_1 R_3}{R_4}}$$
(8-10)

The actual resistances of R_3 and R_4 are not important if their ratio is known, so a potential-divider resistance box could be used as shown in Figure 8-5(b).

Example 8-2

The standard capacitance value in Figure 8-5 is $C_1 = 0.1$ μF, and R_3/R_4 can be set to any ratio between 100:1 and 1:100. Calculate the range of measurements of unknown capacitance C_x.

Solution

Equation 8-10, $\qquad C_x = \dfrac{C_1 R_3}{R_4}$

For $R_3/R_4 = 100:1$:

$$C_x = 0.1 \ \mu F \times \frac{100}{1}$$

$$= 10 \ \mu F$$

For $R_3/R_4 = 1:100$:

$$C_x = 0.1 \ \mu F \times \frac{1}{100}$$

$$= 0.001 \ \mu F$$

The foregoing analysis of the simple capacitance bridge assumes that the capacitors are absolutely pure, with effectively zero leakage current through the dielectric. If a resistance were connected in series or in parallel with C_x in Figure 8-5(a), and the rest of the bridge components remain as shown, balance would be virtually impossible to achieve. This is because i_1 and i_2 could not be brought into phase, and consequently, $i_1 R_3$ and $i_2 R_4$ would not be in phase. As discussed in Section 8-1, the equivalent circuit of a leaky capacitor is a pure capacitance in parallel with a pure resistance. Thus, the simple capacitance bridge is suitable only for measurement of capacitors with high-resistance dielectrics.

Series-Resistance Capacitance Bridge

In the circuit shown in Figure 8-6(a), the unknown capacitance is represented as a pure capacitance C_S in series with a resistance R_s. A standard adjustable resistance R_1 is connected in series with standard capacitor C_1. The voltage drop across R_1 balances the resistive voltage drops in branch Z_2 when the bridge is balanced. The additional resistor in series with C_s increases the total resistive component in Z_2, so that inconveniently small values of R_1 are not required to achieve balance. Bridge balance is most easily achieved when each capacitive branch has a substantial resistive component. To obtain balance, R_1 and either R_3 or R_4 are adjusted alternately. The *series-resistance capacitance bridge* is found to be most suitable for capacitors with a high-resistance dielectric (very low leakage current and low dissipation factor). When the bridge is balanced, Equation 8-9 applies,

$$\frac{Z_1}{Z_3} = \frac{Z_2}{Z_4}$$

giving

$$\frac{R_1 - j1/\omega C_1}{R_3} = \frac{R_s - j1/\omega C_s}{R_4} \qquad (8\text{-}11)$$

Equating the real terms in Equation 8-11,

$$\frac{R_1}{R_3} = \frac{R_s}{R_4}$$

giving

$$R_s = \frac{R_1 R_4}{R_3} \qquad (8\text{-}12)$$

Equating the imaginary terms in Equation 8-11,

$$\frac{1}{\omega C_1 R_3} = \frac{1}{\omega C_s R_4}$$

giving

$$C_s = \frac{C_1 R_3}{R_4} \qquad (8\text{-}13)$$

The phasor diagram for the series-resistance capacitance bridge at balance is drawn in Figure 8-6(b). The voltage drops across Z_3 and Z_4 are $i_1 R_3$ and $i_2 R_4$, respectively. These two voltages must be equal and in phase for the bridge to be balanced. Thus, they are drawn equal and in phase in the phasor diagram. Since R_3 and R_4 are resistive, i_1 is in phase with $i_1 R_3$ and i_2 is in phase with $i_2 R_4$. The impedance of C_1 is purely capacitive, and current leads voltage by 90° in a pure capacitance. Therefore, the capacitor voltage

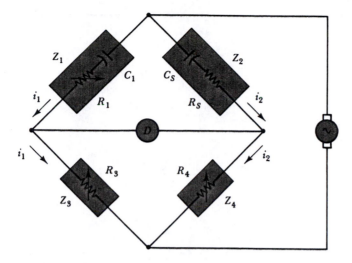

(a) Circuit of series-resistance capacitance bridge

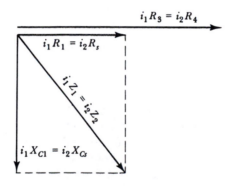

(b) Phasor diagram for balanced bridge

Figure 8-6 The series-resistance capacitance bridge is similar to the simple capacitance bridge, except that an adjustable series resistance (R_1) is included to balance the resistive component (R_s) of Z_2. This bridge is most suitable for measuring capacitors with a high-resistance dielectric.

drop $i_1 X_{C1}$ is drawn 90° lagging i_1. Similarly, the voltage drop across C_s is $i_2 X_{CS}$, and is drawn 90° lagging i_2. The resistive voltage drops $i_1 R_1$ and $i_2 R_s$ are in phase with i_1 and i_2, respectively.

The total voltage drop across Z_1 is the phasor sum of $i_1 R_1$ and $i_1 X_{C1}$, as illustrated in Figure 8-6(b). Also, $i_2 Z_2$ is the phasor sum of $i_2 R_s$ and $i_2 X_{CS}$. Since $i_2 Z_2$ must be equal to and in phase with $i_i Z_1$, $i_1 R_1$ and $i_2 R_s$ are equal, as are $i_1 X_{C1}$ and $i_2 X_{CS}$.

Example 8-3

A series-resistance capacitance bridge [as in Figure 8-6(a)] has a 0.4 µF standard capacitor for C_1, and $R_3 = 10$ kΩ. Balance is achieved with a 100 Hz supply frequency when $R_1 = 125$ Ω and $R_4 = 14.7$ kΩ. Calculate the resistive and capacitive components of the measured capacitor and its dissipation factor.

Solution

Equation 8-13,
$$C_s = \frac{C_1 R_3}{R_4} = \frac{0.1 \ \mu\text{F} \times 10 \ \text{k}\Omega}{14.7 \ \text{k}\Omega}$$

$$= 0.068 \ \mu\text{F}$$

Equation 8-12,
$$R_s = \frac{R_1 R_4}{R_3} = \frac{125 \ \Omega \times 14.7 \ \text{k}\Omega}{10 \ \text{k}\Omega}$$

$$= 183.8 \ \Omega$$

Equation 8-8,
$$D = \omega C_s R_s$$

$$= 2\pi \times 100 \ \text{Hz} \times 0.068 \ \mu\text{F} \times 183.8 \ \Omega$$

$$\approx 0.008$$

Parallel-Resistance Capacitance Bridge

The circuit of a *parallel-resistance capacitance bridge* is illustrated in Figure 8-7. In this case, the unknown capacitance is represented by its parallel equivalent circuit; C_p in parallel with R_p. Z_3 and Z_4 are resistors, as before, either or both of which may be adjustable. Z_2 is balanced by a standard capacitor C_1 in parallel with an adjustable resistor R_1. Bridge balance is achieved by adjustment or R_1 and either R_3 or R_4. The parallel-resistance capacitance bridge is found to be most suitable for capacitors with a low resistance dielectric (relatively high leakage current and high dissipation factor). At balance, Equation 8-9 once again applies:

$$\frac{Z_1}{Z_3} = \frac{Z_2}{Z_4}$$

Also,
$$\frac{1}{Z_1} = \frac{1}{R_1} - \frac{1}{j(1/\omega C_1)}$$

$$= \frac{1}{R_1} + j\omega C_1$$

or
$$Z_1 = \frac{1}{1/R_1 + j\omega C_1}$$

Inductance and Capacitance Measurements   Chap. 8

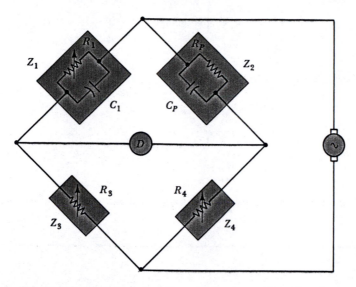

Figure 8-7 The parallel-resistance capacitance bridge uses an adjustable resistance (R_1) connected in parallel with C_1 to balance the resistive component (R_p) of Z_2. This bridge is most suitable for measuring capacitors with a low-resistance dielectric.

and

$$\frac{1}{Z_2} = \frac{1}{R_p} - \frac{1}{j(1/\omega C_p)}$$

$$= \frac{1}{R_p} + j\omega C_p$$

or

$$Z_2 = \frac{1}{1/R_p + j\omega C_p}$$

Substituting into Equation 8-9,

$$\frac{1/(1/R_1 + j\omega C_1)}{R_3} = \frac{1/(1/R_p + j\omega C_p)}{R_4}$$

$$\frac{1}{R_3(1/R_1 + j\omega C_1)} = \frac{1}{R_4(1/R_p + j\omega C_p)}$$

or

$$\boxed{R_3\left(\frac{1}{R_1} + j\omega C_1\right) = R_4\left(\frac{1}{R_p} + j\omega C_p\right)} \qquad (8\text{-}14)$$

Equating the real terms in Equation 8-14,

$$\frac{R_3}{R_1} = \frac{R_4}{R_p}$$

giving

$$R_p = \frac{R_1 R_4}{R_3}$$ (8-15)

Equating the imaginary terms in Equation 8-14,

$$\omega C_1 R_3 = \omega C_p R_4$$

giving

$$C_p = \frac{C_1 R_3}{R_4}$$ (8-16)

Note the similarity between Equations 8-15 and 8-12, and between Equations 8-16 and 8-13.

Example 8-4

A parallel-resistance capacitance bridge (as in Figure 8-7) has a standard capacitance value of $C_1 = 0.1 \ \mu F$ and $R_3 = 10 \ k\Omega$. Balance is achieved at a supply frequency of 100 Hz when $R_1 = 375 \ k\Omega$, $R_3 = 10 \ k\Omega$, and $R_4 = 14.7 \ k\Omega$. Calculate the resistive and capacitive components of the measured capacitor and its dissipation factor.

Solution

Equation 8-16, $$C_p = \frac{C_1 R_3}{R_4} = \frac{0.1 \ \mu F \times 10 \ k\Omega}{14.7 \ k\Omega}$$

$$= 0.068 \ \mu F$$

Equation 8-15, $$R_p = \frac{R_1 R_4}{R_3} = \frac{375 \ k\Omega \times 14.7 \ k\Omega}{10 \ k\Omega}$$

$$= 551.3 \ k\Omega$$

Equation 8-7, $$D = \frac{1}{\omega C_p R_p} = \frac{1}{2\pi \times 100 \ Hz \times 0.068 \ \mu F \times 551.3 \ k\Omega}$$

$$= 42.5 \times 10^{-3}$$

Example 8-5

Calculate the parallel equivalent circuit for the C_s and R_s values determined in Example 8-3. Also determine the component values of R_1 and R_4 required to balance the calculated C_p and R_p values in a parallel-resistance capacitance bridge. Assume that R_3 remains 10 $k\Omega$.

Inductance and Capacitance Measurements   Chap. 8

Solution

$$X_s = \frac{1}{2\pi f C_s} = \frac{1}{2\pi \times 100 \text{ Hz} \times 0.068 \text{ } \mu\text{F}}$$

$$= 23.4 \text{ k}\Omega$$

Equation 8-1,

$$R_p = \frac{R_s^2 + X_s^2}{R_s} = \frac{(183.8 \text{ } \Omega)^2 + (23.4 \text{ k}\Omega)^2}{183.8 \text{ } \Omega}$$

$$= 2.98 \text{ M}\Omega$$

Equation 8-2,

$$X_p = \frac{R_s^2 + X_s^2}{X_s} = \frac{(183.8 \text{ } \Omega)^2 + (23.4 \text{ k}\Omega)^2}{23.4 \text{ k}\Omega}$$

$$= 23.4 \text{ k}\Omega$$

$$C_p = \frac{1}{2\pi f X_p} = \frac{1}{2\pi \times 100 \text{ Hz} \times 23.4 \text{ k}\Omega}$$

$$\approx 0.068 \text{ } \mu\text{F}$$

From Equation 8-16,

$$R_4 = \frac{C_1 R_3}{C_p} = \frac{0.1 \text{ } \mu\text{F} \times 10 \text{ k}\Omega}{0.068 \text{ } \mu\text{F}}$$

$$= 14.7 \text{ k}\Omega$$

From Equation 8-15,

$$R_1 = \frac{R_3 R_p}{R_4} = \frac{10 \text{ k}\Omega \times 2.98 \text{ M}\Omega}{14.7 \text{ k}\Omega}$$

$$= 2.03 \text{ M}\Omega$$

The capacitor, which was determined in Example 8-3 as having a series equivalent circuit of 0.068 μF and 183.8 Ω, was shown in Example 8-5 to have a parallel equivalent circuit of 0.068 μF and 2.98 MΩ. It was also shown that to measure the capacitor on a parallel-resistance capacitance bridge, R_1 (in Figure 8-7) would have to be 2.03 MΩ. This is an inconveniently large value for a precision adjustable resistor. So a capacitor with a high leakage resistance (low D factor) is best measured in terms of its series RC equivalent circuit.

The capacitor in Example 8-4 has a parallel RC equivalent circuit of 0.068 μF and 551.3 kΩ. Conversion to the series equivalent circuit would demonstrate that this capacitor is not conveniently measured as a series RC circuit. Thus, a capacitor with a low leakage resistance (high D factor) is best measured as a parallel RC equivalent circuit.

Capacitors with a very high leakage resistance should be measured as series RC circuits. Capacitors with a low leakage resistance should be measured as parallel RC circuits. Capacitors that have neither a very high nor a very low leakage resistance are best measured as a parallel *RC* circuit, because this gives a direct indication of the capacitor leakage resistance.

8-4 INDUCTANCE BRIDGES

Inductance Comparison Bridge

The circuit of the *inductance comparison bridge* shown in Figure 8-8 is similar to the series-resistance capacitance bridge except that inductors are involved instead of capacitors. The unknown inductance, represented by its (series equivalent circuit) inductance L_s and R_s, is measured in terms of a precise standard value inductor. L_1 is the standard inductor, R_1 is a variable standard resistor to balance R_s, R_3 and R_4 are standard resistors. Balance of the bridge is achieved by alternately adjusting R_1 and either R_3 or R_4. At balance, Equation 8-9 once again applies:

$$\frac{Z_1}{Z_3} = \frac{Z_2}{Z_4}$$

$$= \frac{R_1 + j\omega L_1}{R_3} \quad \frac{R_s + j\omega L_s}{R_4}$$

or

$$\boxed{\frac{R_1}{R_3} + j\frac{\omega L_1}{R_3} = \frac{R_s}{R_4} + j\frac{\omega L_s}{R_4}}$$

(8-17)

Equating the real components in Equation 8-17,

$$\frac{R_1}{R_3} = \frac{R_s}{R_4}$$

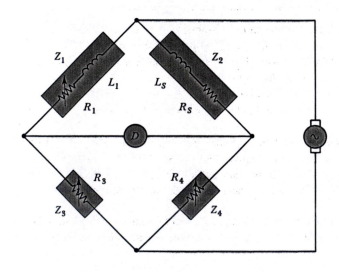

Figure 8-8 The inductance comparison bridge uses a standard inductor L_1 together with adjustable precision resistors R_1, R_3, and R_4 to measure an unknown inductor in terms of its series equivalent circuit L_s and R_s.

Inductance and Capacitance Measurements   Chap. 8

giving
$$R_s = \frac{R_1 R_4}{R_3}$$
(8-18)

Equating the imaginary components in Equation 8-17,

$$\frac{\omega L_1}{R_3} = \frac{\omega L_s}{R_4}$$

giving
$$L_s = \frac{L_1 R_4}{R_3}$$
(8-19)

Example 8-6

An inductor that is marked as 500 mH is to be measured on an inductance comparison bridge. The bridge uses a 100 mH standard inductor for L_1, and a 5 kΩ standard resistor for R_4. If the coil resistance of the 500 mH inductor is measured as 270 Ω, determine the resistances of R_1 and R_3 (in Figure 8-8) at which balance is likely to occur.

Solution

From Equation 8-19,
$$R_3 = \frac{R_4 L_1}{L_s} = \frac{5 \text{ k}\Omega \times 100 \text{ mH}}{500 \text{ mH}}$$

$$= 1 \text{ k}\Omega$$

From Equation 8-18,
$$R_1 = \frac{R_s R_3}{R_4} = \frac{270 \ \Omega \times 1 \text{ k}\Omega}{5 \text{ k}\Omega}$$

$$= 54 \ \Omega$$

Maxwell Bridge

Accurate pure standard capacitors are more easily constructed than standard inductors. Consequently, it is desirable to be able to measure inductance in a bridge that uses a capacitance standard rather than an inductance standard. The *Maxwell bridge* (also known as the *Maxwell-Wein bridge*) is shown in Figure 8-9. In this circuit, the standard capacitor C_3 is connected in parallel with adjustable resistor R_3. R_1 is again an adjustable standard resistor, and R_4 may also be made adjustable. L_s and R_s represent the inductor to be measured.

The Maxwell bridge is found to be most suitable for measuring coils with a low Q factor (i.e., where ωL_s is *not* much larger than R_s). To determine the expression for Z_2 and Z_3,

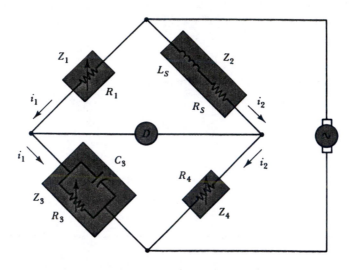

Figure 8-9 The Maxwell bridge uses a standard capacitor C_3 and three adjustable precision resistors to measure an unknown inductor in terms of its series equivalent circuit, L_s and R_s. This bridge is most suitable for measuring coils with a low Q factor.

$$\frac{1}{Z_3} = \frac{1}{R_3} - \frac{1}{j1/\omega C_3} = \frac{1}{R_3} + j\omega C_3$$

or

$$Z_3 = \frac{1}{1/R_3 + j\omega C_3}$$

and

$$Z_2 = R_s + j\omega L_s$$

Substituting for all components in Equation 8-9,

$$\frac{R_1}{1/(1/R_3 + j\omega C_3)} = \frac{R_s + j\omega L_s}{R_4}$$

or

$$\boxed{\frac{R_1}{R_3} + j\omega C_3 R_1 = \frac{R_s}{R_4} + j\frac{\omega L_s}{R_4}}$$ (8-20)

Equating the real components in Equation 8-20,

$$\frac{R_1}{R_3} = \frac{R_s}{R_4}$$

or

$$\boxed{R_s = \frac{R_1 R_4}{R_3}}$$ (8-21)

Equating the imaginary components in Equation 8-20,

$$\omega C_3 R_1 = \frac{\omega L_s}{R_4}$$

giving

$$\boxed{L_s = C_3 R_1 R_4}$$

(8-22)

Example 8-7

A Maxwell inductance bridge uses a standard capacitor of $C_3 = 0.1\ \mu F$ and operates at a supply frequency of 100 Hz. Balance is achieved when $R_1 = 1.26\ k\Omega$, $R_3 = 470\ \Omega$, and $R_4 = 500\ \Omega$. Calculate the inductance and resistance of the measured inductor, and determine its Q factor.

Solution

Equation 8-22,
$$L_s = C_3 R_1 R_4$$
$$= 0.1\ \mu F \times 1.26\ k\Omega \times 500\ \Omega$$
$$= 63\ mH$$

Equation 8-21,
$$R_s = \frac{R_1 R_4}{R_3} = \frac{1.26\ k\Omega \times 500\ \Omega}{470\ \Omega}$$
$$= 1.34\ k\Omega$$

Equation 8-5,
$$Q = \frac{\omega L_s}{R_s} = \frac{2\pi \times 100\ Hz \times 63\ mH}{1.34\ k\Omega}$$
$$= 0.03$$

Hay Inductance Bridge

The *Hay bridge* circuit in Figure 8-10 is similar to the Maxwell bridge, except that R_3 and C_3 are connected in series instead of parallel, and the unknown inductance is represented as a parallel LR circuit instead of a series circuit. The balance equations are found to be exactly the same as those for the Maxwell bridge. It must be remembered, however, that the measured L_p and R_p are a parallel equivalent circuit. The equivalent series RL circuit can be determined by substitution into Equations 8-3 and 8-4.

When the bridge in Figure 8-10 is balanced,

$$\frac{Z_1}{Z_3} = \frac{Z_2}{Z_4}$$

giving

$$\boxed{\frac{R_4}{R_P} - j\frac{R_4}{\omega L_P} = \frac{R_3}{R_1} - j\frac{1}{\omega C_3 R_1}}$$

(8-23)

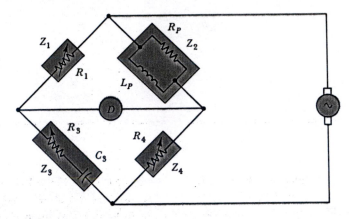

Figure 8-10 The Hay bridge uses a standard capacitor C_3 and three adjustable precision resistors to measure an unknown inductor in terms of its parallel equivalent circuit, L_p and R_p. This circuit is most suitable for inductors with a high Q factor.

Equating the real components in Equation 8-23,

$$\frac{R_4}{R_P} = \frac{R_3}{R_1}$$

or

$$R_P = \frac{R_1 R_4}{R_3} \qquad (8\text{-}24)$$

Equating the imaginary components in Equation 8-23,

$$\frac{R_4}{\omega L_P} = \frac{1}{\omega C_3 R_1}$$

giving

$$L_P = C_3 R_1 R_4 \qquad (8\text{-}25)$$

Example 8-8

A Hay bridge operating at a supply frequency of 100 Hz is balanced when the components are $C_3 = 0.1\ \mu\text{F}$, $R_1 = 1.26\ \text{k}\Omega$, $R_3 = 75\ \Omega$, and $R_4 = 500\ \Omega$. Calculate the inductance and resistance of the measured inductor. Also, determine the Q factor of the coil.

Inductance and Capacitance Measurements   Chap. 8

Solution

Equation 8-25,
$$L_P = C_3 R_1 R_4$$
$$= 0.1\ \mu F \times 1.26\ k\Omega \times 500\ \Omega$$
$$= 63\ mH$$

Equation 8-24,
$$R_P = \frac{R_1 R_4}{R_3} = \frac{1.26\ k\Omega \times 500\ \Omega}{75\ \Omega}$$
$$= 8.4\ k\Omega$$

Equation 8-6,
$$Q = \frac{R_P}{\omega L_P}$$
$$= \frac{8.4\ k\Omega}{2\pi \times 100\ Hz \times 63\ mH}$$
$$= 212$$

Example 8-9

(a) Calculate the series equivalent circuit for the L_P and R_P values determined in Example 8-8.

(b) Determine the component values of R_1 and R_3 required to balance the calculated L_s and R_s values in the Maxwell bridge. Assume that R_4 remains 500 Ω.

Solution

(a)
$$X_p = 2\pi f L_p = 2\pi \times 100\ Hz \times 63\ mH$$
$$= 39.6\ \Omega$$

Equation 8-3,
$$R_s = \frac{R_p X_p^2}{X_p^2 + R_p^2} = \frac{8.4\ k\Omega \times (39.6\ \Omega)^2}{(39.6\ \Omega)^2 + (8.4\ k\Omega)^2}$$
$$= 0.187\ \Omega$$

Equation 8-4,
$$X_s = \frac{R_p^2 X_p}{X_p^2 + R_p^2} = \frac{(8.4\ k\Omega)^2 \times 39.6\ \Omega}{(39.6\ \Omega)^2 + (8.4\ k\Omega)^2}$$
$$= 39.6\ \Omega$$

$$L_s = \frac{X_s}{2\pi f} = \frac{39.6\ \Omega}{2\pi \times 100\ Hz}$$
$$= 63\ mH$$

(b)   From Equation 8-22,
$$R_1 = \frac{L_s}{C_3 R_4} = \frac{63\ mH}{0.1\ \mu F \times 500\ \Omega}$$
$$= 1.26\ k\Omega$$

$$R_3 = \frac{R_1 R_4}{R_s} = \frac{1.26 \text{ k}\Omega \times 500 \text{ }\Omega}{0.187 \text{ }\Omega}$$

$$= 3.37 \text{ M}\Omega$$

Example 8-9 demonstrates that the inductor parallel equivalent circuit determined in Example 8-8 actually represents a coil that has an inductance of 63 mH and a coil resistance of 0.187 Ω. The series equivalent circuit more correctly represents the measurable resistance and inductance of a coil. Conversely, the parallel *CR* equivalent circuit represents the measurable dielectric resistance and capacitance of a capacitor more correctly than a series *CR* equivalent circuit.

The (high) calculated value of R_3 in Example 8-9 shows that the low-resistance (high-Q) coil cannot be conveniently measured on a Maxwell bridge. Thus, the Hay bridge is best for measurement of inductances with high Q. Similarly, it can be demonstrated that the Maxwell bridge is best for measurement of low-Q inductances, and that the Hay bridge is not suited to low-Q inductance measurements.

Some inductors which have neither very low nor very high Q factors may easily be measured on either type of bridge. In this case it is best to use the Maxwell circuit, because the inductor is then measured directly in terms of its (preferable) series equivalent circuit.

8-5 MULTIFUNCTION IMPEDANCE BRIDGE

All but one of the capacitance and inductance bridges discussed in the preceding sections can be constructed using a standard capacitor and three adjustable standard resistors. The single exception is the inductance comparison bridge (Figure 8-8).

Figure 8-11 shows the circuits of five different bridges constructed from the four basic components. These are a Wheatstone bridge, a series-resistance capacitance bridge, a parallel-resistance capacitance bridge, a Maxwell bridge, and a Hay bridge. All five circuits are normally provided in commercial impedance bridges. Such instruments contain the four basic components and appropriate switches to set the components into any one of the five configurations. A null detector and internal ac and dc supplies are also usually included.

8-6 MEASURING SMALL *C*, *R*, AND *L* QUANTITIES

When measuring very small quantities of *C*, *L*, or *R*, the *stray* capacitance, inductance, and resistance of connecting leads can introduce considerable errors. This is minimized by connecting the unknown component directly to the bridge terminal or by means of very short connecting leads. Even when such precautions are observed, there are still small internal *L*, *C*, and *R* quantities in all instruments. These are termed *residuals,* and

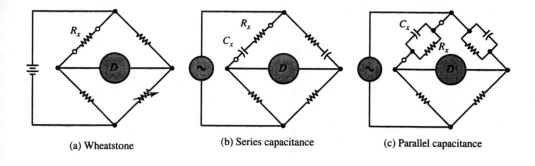

(a) Wheatstone   (b) Series capacitance   (c) Parallel capacitance

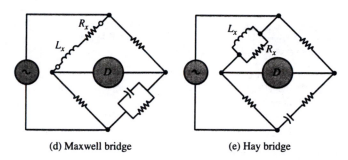

(d) Maxwell bridge   (e) Hay bridge

Figure 8-11 The standard capacitor and three precision resistors typically contained in a commercial impedance bridge can be connected to function as a series-resistance capacitance bridge, a parallel-resistance capacitance, a Wheatstone bridge, a Maxwell inductance bridge, or a Hay inductance bridge.

instrument manufacturers normally list the residuals on the specification. A typical impedance bridge has residuals of $R = 1 \times 10^{-3}$ Ω, $C = 0.5$ pF, and $L = 0.2$ μH. Obviously, these quantities can introduce serious errors if they are a substantial percentage of any measured quantity.

The errors introduced by strays and residuals can be eliminated by a *substitution technique* (see Figure 8-12). In the case of a capacitance measurement, the bridge is first balanced with a larger capacitor connected in place of the small capacitor to be measured. The small capacitor is then connected in parallel with the larger capacitor, and the bridge is readjusted for balance. The first measurement is the large capacitance C_1 plus the stray and residual capacitance C_s. So the measured capacitance is $C_1 + C_s$. When the small capacitor C_x is connected, the measured capacitance is $C_1 + C_s + C_x$. C_x is found by subtracting the first measurement from the second.

A similar approach is used for measurements of low value inductance and resistance, except that in this case the low value component must be connected in series with the larger L or R quantity. The substitution technique can also be applied to other (non-bridge) measurement methods.

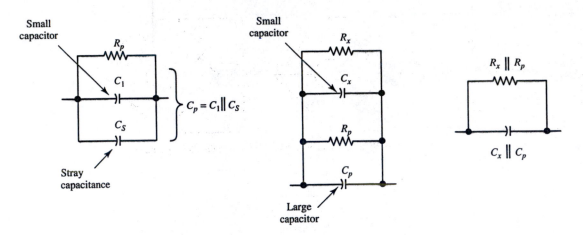

(a) Stray capacitance affects measurement accuracy

(b) Small capacitor connected in parallel with large capacitor for measurement

(c) Measurement gives $C_x \| C_p$ and $R_x \| R_p$

Figure 8-12 Stray capacitance can seriously affect the accuracy of measurement of a small capacitor. For best accuracy, the unknown small capacitor (C_x) should be connected in parallel with a larger capacitor. C_x can then be determined from the measured value of $C_x \| C_p$.

Example 8-10

On the bridge in Example 8-4 a new balance is obtained when a small capacitor (C_x) is connected in parallel with the measured capacitor C_p. The new component values for balance are $R_1 = 369.3$ kΩ, $R_3 = 10$ kΩ, and $R_4 = 14.66$ kΩ. Determine the value of C_x and its parallel resistive component R_x.

Solution

$$C_x \| C_p = C_x + C_p = \frac{C_1 R_3}{R_4}$$

$$= \frac{0.1\ \mu F \times 10\ k\Omega}{14.66\ k\Omega} = 0.682\ \mu F$$

$$C_x = 0.0682\ \mu F - C_p = 0.0682\ \mu F - 0.068\ \mu F$$

$$= 200\ pF$$

and

$$R_x \| R_p = \frac{R_1 R_4}{R_3} = \frac{369.3\ k\Omega \times 14.66\ k\Omega}{10\ k\Omega}$$

$$= 541.4\ k\Omega$$

$$\frac{1}{R_x} + \frac{1}{R_p} = \frac{1}{R_x \| R_p}$$

$$or \qquad R_x = \frac{1}{1/(R_x \| R_p) - 1/R_p}$$

From Example 8-4, $\qquad R_p = 553.1 \text{ k}\Omega$

$$so \qquad R_x = \frac{1}{1/541.4 \text{ k}\Omega - 1/553.1 \text{ k}\Omega}$$

$$= 30 \text{ M}\Omega$$

8-7 DIGITAL L, C, AND R MEASUREMENTS

Inductance Measurement

Inductance and capacitance must be first converted into voltages before any measurement can be made by digital techniques. Figure 8-13 illustrates the basic method.

In Figure 8-13(a) an ac voltage is applied to the noninverting input terminal of an operational amplifier. The input voltage is developed across resistor R_1 to give a current: $I = V_i/R_1$. This current also flows through the inductor giving a voltage drop: $V_L = IX_L$. If $V_i = 1.592$ Vrms, $f = 1$ kHz, $R_1 = 1$ kΩ, and $L = 100$ mH:

$$I = \frac{V_i}{R_i} = \frac{1.592 \text{ V}}{1 \text{ k}\Omega} = 1.592 \text{ mA}$$

and $\qquad V_L = I(2\pi f L) = 1.592 \text{ mA} \times 2\pi \times 1 \text{ kHz} \times 100 \text{ mH}$

$$= 1 \text{ V(rms)}$$

when $\qquad L = 200$ mH, $V_L = 2$ V; when $L = 300$ mH, $V_L = 3$ V; and so on.

It is seen that the voltage developed across L is directly proportional to the inductive impedance. A *phase-sensitive detector* [Figure 8-13(a)] is employed to resolve the inductor voltage into quadrature and in-phase voltages. These two components represent the series equivalent circuit of the measured inductor. The voltages are fed to digital measuring circuits to display the series equivalent circuit inductance L_s, the dissipation factor ($D = 1/Q$), and/or the Q factor.

Capacitance Measurement

Capacitive impedance is treated in a similar way to inductive impedance, except that the input voltage is developed across the capacitor and the output voltage is measured across the resistor [see Figure 8-13(b)].

In this case $I = V_i/X_c$, and $V_R = IR$. With $V_i = 1.592$ Vrms, $f = 1$ kHz, $R_1 = 1$ kΩ, and $C = 0.1$ μF:

$$I = \frac{V_i}{X_c} = V_i(2\pi f C)$$

$$= 1.592 \text{ V} \times 2\pi \times 1 \text{ kHz} \times 0.1 \text{ }\mu\text{F}$$

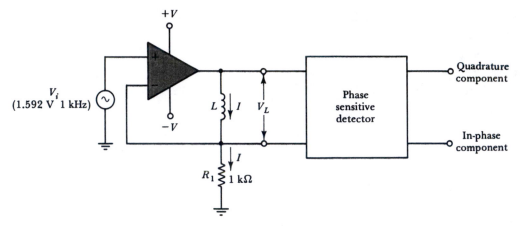

(a) Linear conversion of inductive
impedance into voltage

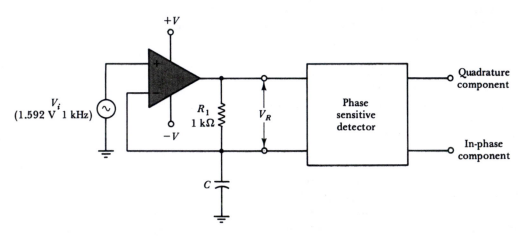

(b) Linear conversion of capacitive
impedance into voltage

Figure 8-13 Basic circuits for converting inductive and capacitive impedances into voltage components for electronic measurement. The voltages are resolved into in-phase and quadrature components for determination of the D and Q factors.

$$= 1 \text{ mA}$$

and
$$V_R = IR = 1 \text{ mA} \times 1 \text{ k}\Omega$$

$$= 1 \text{ V (rms)}$$

when $C = 0.2 \ \mu\text{F}$, $V_R = 2$ V; when $C = 0.3 \ \mu\text{F}$, $V_R = 3$ V; and so on.

The voltage developed across R is directly proportional to the capacitive impedance. The phase sensitive detector [Figure 8-13(b)] resolves the resistor voltage into quadrature and in-phase components, which in this case are proportional to the capacitor current. The displayed capacitance measurement is that of the parallel equivalent circuit (C_p). The dissipation factor (D) of the capacitor is also displayed.

Capacitance Measurement on Digital Multimeters

Some digital multimeters have a facility for measuring capacitance. This normally involves charging the capacitor at a constant rate, and monitoring the time taken to arrive at a given terminal voltage. In the ramp generator digital voltmeter system in Figure 6-1, the ramp is produced by using a constant current to charge a capacitor. Figure 8-14 shows the basic method. Transistor Q_1, together with resistors R_1, R_2, and R_3, produce the constant charging current to capacitor C_1 when Q_2 is *off*. C_1 is discharged when Q_2 switches *on*. (A similar circuit is treated in more detail in Section 9-4.)

As already explained for the digital voltmeter, a ramp time (t_1) of 1 s and a clock generator frequency of 1 kHz result in a count of 1000 clock pulses, which is then read as a voltage. If V_i remains fixed at 1 V, the display could be read as a measure of the capacitor in the ramp generator. A 1 μF capacitor might produce the 1 s counting time, so that the display is read as 1.000 μF. A change of capacitance to 0.5 μF would give a 0.5 s counting time and a display of 0.500 μF. Similarly, a capacitance increase to 1.5 μF would produce a 1.5 s counting time and a 1.500 μF display. In this way, the digital voltmeter is readily converted into a digital capacitance meter.

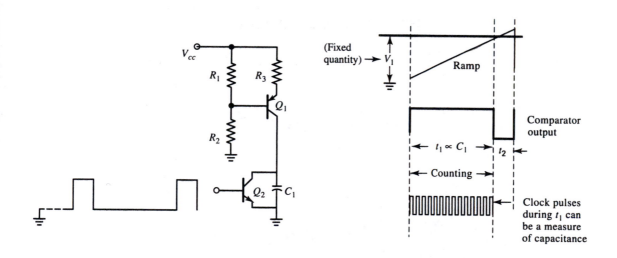

(a) Ramp generator circuit

(b) Waveforms

Figure 8-14 Basic ramp generator circuit and waveforms for a digital voltmeter. If V_i is a fixed quantity, time t_1 is directly proportional to capacitor C_1, and the digital output can be read as a measure of the capacitance.

Sec. 8-7   Digital *L, C,* and *R* Instruments

217

8-8 DIGITAL *RCL* METER

The digital *RCL* meter shown in Figure 8-15 can measure *inductance, capacitance, resistance, conductance,* and *dissipation factor.* The desired function is selected by pushbutton. The range switch is normally set to the automatic (AUTO) position for convenience. However, when a number of similar measurements are to be made, it is faster to use the appropriate range instead of the automatic range selection. The numerical value of the measurement is indicated on the $3\frac{1}{2}$-digit display, and the multiplier and measured quantity are identified by LED indicating lamps.

Four (*current* and *potential*) terminals are provided for connection of the component to be measured. (See Section 7-4 for four-terminal resistors.) For general use each pair of current and voltage terminals are joined together at two spring clips (known as *Kelvin clips*) which facilitate quick connection of components. A ground terminal for guard-ring measurements (see Section 7-6) is provided at the rear of the instrument. The ground terminal together with the other four terminals is said to give the instrument *five-terminal measurement* capability. *Bias terminals* are also available at the rear of the instrument, so that a bias current can be passed through an inductor or a bias voltage applied to a capacitor during measurement.

For *R, L, C,* and *G,* typical measurement accuracies are $\pm[0.25\% + (1 + 0.002\ R, L, C,$ or $G)$ digits]; for *D,* the measurement accuracy is $\pm(2\% + 0.010)$.

Resistance measurements may be made directly on the digital LCR instrument in Figure 8-15 over a range of 2 Ω to 2 MΩ. Conductance is measured directly over a range

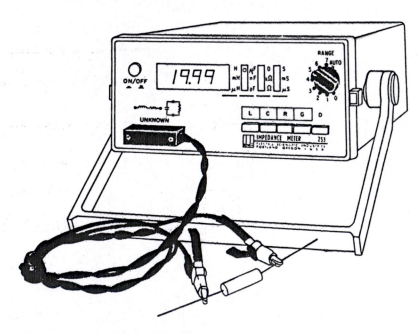

Figure 8-15 Digital impedance meter that can measure inductance, capacitance, resistance, conductance, and dissipation factor. (Courtesy of Electro Scientific Industries, Inc.)

of 2 μS to 20 S. Resistances between 2 MΩ and 1000 MΩ can be measured as conductances, and the resistance calculated: $R = 1/G$. For example, a resistance of 10 MΩ is measured as 0.100 μS.

Inductance and capacitance measurements may be made directly over a range of 200 μH to 200 H, and 200 pF to 2000 μF, respectively. The dissipation factor D is determined by pressing and holding in the D button while the L or C button is still selected. The directly measured inductance is the series equivalent circuit quantity L_s. The Q factor of the inductor is calculated as the reciprocal of D:

$$Q = \frac{\omega L_s}{R_s} = \frac{1}{D} \qquad \text{(see Section 8-1)}$$

Direct capacitance measurements give the parallel equivalent circuit quantity C_p. In this case D (for the parallel equivalent CR circuit) is

$$D = \frac{1}{\omega C_p R_p} \qquad \text{(see Section 8-1)}$$

When measuring low values of resistance or inductance, the connecting clips should first be shorted together and the residual R or L values noted (as indicated digitally). These values should then be subtracted from the measured value of the component. When measuring low capacitances, the connecting clips should first be placed as close together as the terminals of the component to be measured (i.e., without connecting the component). The indicated residual capacitance is noted and then subtracted from the measured component capacitance.

Return to Figure 8-13(a) and assume that a capacitor is connected in place of the inductor. The measured quantity is displayed as an inductance prefixed by a negative sign on the RLC meter in Figure 8-15. The capacitive impedance is equivalent to the impedance of the indicated inductance:

$$\omega L_s = \frac{1}{\omega C_s}$$

or

$$C_s = \frac{1}{\omega^2 L_s}$$

For an indicated inductance of 100 mH, and a measuring frequency of 1 kHz,

$$C_s = \frac{1}{(2\pi \times 1 \text{ kHz})^2 \times 100 \text{ mH}} = 0.25 \text{ μF}$$

Similarly, inductance can be measured as capacitance when it is convenient to do so.

The digital RCL meter shown in Figure 8-16 displays the measured quantity and the units of measurement. It also displays the equivalent circuit (parallel RC, series RL, etc.) of the measured quantity. In RCL $AUTO$ mode of operation, the dominating component is measured, and its equivalent circuit is displayed. Any one of several parameters (Q, D, R_p, R_s, etc.) may be selected manually for measurement.

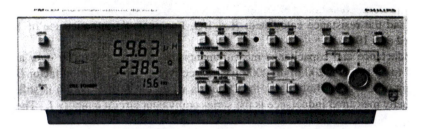

Figure 8-16 Digital *RCL* meter that displays the equivalent circuit of the measured quantity, as well as the numerical value and the units. (© 1991, John Fluke Mfg. Co., Inc. All rights reserved. Reproduced with permission.)

8-9 *Q* METER

Q-Meter Operation

Inductors, capacitors, and resistors which have to operate at radio frequencies (RF) cannot be measured satisfactorily at lower frequencies. Instead, resonance methods are employed in which the unknown component may be tested at or near its normal operating frequency. The *Q meter* is designed for measuring the *Q* factor of a coil and for measuring inductance, capacitance, and resistance at RF.

The basic circuit of a *Q* meter shown in Figure 8-17 consists of a variable calibrated capacitor, a variable-frequency ac voltage source, and the coil to be investigated. All are connected in series. The capacitor voltage (V_C) and the source voltage (E) are monitored by voltmeters. The source is set to the desired measuring frequency, and its voltage is adjusted to a convenient level. Capacitor *C* is adjusted to obtain resonance, as indicated

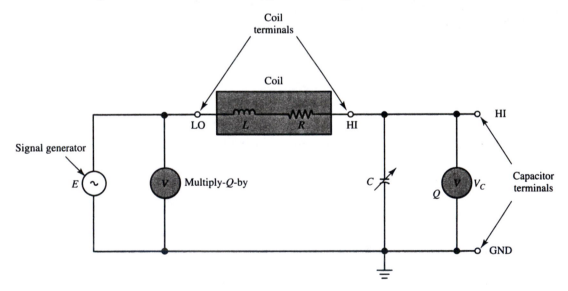

Figure 8-17 A basic *Q* meter circuit consists of a stable ac supply, a variable capacitor, and a voltmeter to monitor the capacitor voltage. When the circuit is in resonance, $V_C = V_L$, and $Q = V_C/E$.

Inductance and Capacitance Measurements   Chap. 8

when the voltage across C is a maximum. If necessary, the source is readjusted to the desired output level when resonance is obtained.

At resonance: $\qquad\qquad\qquad V_C = V_L \qquad$ and $\qquad I = \dfrac{E}{R}$

also

$$Q = \frac{\omega L}{R} = \frac{1}{\omega CR} \qquad\qquad (8\text{-}26)$$

and

$$Q = \frac{V_L}{E} = \frac{V_C}{E} \qquad\qquad (8\text{-}27)$$

Example 8-11

When the circuit in Figure 8-17 is in resonance, $E = 100$ mV, $R = 5$ Ω, and $X_L = X_C = 100$ Ω.
(a) Calculate the coil Q and the voltmeter indication.
(b) Determine the Q factor and voltmeter indication for another coil that has $R = 10$ Ω and $X_L = 100$ Ω at resonance.

Solution

(a) $\qquad\qquad\qquad I = \dfrac{E}{R} = \dfrac{100 \text{ mV}}{5 \text{ }\Omega} = 20 \text{ mA}$

$\qquad\qquad\qquad V_L = V_C = I\,X_C$

$\qquad\qquad\qquad\qquad = 20 \text{ mA} \times 100 \text{ }\Omega$

$\qquad\qquad\qquad\qquad = 2 \text{ V}$

$\qquad\qquad\qquad Q = \dfrac{V_L}{E} = \dfrac{2 \text{ V}}{100 \text{ mV}} = 20$

(b) *For the second coil:*

$\qquad\qquad\qquad I = \dfrac{E}{R} = \dfrac{100 \text{ mV}}{10 \text{ }\Omega} = 10 \text{ mA}$

$\qquad\qquad\qquad V_L = V_C = I\,X_L$

$\qquad\qquad\qquad\qquad = 10 \text{ mA} \times 100 \text{ }\Omega$

$\qquad\qquad\qquad\qquad = 1 \text{ V}$

$\qquad\qquad\qquad Q = \dfrac{V_L}{E} = \dfrac{1 \text{ V}}{100 \text{ mV}} = 10$

Q Meter Controls

Example 8-11 shows that when $Q = 20$ the capacitor voltmeter indicates 2 V, and when $Q = 10$ the voltmeter indicates 1 V. Clearly, the voltmeter can be calibrated to indicate the coil Q directly [see Figure 8-18(a)].

If the ac supply voltage in Example 8-11 is halved, the circuit current is also halved. This results in V_C and V_L becoming half of the values calculated. Thus, instead of indicating 2 V for a Q of 20, the capacitor voltmeter would indicate only 1 V. The problem of supply voltage stability can be avoided by always setting the signal generator voltage to the correct level or by having the signal generator output voltage precisely stabilized. However, it can sometimes be convenient to adjust the supply to other voltage levels. If the 100 mV position on the supply voltmeter is marked as *1*, and the 50 mV position is marked as *2*, and so on, the supply voltmeter becomes a *multiply-Q-by* meter [Figure 8-18(b)]. When E is set to give a *1* indication, all Q values measured on the capacitor voltmeter are correct. If E is set to the *2* position, measured Q values must be multiplied by 2. Instruments that have a signal generator with a stabilized output do not use a meter for monitoring the source voltage (i.e., there is no multiply-Q-by meter). In this case, the voltage level of the supply is selected by means of a switch, and this switch becomes a Q-meter range control.

If the adjustable capacitor in the Q meter circuit is calibrated and its capacitance indicated on a dial, it can be used to measure the coil inductance. From Equation 8-26,

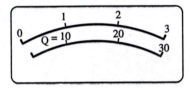

(a) Capacitor voltmeter calibrated to monitor Q

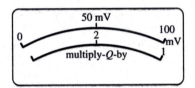

(b) Supply voltmeter calibrated as a multiply-Q-by meter

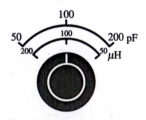

(c) Capacitance dial calibrated to indicate coil inductance

Figure 8-18 With the Q-meter supply voltage *(E)* set to a convenient level, the capacitor voltmeter can directly indicate Q, the supply voltmeter can function as a multiply-Q- by meter, and the capacitance dial can indicate coil inductance as well as capacitance.

$$L = \frac{1}{\omega^2 C} = \frac{1}{(2\pi f)^2 C}$$

Suppose that $f = 1.592$ MHz, and resonance is obtained with $C = 100$ pF.

$$L = \frac{1}{(2\pi \times 1.592 \text{ MHz})^2 \times 100 \text{ pF}}$$

$$\approx 100 \ \mu\text{H}$$

When resonance is obtained at the same frequency with $C = 200$ pF, $L \approx 50 \ \mu\text{H}$. Also, if $C = 50$ pF at 1.592 MHz, L is calculated as $200 \ \mu\text{H}$. It is seen that the capacitance dial can be calibrated to indicate the coil inductance directly (in addition to capacitance) [Figure 8-18(c)].

If the capacitor dial is calibrated to indicate inductance when $f = 1.592$ MHz, any change in f changes the inductance scale. For $f = 15.92$ MHz and $C = 100$ pF,

$$L = \frac{1}{(2\pi \times 15.92 \text{ MHz})^2 \times 100 \text{ pF}}$$

$$= 1 \ \mu\text{H}$$

With $C = 200$ pF and 50 pF, L becomes $0.5 \ \mu\text{H}$ and $2 \ \mu\text{H}$, respectively. Therefore, if the frequency is changed in multiples of 10, the inductance scale can still be used with an appropriate multiplying factor.

As an alternative to using a fixed frequency and adjusting the capacitor, it is sometimes convenient to leave C fixed and adjust f to obtain resonance. In this case, the inductance scale on the capacitor dial is no longer usable. However, Equation 8-26 still applies, so L can be calculated from the C and f values.

Residuals

Residual resistance and inductance in the Q meter circuit can be an important source of error when the signal generator voltage is not metered. If the signal generator has a source resistance R_E, the circuit current at resonance is

$$I = \frac{E}{R_E + R} \qquad \text{instead of} \qquad I = \frac{E}{R}$$

Also, the indicated Q factor of the coil is

$$Q = \frac{\omega L}{R_E + R}$$

instead of the actual coil Q, which is

$$Q = \frac{\omega L}{R}$$

Obviously, R_E must be much smaller than the resistance of any coil to be investigated. Similarly, residual inductance must be held to a minimum to avoid measurement errors.

In a practical Q meter, the output resistance of the signal generator is around 0.02 Ω, and the residual inductance may typically be 0.015 μH.

Commercial Q Meter

The Q meter shown in Figure 8-19 has a meter for indicating circuit Q and a Q *LIMIT* (range) switch. A frequency dial with a window is included, and controls are provided for frequency range selection and for continuous adjustment of frequency. The L/C dial indicates the circuit L and C and is adjusted by the series capacitor control identified as *L/C*. The ΔC control (alongside the L/C control) provides fine adjustment of the series capacitor. Its dial indicates the capacitance as a plus (+) or minus (−) quantity. The total resonating capacitance is the sum or difference of that indicated on the two capacitance dials. ΔQ ZERO *COARSE* and *FINE* controls are situated to the right of the Q indicating meter. These are used to measure the difference in Q between two or more coils that have closely equal Q factors.

Measuring Procedures

Medium-range inductance measurement (direct connection). Coils with inductances of up to about 100 mH can be connected directly to the inductance terminals, as explained earlier. The signal generator is set to the desired frequency, and its output level is adjusted to give a convenient Q-factor range. With the ΔC capacitor dial set to zero, the Q capacitor control is adjusted to give maximum deflection on the Q meter. The Q factor of the coil is now read directly from the meter. The coil inductance may also be read from the C/L dial if the signal generator is set to a specified frequency. When some other frequency is employed, the inductance can be calculated from f and C (Equation 8-26). With the coil Q and L known, its resistance can also be calculated.

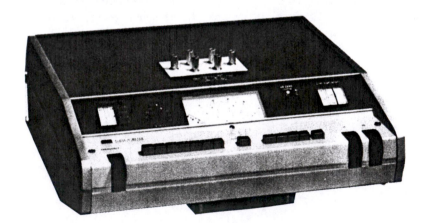

Figure 8-19 HP4342A Q meter has a deflection meter for indicating Q, a frequency dial, and an *L/C* dial. (Courtesy of Hewlett-Packard.)

Example 8-12

With the signal generator frequency of a Q meter set to 1.25 MHz, the Q of a coil is measured as 98 when $C = 147$ pF. Determine the coil inductance and resistance.

Solution

From Equation 8-26,

$$L = \frac{1}{\omega^2 C} = \frac{1}{(2\pi f)^2 C}$$

$$= \frac{1}{(2\pi \times 1.25 \text{ MHz})^2 \times 147 \text{ pF}}$$

$$= 110 \text{ } \mu\text{H}$$

and

$$Q = \frac{\omega L}{R}$$

or

$$R = \frac{2\pi f L}{Q} = \frac{2\pi \times 1.25 \text{ MHz} \times 110 \text{ } \mu\text{H}}{98}$$

$$= 8.8 \text{ } \Omega$$

High-impedance measurements (parallel connection). Inductances greater than 100 mH, capacitances smaller than 400 pF, and high-value resistances are best measured by connecting them in parallel with the capacitor terminals.

For measurement of parallel-connected inductance (L_P), the circuit is first resonated using a *reference inductor* (or *work coil*). The values of C and Q are recorded as C_1 and Q_1. L_P is now connected, and the circuit is readjusted for resonance to obtain C_2 and Q_2. The parameters of the unknown inductance are now determined from the following equations:

$$L_P = \frac{1}{\omega^2 (C_2 - C_1)} \qquad (8\text{-}28)$$

$$Q = \frac{Q_1 Q_2 (C_2 - C_1)}{C_1 (Q_2 - Q_1)} \qquad (8\text{-}29)$$

To measure a parallel-connected capacitance (C_P), the circuit is first resonated using a reference inductor, as before. The values of C_1 and Q_1 are noted. Then the capacitor is connected. Resonance is again found by adjusting the resonating capacitor to give a value C_2. Normally, the circuit Q is not affected. The unknown capacitance is

$$C_P = C_1 - C_2 \tag{8-30}$$

Large-value resistors (R_P) connected in parallel with the resonating capacitor alter the circuit Q, but no capacitance adjustment is necessary (unless R_P also has capacitance or inductance). Once again, the circuit is first resonated using a reference inductor. Then R_P is connected, and the change in Q factor (ΔQ) is measured. The unknown resistance is calculated from

$$R_P = \frac{Q_1 Q_2}{\omega C_1 \, \Delta Q} \tag{8-31}$$

Low-impedance measurements (series connection). Small values of resistance, small inductors, and large capacitors can be measured by placing them in series with the reference inductor. The component to be measured is connected between the LO terminal of the Q meter and the low potential terminal of the reference inductor. The other end of the reference inductor is connected to the HI terminal of the Q meter. Initially, a low-resistance *shorting strap* is connected to short-out the unknown component. The circuit is now tuned for resonance (using an internal coil), and the values of Q_1 and C_1 are noted. The shorting strap is removed, and the circuit is retuned for resonance.

When a pure resistance is involved, circuit resonance should not be affected by removal of the shorting strap. However, the circuit Q should be reduced. The change to Q_2 is measured as ΔQ. The series-connected resistance is now calculated as

$$R_S = \frac{\Delta Q}{\omega C_1 Q_1 Q_2} \tag{8-32}$$

A small series-connected inductance (L_s) affects both the Q factor and the circuit resonance. The circuit is initially resonated with L_s shorted, and the capacitor value (C_1) is noted. The shorting strap is removed, and the capacitor is readjusted for resonance and its new value (C_2) is recorded. The inductance is now calculated as

$$L_s = \frac{C_1 - C_2}{\omega^2 C_1 C_2} \tag{8-33}$$

With a large series-connected capacitor (C_s), the circuit is first resonated with a shorting strap across the capacitor terminals. The strap is removed, and the circuit capacitor is readjusted for resonance. In this case, the Q of the circuit should be largely unaffected. The series-connected capacitance is

$$C_s = \frac{C_1 C_2}{C_2 - C_1} \tag{8-34}$$

REVIEW QUESTIONS

8-1 Sketch *RC* series and parallel equivalent circuits for a capacitor. Discuss the capacitor types best represented by each circuit.

8-2 Derive equations for converting a series *RC* circuit into its equivalent parallel circuit.

8-3 Sketch *RL* series and parallel equivalent circuits for an inductor. Explain which of the two equivalent circuits best represents an inductor.

8-4 Derive equations for converting a parallel *RL* circuit into its equivalent series circuit.

8-5 Define the *Q* factor of an inductor. Write the equations for inductor *Q* factor with *RL* series and parallel equivalent circuits.

8-6 Define the *D* factor of a capacitor. Write the equations for capacitor *D* factor with *RC* series and parallel equivalent circuits.

8-7 Sketch the basic circuit for an ac bridge and explain its operation. Discuss the adjustment procedure for obtaining bridge balance, and derive the balance equations.

8-8 Draw the circuit diagram of a simple capacitance bridge. Derive the balance equation, and discuss the limitations of the bridge.

8-9 Sketch the circuit diagram of a series-resistance capacitance bridge. Derive the equations for the measured capacitance and its resistive component.

8-10 Draw the phasor diagram for a series-resistance capacitance bridge at balance. Explain.

8-11 Sketch the circuit diagram of a parallel-resistance capacitance bridge. Derive the equations for the measured capacitance and its resistive component. Discuss the different applications of series *RC* and parallel *RC* bridges.

8-12 Sketch the circuit diagram of an inductance comparison bridge. Derive the equations for the resistive and inductive components of the measured inductor.

8-13 Sketch the circuit diagram of a Maxwell bridge. Derive the equations for the resistive and inductive components of the measured inductor.

8-14 Sketch the circuit diagram of a Hay inductance bridge. Derive the equations for the resistive and inductive components of the measured inductor. Discuss the various applications of the Maxwell and Hay bridges.

8-15 Sketch ac bridge circuit diagrams showing how a standard capacitor and three adjustable standard resistors may be used to measure capacitance as a series *RC* circuit, capacitance as a parallel *RC* circuit, inductance as a series *RL* circuit, and inductance as a parallel *RL* circuit.

8-16 Discuss the problems involved in measuring small *C, R,* and *L* quantities, and explain suitable measuring techniques.

8-17 Sketch the basic circuits for converting inductance and capacitance into voltages for digital measurements. Explain the operation of each circuit.

8-18 Draw a circuit and waveforms to show how capacitance can be measured on a digital multimeter. Explain.

8-19 Draw the basic circuit diagram for a Q meter, explain its operation, and write the equation for Q factor.

8-20 Draw a practical Q-meter circuit, and discuss the various controls involved in Q-meter measurements.

8-21 Discuss the various methods of connecting components to a Q meter for measurement. Explain briefly.

PROBLEMS

8-1 A circuit behaves as a 0.01 μF capacitor in series with a 15 kΩ resistance when measured at a frequency of 1 kHz. If the terminal resistance is measured as 31.9 kΩ, determine the circuit components and the connection method.

8-2 When measured at a frequency of 100 kHz, an unknown circuit behaves as a 1000 pF capacitor and a 1.8 kΩ resistor connected in series. The terminal resistance is measured as greater than 10 MΩ. Determine the actual circuit components and the connection method.

8-3 A simple capacitance bridge, as in Figure 8-5, uses a 0.1 μF standard capacitor and two standard resistors each of which is adjustable from 1 kΩ to 200 kΩ. Determine the minimum and maximum capacitance values that can be measured on the bridge.

8-4 A series-resistance capacitance bridge, as in Figure 8-6, has a 1 kHz supply frequency. The bridge components at balance are $C_1 = 0.1$ μF, $R_1 = 109.5$ Ω, $R_3 = 1$ kΩ, and $R_4 = 2.1$ kΩ. Calculate the resistive and capacitive components of the measured capacitor, and determine the capacitor dissipation factor.

8-5 A parallel-resistance capacitance bridge (Figure 8-7) uses a 0.1 μF capacitor for C_1, and the supply frequency is 1 kHz. At balance, $R_1 = 547$ Ω, $R_3 = 1$ kΩ, and $R_4 = 666$ Ω. Determine the parallel RC components of the measured capacitor, and calculate the capacitor dissipation factor.

8-6 Calculate the parallel equivalent circuit components (C_p and R_p) for the measured capacitor in Problem 8-4. Also, determine the values of R_1 and R_4 required to balance C_p and R_p when the bridge is operated as a parallel-resistance capacitor bridge. Assume that R_3 remains 1 kΩ.

8-7 An inductance comparison bridge (Figure 8-8) has $L_1 = 100$ μH and $R_4 = 10$ kΩ. When measuring an unknown inductance, null is detected with $R_1 = 37.1$ Ω and $R_3 = 27.93$ kΩ. The supply frequency is 1 MHz. Calculate the measured inductance and its resistive component. Also, determine the Q factor of the inductor.

8-8 An inductor with a marked value of 100 mH and a Q of 21 at 1 kHz is to be measured on a Maxwell bridge (Figure 8-9). The bridge uses a 0.1 μF standard capacitor and a 1 kΩ standard resistor for R_1. Calculate the resistance values of R_3 and R_4 at which balance is likely to be achieved.

8-9 A Maxwell bridge with a 10 kHz supply frequency has a 0.1 μF standard capacitor and a 100 Ω standard resistor for R_1. Resistors R_3 and R_4 can each be adjusted from 100 Ω to 1 kΩ. Calculate the range of inductances and Q factors that can be measured on the bridge.

8-10 A Hay bridge (Figure 8-10) with a 500 Hz supply frequency has $C_3 = 0.5$ µF and $R_4 = 900$ Ω. If balance is achieved when $R_1 = 466$ Ω and $R_3 = 46.1$ Ω, calculate the inductance, resistance, and Q factor of the measured inductor.

8-11 Calculate the series equivalent circuit components L_s and R_s for the L_p and R_p quantities determined in Problem 8-10. Also, determine the resistances of R_1 and R_3 required to balance L_s and R_s when the circuit components are connected as a Maxwell bridge. Assume that R_4 and C_3 remain 900 Ω and 0.5 µF, respectively.

8-12 The Q-meter circuit in Figure 8-17 is in resonance when $E = 200$ mV, $R = 3$ Ω, and $X_L = X_C = 95$ Ω. Calculate the coil Q and the voltmeter indication.

8-13 The voltmeter in the Q-meter circuit in Figure 8-17 indicates 5 V when a coil is in resonance. If the coil has $R = 3.3$ Ω and $X_L = 66$ Ω at resonance, calculate the coil Q and the supply voltage.

Cathode-Ray Oscilloscopes

9

Objectives

You will be able to:

1. Sketch the basic construction of a cathode-ray tube (CRT) and explain its operation. Also, explain electrostatic deflection.

2. Explain the basic circuit diagram of an oscilloscope deflection amplifier, and discuss the various amplifier controls.

3. Sketch illustrations to show how waveforms applied to the vertical deflecting plates of a CRT are displayed, and determine the displayed waveform for given vertical and horizontal inputs.

4. Draw the circuits and block diagrams for basic and automatic oscilloscope time bases. Explain time base operation and discuss the various controls.

5. Explain the operation of dual-trace oscilloscopes, and explain the function of each of the various controls on the front panel of a basic dual-trace oscilloscope.

6. Use an oscilloscope to measure voltage levels and times on a variety of displayed waveforms.

7. Explain how the oscilloscope input circuit may produce distortion on pulse waveforms, and explain the use of oscilloscope probes.

8. Discuss typical oscilloscope specifications.

Introduction

The oscilloscope is the basic instrument for the study of all types of waveforms. It can be employed to measure such quantities as peak voltage, frequency, phase difference, pulse width, delay time, rise time, and fall time. An oscilloscope consists of a cathode-ray tube

(CRT) and its associated control and input circuitry. In the CRT, electrons generated at a heated cathode are shaped into a fine beam and accelerated toward a fluorescent screen. The screen glows at the point at which the electrons strike. The electron beam is easily deflected vertically and horizontally by voltages applied to deflecting plates. Usually, the beam is swept horizontally across the screen by a ramp voltage generated by a time base circuit, and the waveform to be investigated is applied to deflect the beam vertically. This results in a voltage-versus-time display of the input waveform. Most oscilloscopes are dual-trace (or multitrace) instruments capable of displaying two or more waveforms simultaneously.

9-1 CATHODE-RAY TUBE

Operation

The basic construction and biasing of a cathode-ray tube are shown in Figure 9-1. The system of electrodes is contained in an evacuated glass tube with a viewing screen at one end. A beam of electrons is generated by the cathode and directed to the screen, causing the phosphor coating on the screen to glow where the electrons strike. The electron beam is deflected vertically and horizontally by externally applied voltages.

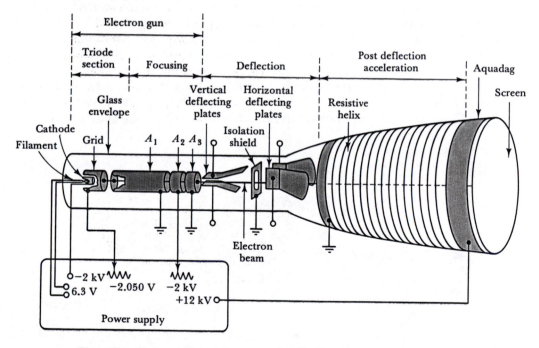

Figure 9-1 Basic cathode-ray tube construction. The filament heats the cathode, which emits electrons. The grid voltage controls the flow of electrons to anodes A_1, A_2, and A_3, which focus the beam to a fine point at the screen. Voltages applied to the vertical and horizontal deflecting plates move the electron beam around the screen.

Triode Section

The triode section of the tube consists of a *cathode,* a *grid,* and an *anode.* The grid, which is a nickel cup with a hole in it (see Figure 9-1), almost completely encloses the cathode. The cathode, also made of nickel, is cylinder shaped with a flat, oxide-coated, electron-emitting surface directed toward the hole in the grid. Cathode heating is provided by an inside filament. The cathode is typically held at approximately −2 kV, and the grid potential is adjustable from approximately −2000 V to −2050 V. The grid-cathode potential controls the electron flow from the cathode and thus controls the number of electrons directed to the screen. A large number of electrons striking one point will cause the screen to glow brightly; a small number will produce a dim glow. Therefore, the grid potential control is a *brightness control.*

The first anode (A_1) is cylinder shaped, open at one end and closed at the other end, with a hole at the center of the closed end. Since A_1 is grounded and the cathode is at a high negative potential, A_1 is highly positive with respect to the cathode. This causes electrons to be accelerated from the cathode through the holes in the grid and anode to the focusing section of the tube.

Focusing Section

The focusing electrodes A_1, A_2, and A_3 are sometimes referred to as an *electron lens.* The function of the electron lens is to focus the electrons to a fine point on the screen of the tube. A_1 provides the accelerating field to draw the electrons from the cathode, and the hole in A_1 limits the initial cross section of the electron beam. A_3 and A_1 are held at ground potential while the A_2 potential is adjustable around −2 kV. The result of the potential difference between anodes is that *equipotential lines* are set up as shown in Figure 9-2. These are lines along which the potential is constant. Line 1, for example, might have a potential of −700 V along its entire length, while the potential of line 2 might be −500 V over its whole length. The electrons enter A_1 as a divergent beam. On crossing the equipotential lines, however, the electrons experience a force that changes their direc-

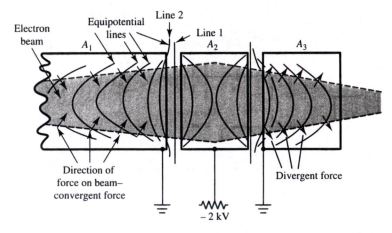

Figure 9-2 Electrostatic focusing for a cathode-ray tube. The potential difference between electrodes A_1, A_2, and A_3 set up equipotential lines that exert convergent and divergent forces on the electron beam, thus focusing the beam at the screen.

Cathode-Ray Oscilloscopes   Chap. 9

tion of travel toward right angles with respect to the equipotential lines. The shape of the lines within A_1 produces a convergent force on the electron beam, and those within A_3 produce a divergent force on the beam. The convergent and divergent forces can be altered by adjusting the potential on A_2. This adjusts the point at which the beam is focused. A_2 is sometimes referred to as the *focus ring*. Figure 9-2 is a two-dimensional representation of a three-dimensional apparatus, so there are really equipotential *planes* rather than equipotential *lines*.

The negative potential on A_2 tends to slow down the electrons, but they are accelerated again by A_3, so that the beam speed leaving A_3 is the same as when entering A_1. The electrons are traveling at a constant velocity as they pass between the deflecting plates.

Deflection Section

If the horizontal and vertical deflecting plates were grounded, or left unconnected, the beam of electrons would pass between each pair of plates and strike the center of the oscilloscope screen. There they would produce a bright glowing point. When one plate of a pair of deflecting plates has a positive voltage applied to it, and the other one has a negative potential, the electrons in the beam are attracted toward the positive plate and repelled from the negative plate. The electrons are actually accelerated in the direction of the positive plate. However, since they are traveling axially between the plates, no electrons ever strike a deflecting plate. Instead, the beam is deflected so that the electrons strike the screen at a new position.

Consider the electrostatic deflection illustration in Figure 9-3. When the potential on each plate is zero, the electrons passing between the plates do not experience any deflecting force. When the upper plate potential is $+E/2$ volts and the lower potential is $-E/2$, the potential difference between the plates is E volts. The (negatively charged) electrons are attracted toward the positive plate and repelled from the negative plate. The tube sensitivity to deflecting voltages can be expressed in two ways. The voltage required to produce one division of deflection at the screen (V/cm) is referred to as the *deflection factor* of the tube. The deflection produced by 1 V (cm/V) is termed the *deflection sensitivity*.

When an alternating voltage is applied to the deflecting plates, the beam is deflected first in one direction, and then in the other. This produces a horizontal line when the ac voltage is applied to the horizontal plates, and a vertical line when the vertical deflection plates are involved. In Section 9-3 it is explained that the waveform to be displayed is ap-

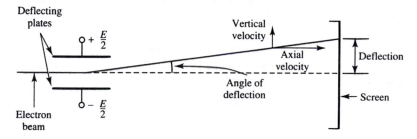

Figure 9-3 Electrostatic deflection. Electrons, traveling from the electron gun to the screen of the cathode-ray tube, are attracted toward the positive deflecting plate and repeled from the negative plate. The electrons do not strike the plate, but the electron beam is deflected.

plied via an amplifier to the vertical deflecting plates, and that the horizontal plates normally have a ramp waveform.

A grounded *isolation shield* is situated between the vertical and horizontal deflecting plates (see Figure 9-1). This prevents the electric fields of one set of plates from influencing the other pair of plates.

Screen

The screen of a CRT is formed by depositing a coating of phosphor materials on the inside of the tube face. When the electron beam strikes the screen, electrons within the screen material are raised to a higher energy level and emit light as they return to their normal levels. The glow may persist for a few milliseconds, for several seconds, or even longer. Depending on the materials employed, the color of the glow produced at the screen may be blue, red, green, or white.

The phosphors used on the screen are insulators, and, but for secondary emission, the screen would develop a negative potential as the primary electrons accumulate. The negative potential would eventually become so great that it would repel the electron beam. The secondary electrons are collected by a graphite coating termed *aquadag* around the neck of the tube (see Figure 9-1), so that the negative potential does not accumulate on the screen. In another type of tube, the screen has a fine film of aluminum deposited on the surface at which the electrons strike. This permits the electron beam to pass through, but collects the secondary electrons and conducts them to ground. The aluminum film also improves the brightness of the glow by reflecting the emitted light toward the glass. A further advantage of the film is that it acts as a *heat sink,* conducting away heat that might otherwise damage the screen.

Brightness of Display

As has already been explained, the brightness of the glow produced at the screen is dependent on the number of electrons making up the beam. Since the grid controls the electron emission from the cathode, the grid voltage control is a brightness control. Brightness also depends on beam speed; so for maximum brightness the electrons should be accelerated to the greatest possible velocity. However, if the electron velocity is very high when passing through the deflection plates, the deflecting voltages will have a reduced influence, and the deflection sensitivity will be poor. It is for this reason that *post deflection acceleration* is provided; that is, the electrons are accelerated again after they pass between the deflecting plates. A helix of resistive material is deposited on the inside of the tube from the deflecting plates to the screen (Figure 9-1). The potential at the screen end of the helix might be typically +12 kV and at the other end zero. Thus, the electrons leaving the deflecting plates experience a continuous accelerating force all the way to the screen, where they strike with high energy.

9-2 DEFLECTION AMPLIFIERS

Any voltage that is to produce deflection of the electron beam must be converted into two equal and opposite voltages, $+E/2$ and $-E/2$, as explained in Section 9-1. This requires an amplifier that accepts an (ac or dc) input and provides *differential* outputs. The basic cir-

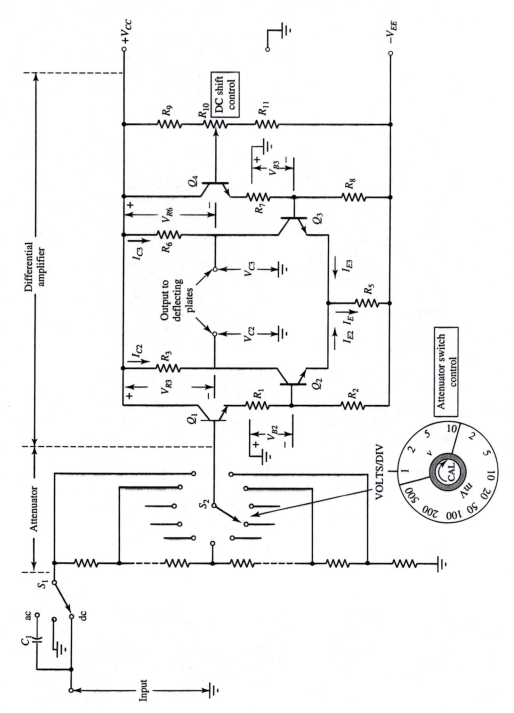

Figure 9-4 Basic circuits of an oscilloscope deflection amplifier and its input attenuator. A positive input applied to the base of transistor Q_1 produces a negative output at the Q_2 collector terminal and a positive output at the collector of Q_3.

235

cuit of such an amplifier is illustrated in Figure 9-4. Transistors Q_2 and Q_3 form an emitter-coupled amplifier (already discussed in Section 4-1). Q_1 and Q_4 are emitter followers to provide high input resistance.

When the input voltage to the attenuator is zero, the base of Q_1 is at ground level. If Q_4 base is also adjusted to ground level, Q_2 and Q_3 bases are both at the same negative potential with respect to ground $(-V_{B2} = -V_{B3})$. Also, $I_{C2} = I_{C3}$, and the voltage drops across R_3 and R_6 set the collectors of Q_2 and Q_3 at ground level. These collectors are the amplifier outputs, and they are connected directly to the deflection plates.

An input voltage that is to produce vertical deflection is coupled to the *attenuator* of the amplifier feeding the vertical plates (see Figure 9-4). The attenuated voltage appears at the base of transistor Q_1, where it is further attenuated (via R_1 and R_2) and then applied to Q_2 base. A positive-going input produces a positive-going voltage at Q_2 base, and causes I_{C2} to increase and I_{C3} to decrease. The I_{C2} increase causes output V_{C2} to fall below its normal ground level, and the I_{C3} decrease makes V_{C3} rise above ground. If the change in V_{C2} is $\Delta V_{C2} = -1$ V, then $\Delta V_{C3} = +1$ V. When the input to the attenuator is a negative-going quantity, I_{C2} decreases and I_{C3} increases. Now ΔV_{C2} is positive and ΔV_{C3} is an equal and opposite negative voltage.

Where the amplifier is used with the vertical deflection plates, the attenuator resistors are selected to produce a wide range of deflection sensitivity. At the most sensitive position of the attenuator switch (least attenuation), a 2 mV input typically produces one division of deflection on the screen. At the greatest attenuation position of the switch, a 10 V input is required to produce one division of deflection. The input to the attenuator may be capacitor coupled (ac) or direct coupled (dc), according to the selected position of switch S_1 in Figure 9-4.

Potentiometer R_{10} in Figure 9-4 is a dc shift control that serves to adjust the voltage at the base of Q_4. When the moving contact is centralized, Q_4 base is at ground level. By adjusting the moving contact of R_{10}, either a positive or a negative dc potential is applied to the base of Q_4. When V_{B4} is positive, Q_3 base voltage is raised positively, so that I_{C3} increases and I_{C2} decreases. This causes V_{C3} to fall and V_{C2} to rise. A differential dc voltage is thus applied to the deflecting plates to deflect (or *shift*) the electron beam above the center of the screen. When R_{10} is adjusted to produce a negative voltage at Q_4 base, the electron beam is adjusted below the screen center. This dc shift does not affect a waveform to be displayed, which is applied to the attenuator input. However, R_{10} adjustment shifts the displayed waveform up or down on the screen as desired by the operator.

9-3 WAVEFORM DISPLAY

When an alternating voltage is applied to the vertical deflecting plates and no input is applied to the horizontal plates, the spot on the tube face moves up and down continuously. If a constantly increasing voltage is also applied to the horizontal deflecting plates, then, as well as moving vertically, the spot on the tube face moves horizontally. Consider Figure 9-5, in which a sine wave is applied to the vertical deflecting plates and a *sawtooth* (or repetitive ramp) is applied to the horizontal plates. If the wave-

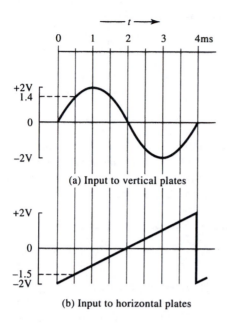

(a) Input to vertical plates

(b) Input to horizontal plates

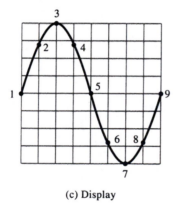

(c) Display

Figure 9-5 A ramp waveform applied to the horizontal deflecting plates of an oscilloscope causes the electron beam to be deflected horizontally across the screen. Another waveform, synchronized with the ramp and applied to the vertical deflecting plates, is displayed on the screen.

forms are perfectly synchronized, then at time $t = 0$ the vertical deflecting voltage is zero and the horizontal deflecting voltage is −2 V. Therefore, assuming a deflecting sensitivity of 2 cm/V, the vertical deflection is zero and the horizontal deflection is 4 cm left from the center of the screen [point 1 on Figure 9-5(c)]. When $t = 0.5$ ms, the horizontal deflecting voltage has become −1.5 V; therefore, the horizontal deflection is 3 cm left from the screen center. The vertical deflecting voltage has now become +1.4 V, and this causes a vertical deflection of +2.8 cm above the center of the screen. The spot is now 2.8 cm up and 3 cm left from the screen center, point 2 on Figure 9-5(c).

The following table gives deflection voltages and deflections at various instants for the waveforms in Figure 9-5:

t (ms)	1	1.5	2	2.5	3	3.5	4
Vertical voltage (V)	+2	+1.4	0	−1.4	−2	−1.4	0
Vertical deflection (cm)	+4	+2.8	0	−2.8	−4	−2.8	0
Horizontal voltage (V)	−1	−0.5	0	+0.5	+1	+1.5	+2
Horizontal deflection (cm)	−2	−1	0	+1	+2	+3	+4
Point	3	4	5	6	7	8	9

At point 9 the horizontal deflecting voltage rapidly goes to −2 V again, so the beam returns to the left side of the screen. From here it is ready to repeat the waveform trace. It is seen that with a sawtooth applied to the horizontal deflecting plates, any waveform applied to the vertical plates will be displayed on the screen of the CRT.

Example 9-1

A 500 Hz triangular wave with a peak amplitude of 40 V is applied to the vertical deflecting plates of a CRT. A 250 Hz sawtooth wave with a peak amplitude of 50 V is applied to the horizontal deflecting plates. The CRT has a vertical deflection sensitivity of 0.1 cm/V and a horizontal deflection sensitivity of 0.08 cm/V. Assuming that the two inputs are synchronized, determine the waveform displayed on the screen.

Solution

For the triangular wave,

$$T = \frac{1}{f} = \frac{1}{500 \text{ Hz}} = 2 \text{ ms}$$

For the sawtooth wave,

$$T = \frac{1}{250 \text{ Hz}} = 4 \text{ ms}$$

The two waveforms are shown in Figure 9-6(a) and (b).
At t = 0:

$$\text{vertical voltage} = 0$$

$$\text{horizontal voltage} = -50 \text{ V}$$

$$\text{horizontal deflection} = \text{voltage} \times (\text{deflection sensitivity})$$

$$= -50 \times 0.08 \text{ cm}$$

$$= -4 \text{ cm (i.e., 4 cm left from center)}$$

Point 1 on the CRT screen [Figure 9-6(c)] is at

$$\text{vertical deflection} = 0,$$

$$\text{horizontal deflection} = 4 \text{ cm left of center}$$

Cathode-Ray Oscilloscopes   Chap. 9

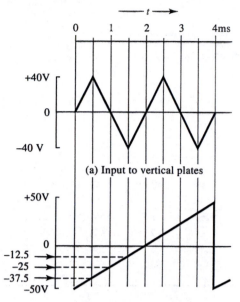

(a) Input to vertical plates

(b) Input to horizontal plates

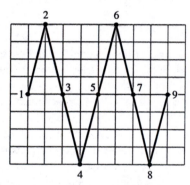

(c) Display at CRT screen

Figure 9-6 Two cycles of triangular wave applied to the vertical deflecting plates of an oscilloscope are displayed on the screen if the triangular wave is synchronized with the ramp wave at the horizontal deflecting plates.

At t = 0.5 ms:

$$\text{vertical voltage} = +40 \text{ V}$$

$$\text{horizontal voltage} = -37.5 \text{ V}$$

Therefore, at point 2 on the CRT screen,

$$\text{vertical deflection} = +40 \times 0.1 \text{ cm} = +4 \text{ cm}$$

$$\text{horizontal deflection} = -37.5 \times 0.08 \text{ cm} = -3 \text{ cm}$$

At t = 1 ms (point 3):

$$\text{vertical deflection} = 0$$

$$\text{horizontal deflection} = -25 \times 0.08 \text{ cm} = -2 \text{ cm}$$

At t = 1.5 ms (point 4):

$$\text{vertical deflection} = -40 \times 0.1 \text{ cm} = -4 \text{ cm}$$

$$\text{horizontal deflection} = -12.5 \times 0.08 \text{ cm} = -1 \text{ cm}$$

At t = 2 ms (point 5):

$$\text{vertical deflection} = 0$$

$$\text{horizontal deflection} = 0$$

t *(ms)*	2.5	3	3.5	4
Vertical voltage *(V)*	+40	0	−40	0
Vertical deflection *(cm)*	+4	0	−4	0
Horizontal voltage *(V)*	+12.5	+25	+37.5	+50
Horizontal deflection *(cm)*	+1	+2	+3	+4
Point	6	7	8	9

9-4 OSCILLOSCOPE TIME BASE

Horizontal Sweep Generator

In Section 9-3 it is explained that a waveform applied to the vertical deflecting plates is displayed on the oscilloscope screen if a sawtooth (or repetitive ramp) voltage is simultaneously applied to the horizontal deflecting plates. The sawtooth voltage is produced by a *sweep generator circuit,* which may be of the type illustrated in Figure 9-7.

The sweep generator shown consists of two major components: a ramp generator and a *noninverting Schmitt trigger circuit.* Note that the inverting input terminal of the operational amplifier is grounded via resistor R_7. Ignoring C_2, the input voltage to the Schmitt is the ramp generator output (V_1), applied via resistor R_6. Because the op-amp has a very large voltage gain (typically 200 000), a very small difference between the inverting and noninverting terminals causes the Schmitt output to be *saturated.* This means that the output voltage is very close to either the positive or the negative supply voltages. Typically, the saturated output voltage is $+(V_{CC} - 1 \text{ V})$, or $-(V_{EE} - 1 \text{ V})$.

Assume that the Schmitt output is negative, and that the ramp input to the Schmitt is at its minimum level. The voltages at both ends of potential divider $R_5 + R_6$ are negative, so the junction of R_5 and R_6 must also be negative. Thus, the op-amp noninverting terminal voltage is below the level of the (grounded) inverting terminal, and the op-amp output remains saturated in a negative direction. This keeps Q_2 biased off.

As the ramp voltage grows, the voltage at the junction of R_5 and R_6 rises toward ground. When the ramp reaches a high enough positive level, the noninverting input terminal is eventually raised slightly above ground. This causes the op-amp output to switch rapidly from the negative saturated level to saturation in the positive direction.

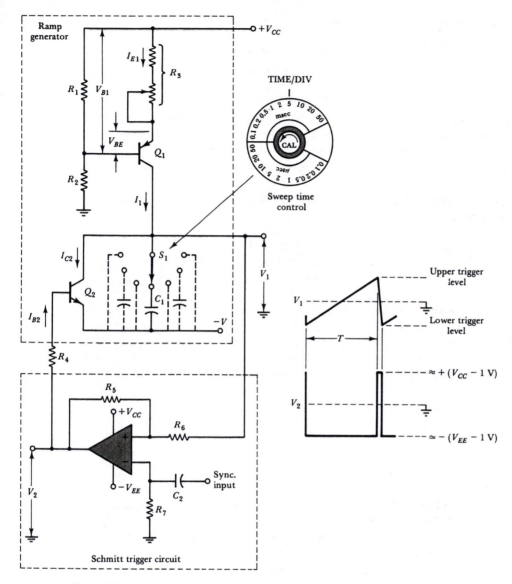

Figure 9-7 Basic circuit of an oscilloscope sweep generator. Transistor Q_1 provides a constant charging current to capacitor C_1, thus generating a ramp waveform. The Schmitt trigger circuit switches Q_2 *on* to discharge C_1 when the ramp reaches its upper peak level, and turns Q_2 *off* again when C_1 is discharged to the desired lower peak of the ramp.

The level of the input voltage (to R_6) that causes the output to switch positively is known as the *upper trigger point (UTP)*. For the output to go negative once more, the input to R_6 must fall to a negative level, which will pull the noninverting terminal below ground. The negative input voltage level required to trigger the output to negative saturation is termed the *lower trigger point (LTP)*, and it is numerically equal to the upper trigger point.

The trigger voltage levels can be shown to be,

$$\text{UTP, LTP} = \pm (V_{CC} - 1 \text{ V}) \frac{R_6}{R_5} \qquad (9\text{-}1)$$

In the ramp generator portion of Figure 9-7 transistor Q_1 and its associated components constitute a *constant current source*. Resistors R_1 and R_2 potentially divide the positive supply voltage (V_{CC}) to provide a constant bias voltage (V_{B1}) to the base of *pnp* transistor Q_1. This makes the voltage drop across emitter resistor R_3 a constant quantity, and maintains I_E constant.

Since the transistor collector current is approximately equal to its emitter current, I_1 is a constant current:

$$I_1 \simeq \frac{V_{B1} - V_{BE}}{R_3} \qquad (9\text{-}2)$$

Alteration of I_1 is possible by means of the adjustable portion of resistor R_3.

When the Schmitt trigger circuit output is positive, a base current (I_{B2}) flows into *npn* transistor Q_2 via R_4. This switches Q_2 into saturation and short-circuits capacitor C_1. When the trigger circuit output goes negative, Q_2 is switched *off* and current I_1 flows into capacitor C_1, charging it linearly and producing a ramp output. The equation for the ramp voltage is

$$\Delta V_1 = \frac{I_1 T}{C_1} \qquad (9\text{-}3)$$

where ΔV_1 is the capacitor voltage change during time T and C_1 is the capacitance.

The capacitor voltage continues to grow linearly until it arrives at the upper trigger level for the Schmitt. Then the Schmitt output goes positive once again, turning Q_2 *on* and rapidly discharging C_1. When the voltage across C_1 falls to the lower trigger level of the Schmitt, the trigger circuit output switches to negative again. Q_2 is once again switched *off*, and the voltage across C_1 commences to grow linearly once more.

The process described above is repeated continuously, producing a repetitive ramp wave (V_1 in Figure 9-7). A pulse waveform (V_2) is also generated at the output of the Schmitt trigger circuit. The complete circuit is termed a *free-running ramp generator*.

The amplitude of the ramp waveform is obviously dictated by the upper and lower trigger points of the Schmitt trigger circuit. The time period (T) of the ramp depends on charging current I_1 and on the capacitance of C_1. I_1 can be altered by the adjustable component of R_3. As illustrated, any one of several capacitors can be switched into the circuit, thus altering time period T. Individual capacitors can be selected to generate ramps that sweep the electron beam across the screen of the CRT at 1 div/10 ms, 1 div/50 ms, or at any one of several other rates. The capacitor selection switch on the oscilloscope front panel is identified as TIME/DIV.

Example 9-2

The sweep generator circuit in Figure 9-7 has $R_3 = 4.2$ kΩ, $C_1 = 0.25$ μF, $V_{B1} = 4.9$ V, and the trigger levels for the Schmitt are ±2 V. Calculate the peak-to-peak amplitude and time period of the ramp waveform.

Solution

$$\Delta V_1 = 2 \times (\text{UTP, LTP}) = 2 \times 2 \text{ V}$$

$$= 4 \text{ V p-to-p}$$

Equation 9-2,

$$I_{C1} \simeq \frac{V_{B1} - V_{BE}}{R_3} = \frac{4.9 \text{ V} - 0.7 \text{ V}}{4.2 \text{ k}\Omega}$$

$$= 1 \text{ mA}$$

From Equation 9-3,

$$T = \frac{\Delta V_1 \times C_1}{I_{C1}} = \frac{4 \text{ V} \times 0.25 \text{ μF}}{1 \text{ mA}}$$

$$= 1 \text{ ms} \qquad (\text{see Figure 9-8})$$

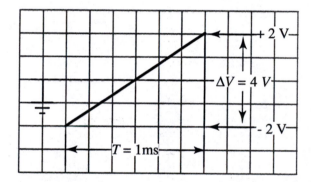

Figure 9-8 Ramp waveform for Example 9-2.

Automatic Time Base

For a waveform to be displayed correctly on an oscilloscope, it is important that the ramp voltage producing the horizontal sweep begin the instant the displayed waveform goes positive. Alternatively, it may be made to commence as the displayed wave goes negative. The ramp wave must be *synchronized* with the input waveform. If the input and ramp waveforms are not synchronized, the displayed wave will appear to continuously slide off to one side of the screen. Synchronization is accomplished by means of the *sync input* to the Schmitt trigger in Figure 9-7, and by the other components of the *automatic time base* in Figure 9-9.

The voltage waveform to be displayed (V_i) is applied to the *vertical amplifier* and to the time base *triggering amplifier* (see Figure 9-9). Like the vertical amplifier, the triggering amplifier has differential outputs (see Section 9-2). These provide two identical but antiphase voltage waveforms (V_{o1} and V_{o2}). In the triggering amplifier the input is ampli-

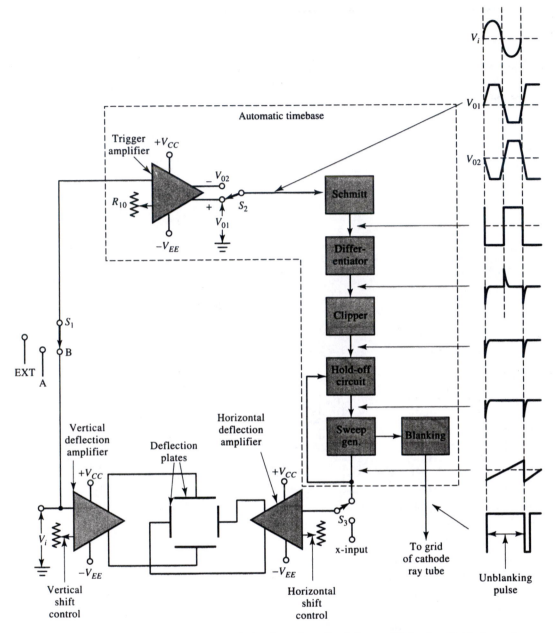

Figure 9-9 Oscilloscope automatic time base. As well as being amplified and applied to the vertical deflecting plates, input signal input V_i is routed to a Schmitt trigger circuit to generate a square wave. This is differentiated and clipped to produce a train of negative spikes that occur precisely at the beginning of the V_i cycle. The spikes are used to synchronize the sweep generator to the input waveform.

Cathode-Ray Oscilloscopes   Chap. 9

fied so much that its peaks are cut off by saturation of the amplifier output stage. So the output waveforms are almost square (Figure 9-9). One of these waveforms is passed via switch S_2 to the input of an inverting Schmitt trigger circuit. The Schmitt is designed to have upper and lower trigger points slightly above and below ground. (With this condition, it is often called a *zero-crossing detector.*) The Schmitt output rapidly goes negative as the input passes the upper trigger point, and positive as the input passes the lower trigger point (see the waveforms in Figure 9-9).

The output from the Schmitt circuit is a square waveform exactly in antiphase with the input wave to be displayed. This square wave is applied to a *differentiating circuit.* The output produced by the differentiator is proportional to the rate of change of the square wave. During the times that the square wave is at its constant positive level or at its constant negative level, its rate of change is zero. So the differentiator output is zero at these times. At the positive-going edge of the square wave, the rate of change is a large positive quantity. At the negative-going edge, the rate of change is a large negative quantity. Therefore, the differentiated square wave is a series of positive spikes coinciding with the positive-going edges of the square wave, and negative spikes coinciding with the negative-going edges (see Figure 9-9).

The spike waveform is now fed to a *positive clipper circuit.* This is essentially a rectifier circuit that passes the negative spikes but blocks (or *clips off*) the positive spikes. The negative spikes (which coincide with the commencement of each cycle of the original input) are passed via a *hold-off circuit* to the sync input of the sweep generator. For now, ignore the function of the hold-off circuit.

Return once again to the sweep generator circuit in Figure 9-7. Note that the sync input is capacitor coupled to the inverting input terminal of the Schmitt operational amplifier. Now assume that the ramp voltage from the integrator is approaching the upper trigger point of the Schmitt but that it will not get there before commencement of another cycle of input. The situation is illustrated in Figure 9-10(a). The negative spike from the clipper (Figure 9-9) arrives at the op-amp inverting terminal via capacitor C_2. The inverting terminal is driven below ground level and below the level of the voltage at the noninverting terminal. The noninverting terminal is now positive with respect to the inverting terminal, and the Schmitt output rapidly switches to positive saturation. Q_2 is switched *on,* and the ramp output falls rapidly to the Schmitt lower trigger point and then begins to grow linearly again.

It is seen that the train of negative spikes causes the ramp output of the sweep generator to be synchronized with the input waveform that is to be displayed. The ramp commences at the beginning of each positive half-cycle of the input. The ramp output from the sweep generator is fed to the horizontal deflection amplifier (Figure 9-9). This amplifier provides differential (antiphase) ramp outputs to the horizontal deflecting plates, and causes the electron beam to sweep from left to right on the oscilloscope screen. The input waveform is now displayed exactly as explained in Section 9-3.

It is not desirable that *every* negative spike should trigger the Schmitt and reset the ramp to its starting point. If this occurred, the displayed wave would always consist of only one cycle instead of the two or more cycles that frequently are required. The hold-off circuit in Figure 9-9 is a spike-suppressing circuit controlled by the level of the ramp output from the sweep generator. Once a negative spike has synchronized the time base, no more spikes are allowed to pass to the sync input terminal until the ramp output of the

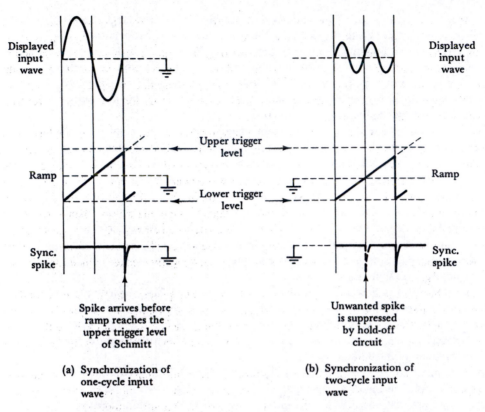

Figure 9-10 Automatic time base synchronization process. The spike produced at the end (beginning) of a cycle of the input wave switches the time base ramp voltage back to its negative peak, thus sweeping the electron beam horizontally back to the required position for tracing another cycle of input.

sweep generator approaches its maximum amplitude [Figure 9-10(b)]. Note that in Figure 9-9, the sweep generator output is fed back to the hold-off circuit as well as to the horizontal amplifier.

With all spikes suppressed until the ramp approaches its peak level, the ramp is *not* reset to zero until the electron beam has been swept horizontally to the right-hand side of the oscilloscope screen. Any number of waveform cycles can now be displayed on the oscilloscope screen.

The exact instant at which the time base is synchronized with the input wave is usually taken as the instant at which the displayed wave commences it positive half-cycle. However, as illustrated in Figure 9-11, synchronization can be made to occur shortly after the wave crosses zero. The triggering amplifier in Figure 9-9 is similar to the deflection amplifier in Figure 9-4. Potentiometer R_{10} in Figure 9-4 is a dc *shift control,* and in the triggering amplifier R_{10} performs a similar function [see Figure 9-11(a)]. By means of this potentiometer, the amplifier dc output levels can be adjusted close to the upper trigger point of the Schmitt in Figure 9-9. In Figure 9-11(b) the triggering amplifier output reaches the upper trigger point approximately at the same instant that the input wave begins its cycle. Therefore, the displayed wave commences at this instant. In Figure 9-11(c) the out-

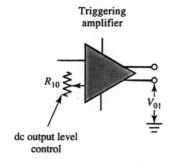

Triggering
amplifier

R_{10}

V_{01}

dc output level
control

(a) R_{10} controls the trigger level

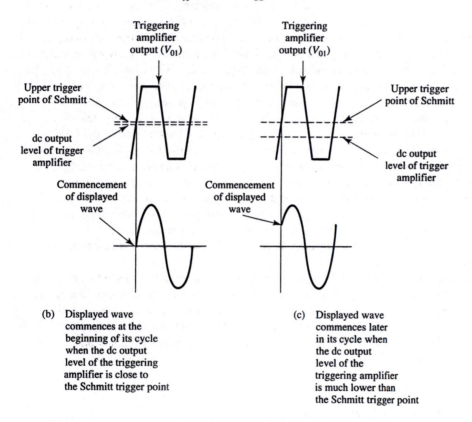

Triggering
amplifier
output (V_{01})

Triggering
amplifier
output (V_{01})

Upper trigger
point of Schmitt

Upper trigger
point of Schmitt

dc output
level of trigger
amplifier

dc output
level of trigger
amplifier

Commencement
of displayed
wave

Commencement
of displayed
wave

(b) Displayed wave
commences at the
beginning of its cycle
when the dc output
level of the triggering
amplifier is close to
the Schmitt trigger point

(c) Displayed wave
commences later
in its cycle when
the dc output
level of the
triggering amplifier
is much lower than
the Schmitt trigger point

Figure 9-11   The dc level control on the oscilloscope triggering amplifier determines the
instant at which a displayed waveform commences.

put of the triggering amplifier does not reach the trigger point until some time after the
input wave has begun its cycle. The displayed wave now commences at this point. It is
seen that potentiometer R_{10} functions as a *trigger level control*, to adjust the instant in
time at which the displayed wave commences on the oscilloscope screen.

The displayed wave may be made to commence at the beginning of its negative
half-cycle instead of its positive half-cycle. This is accomplished simply by switching S_2

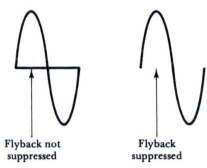

Flyback not suppressed

Flyback suppressed

Figure 9-12 The electron beam in an oscilloscope is normally suppressed during the flyback time from right to left.

in Figure 9-9 to the antiphase output of the triggering amplifier (V_{o2}). The input to the Schmitt is now in phase with the input wave, and the negative spikes are generated at the instant that V_i crosses zero from positive to negative. The horizontal sweep now commences at this instant, and the negative half-cycle of the displayed wave occurs first.

In Figure 9-9 an output from the sweep generator is fed to a *blanking circuit*. This is the pulse waveform output (V_2) from the Schmitt in Figure 9-7. The pulse wave is inverted and capacitor coupled to convert it to a train of positive pulses during the sweep time, and negative pulses that occur each time the ramp wave falls from its maximum positive level to its maximum negative level. These pulses are fed to the grid of the CRT (Figure 9-1). The negative pulses (known as *blanking pulses*) drive the grid sufficiently negative to suppress the electron beam completely. This means that no electrons strike the screen while the ramp is going from its maximum positive to its maximum negative level. If the electrons were not suppressed during this time, every displayed wave would have a horizontal line traced by the electron beam during the *flyback time* from the right-hand side of the screen to the left (see Figure 9-12). The positive pulses are termed *unblanking pulses,* and the bias that these apply to the CRT grid cause electrons to travel from the cathode to the screen.

Switch S_1 in Figure 9-9 permits triggering inputs to be selected from channel *A* or channel *B* in a dual-trace oscilloscope (see Section 9-5). Alternatively, an external (EXT) triggering source can be selected. The time base can be disconnected from the horizontal deflection amplifier by means of switch S_3 in Figure 9-9. An external *x*-input may be applied to produce horizontal deflection of the electron beam. Applications of this facility are discussed in Section 9-10.

9-5 DUAL-TRACE OSCILLOSCOPES

Most oscilloscopes can display two waveforms rather than just one. This allows waveforms to be compared in terms of amplitude and phase or time. Two input terminals and two sets of controls are provided, identified as channel *A* and channel *B*.

The construction of a dual-trace CRT could be exactly as illustrated in Figure 9-1, with the exception that two complete electron guns are contained in a single tube [see Figure 9-13(a)]. In this case the instrument can be termed a *dual-beam oscilloscope* because there are two separate electron beams, one for each waveform trace. In another type of dual-trace CRT, a single electron gun is involved, but the beam is split into two sepa-

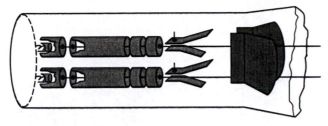

(a) CRT with two electron guns

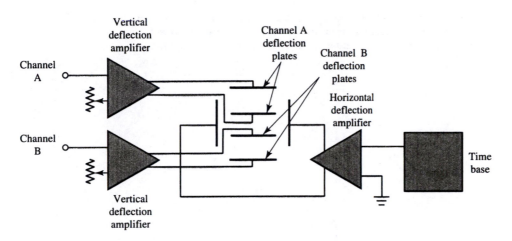

(b) Deflection system for dual-beam oscilloscope

Figure 9-13   A dual-trace oscilloscope may be produced by using two electron guns. Two sets of vertical deflection plates are required together with one set of horizontal deflecting plates.

rate beams before it passes to the deflection plates. This is referred to as a *split-beam* CRT. The dual-beam and split-beam instruments each have only one set of horizontal deflection plates [Figure 9-13(b)]. The sawtooth wave from the time base is applied to the single set of horizontal deflection plates, and both beams are made to sweep across the screen simultaneously. There are two completely separate vertical inputs: channel A and channel B. Each channel has its own deflection amplifier feeding one pair of vertical deflection plates.

Another common type of dual-trace oscilloscope is illustrated in Figure 9-14. A single-beam CRT is shown, with only one set of vertical deflection plates. Two separate (channel A and channel B) input amplifiers are employed, with a single amplifier feeding the vertical deflection plates. The input to this amplifier is alternately switched between channels A and B, and the switching frequency is controlled by the time base circuit.

Consider Figure 9-15, which illustrates the process of alternately displaying first one input wave and then the other. The input to channel A is a sine wave with time period T, and that to channel B is a triangular wave also having a time period T [Figure 9-15(a)].

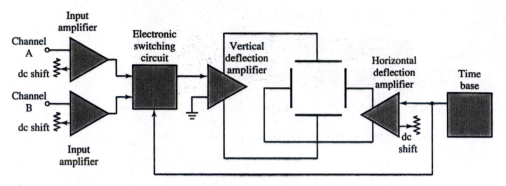

Figure 9-14 Most dual-trace (and multitrace) oscilloscopes have a single electron beam and a single set of vertical deflecting plates. Two (or more) waveforms are displayed by rapidly switching the input of the vertical deflection amplifier between two (or more) input channels.

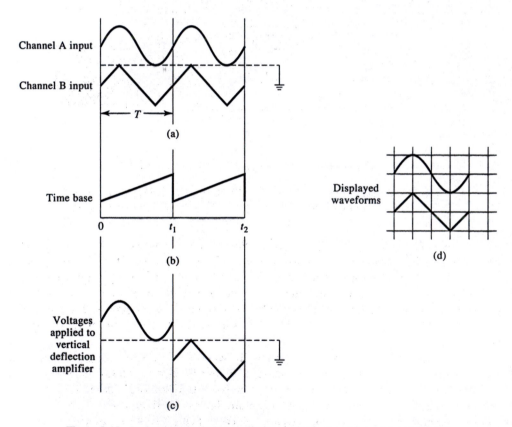

Figure 9-15 In the alternate dual-trace display mode, the two (channel A and channel B) waveforms are alternately switched to the input of the oscilloscope vertical deflection amplifier (one, or more, cycles at a time). Except at very low frequencies, the two waves appear to be displayed on the screen simultaneously.

The two waveforms are in synchronism. Note that channel A input has a dc offset, which puts it above ground level, while channel B is offset below ground. Channel A input is switched to the vertical deflection amplifier and traced on the oscilloscope screen during time 0 to t_1. Channel B input is next applied to the vertical deflection amplifier and traced on the oscilloscope screen during time t_1 to t_2 [Figure 9-15(b), (c)]. The dc offsets on the inputs cause the waveform on channel A to be traced on the top half of the oscilloscope screen and that on channel B to be traced on the bottom half of the screen. During the next cycle of the time base channel A input is again traced on the screen, followed by channel B input once again. Thus, the two inputs are alternately and repeatedly traced on the screen. The repetition frequency is usually so high that the waveforms appear to be displayed simultaneously [Figure 9-15(d)].

When the method described above is employed to display two waveforms, the oscilloscope is said to be operating in *alternate mode*. A similar method which uses a much higher switching frequency is termed *chop mode*.

Figure 9-16 illustrates oscilloscope operation in the chop mode. Channel A input is traced for a short time, t_1, then channel B input is traced for time t_2. Back to channel A input for t_3, then channel B for t_4, and so on, as illustrated. High-frequency waveforms would be displayed as dashed lines, exactly as shown in Figure 9-16. However, the breaks in the traced waveforms are of such short duration that they become invisible when medium- and low-frequency waves are displayed.

Dual-trace oscilloscopes that employ channel switching normally permit selection of operation in either alternate mode or chop mode. For high-frequency inputs the alternate mode is best, because the waveform traces appear continuous rather than broken. When the alternate mode is used with low-frequency inputs, the beam is seen slowly trac-

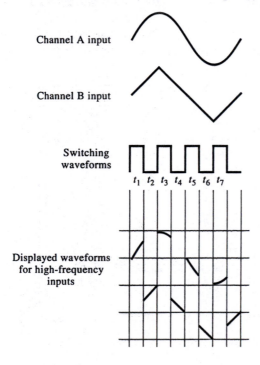

Figure 9-16 In the chop dual-trace display mode, the input of the vertical deflection amplifier is rapidly switched between the channel A and channel B inputs. The waves are displayed as a dashed-line trace with gaps that are virtually invisible at low frequency.

ing out first one wave and then the other. The two waveforms are not displayed continuously, and this makes comparison difficult. Using the chop mode with low-frequency inputs results in both waves being displayed continuously. The breaks in each trace are so short that they cannot be seen.

9-6 OSCILLOSCOPE CONTROLS

The front panel and controls of a representative single-beam dual-trace analog oscilloscope are illustrated in Figure 9-17. The waveforms under investigation are displayed on the screen, which is protected with a calibrated flat piece of hard plastic called a *graticule*. The graticule is used for measuring the amplitude (in vertical divisions) and the time period (in horizontal divisions) of any displayed waveform. Immediately below the screen is a power POWER ON/OFF switch, an INTENSITY control to adjust the brightness of the display, a FOCUS control to focus the display to a fine line, and a pushbutton BEAM FINDER to locate a display that has been shifted off the screen.

To facilitate display of two waveforms, there are two separate sets of VERTICAL controls, identified as CHANNEL A and CHANNEL B. The purpose of the POSITION controls is to move each waveform vertically up or down the screen to set it in the best position for viewing. Each of these is a deflection amplifier dc shift control (see Figure 9-4). The VOLTS/DIV selector switch for each channel determines the sensitivity of the display to input voltages. This is an attenuator selection switch (S_2 in Figure 9-4). When this control is set to 1 V, a signal having a peak-to-peak amplitude of 1 V would occupy 1 division of the screen graticule. A signal that occupies four screen divisions at this setting would have a peak-to-peak amplitude of 4 V. The VOLTS/DIV setting is correct only when the *vernier* knob at its center is in the CAL (calibrated) position. The vernier control provides continuous volts/div adjustment, to reduce the display amplitude typically by a factor of 2.5.

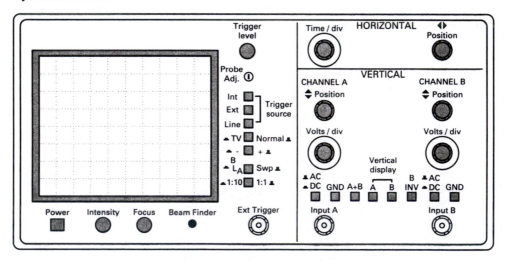

Figure 9-17   Typical front panel of a single-beam, dual-trace oscilloscope.

Below the channel A and channel B VOLTS/DIV switches is a horizontal row of pushbutton switches, and below these are channel A and channel B input (coaxial cable) terminals. The two pushbuttons immediately above each input terminal facilitate ac or dc connection of the input voltage. Sometimes it is necessary to display an alternating voltage and block a dc level that it may be superimposed upon. This is achieved by setting the AC-DC button to AC, where a coupling capacitor passes the ac quantity and blocks the dc. The GND (ground) buttons alongside each AC-DC button disconnects the input signal and grounds the input terminal. This allows each trace to be set to a convenient *zero* position on the screen. When an input voltage is displayed, its dc level may be measured with respect to the zero position. The AC-DC and GND switches perform the same function as switch S_1 in Figure 9-4.

The VERTICAL DISPLAY A and B pushbuttons allow channel A input, channel B input, or both to be displayed on the screen. The B INV button is included to invert the channel B input, and the $A + B$ button permits the channel A and channel B waveforms to be added together and displayed. With the $A + B$ and B INV buttons pressed, the displayed waveform is the *difference* of channel A and channel B input voltages.

The HORIZONTAL TIME/DIV selector switch determines the number of horizontal divisions occupied by each cycle of displayed waveform. This control applies to both waveforms, and it is in fact the time base capacitor selection switch S_1 illustrated in Figure 9-7. With the TIME/DIV switch at 1 ms, a cycle of displayed waveform that occupies exactly 1 horizontal division on the screen has a time period of exactly 1 ms. Similarly, when one cycle of waveform occupies exactly 3.5 division at 1 ms/division its time period is 3.5 ms. The TIME/DIV setting is correct only when the vernier knob at its center is in the CAL position. The vernier control or *expander* provides continuous time/division adjustment, so that one cycle of displayed waveform may be widened as desired up to 10 times the horizontal time/div setting. The HORIZONTAL POSITION knob performs a similar function to the VERTICAL POSITION controls. The displayed waveforms may be moved horizontally about the screen as desired. Once again, this control functions similarly to the dc shift control on the deflection amplifier in Figure 9-4.

The TRIGGER LEVEL knob continuously adjusts the triggering point of the input wave, as explained in Section 9-4. When set to AUTO OFF, the time base circuit no longer operates automatically. Each horizontal sweep of the electron beam must be triggered by an input waveform.

Immediately below the TRIGGER LEVEL control a PROBE ADJ (probe adjust) terminal is situated. A 2 kHz, 0.5 V square-wave output may be taken from this terminal for use in calibrating attenuator probes. The calibration process is explained in Section 9-9.

The first three pushbutton switches in the vertical row below the PROBE ADJ terminal are for trigger source selection. They perform the same function as switch S_1 in Figure 9-9. The INT button selects an *internal* trigger source, which means that the time base is triggered from one of the input waveforms. When both channel A and channel B inputs are displayed, the internal trigger source is channel A. When the EXT (external) button is depressed, the time base is triggered from an external source connected to the EXT TRIGGER terminal. When LINE is selected as a trigger source, the time base is triggered from the line or ac power frequency. The TV-NORM pushbutton is kept in its NORMAL position for most purposes. For studying TV video signals, the button is placed in the TV position. By means of the $-+$ button, displayed waveform may be made to commence at the beginning of the *positive* half-cycle (+) or at the beginning of the *negative* half-cycle (−). This − + button performs the same function as S_2 in Figure 9-9.

The A-B SWP button is usually set in the SWP (sweep) position, where it allows the internal time base to sweep the electron beam horizontally across the tube face for waveform display. When the button is in the A-B position, signals applied to channel *B* input terminal produce vertical deflection, while channel *A* inputs produce horizontal deflection. One application for this facility is the display of electronic device characteristics (see Section 9-10).

9-7 MEASUREMENT OF VOLTAGE, FREQUENCY, AND PHASE

Peak-to-Peak Voltage Measurement

The peak-to-peak amplitude of a displayed waveform is very easily measured on an oscilloscope. Figure 9-18 shows two sine waves with different amplitudes and time periods. Waveform *A* has a peak-to-peak amplitude of 4.6 vertical divisions on the graticule, while wave *B* is measured as 2 vertical divisions peak-to-peak. *It is very important to check that the central vernier knob on the VOLTS/DIV control is in its calibrated (CAL) position before measuring the waveform amplitudes.*

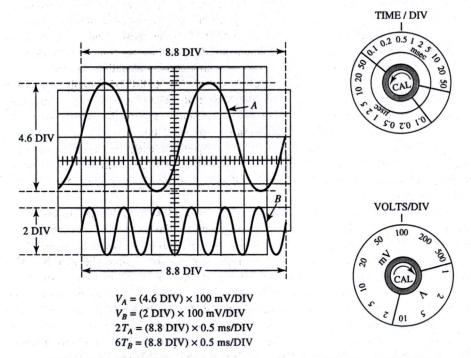

$$V_A = (4.6 \text{ DIV}) \times 100 \text{ mV/DIV}$$
$$V_B = (2 \text{ DIV}) \times 100 \text{ mV/DIV}$$
$$2T_A = (8.8 \text{ DIV}) \times 0.5 \text{ ms/DIV}$$
$$6T_B = (8.8 \text{ DIV}) \times 0.5 \text{ ms/DIV}$$

Figure 9-18 The peak-to-peak voltage of a waveform is measured by multiplying the VOLTS/DIV setting by the peak-to-peak vertical divisions occupied by the waveform. The time period is determined by multiplying the horizontal divisions for one cycle by the TIME/DIV setting.

$$\boxed{V_{\text{p-to-p}} = (\text{vertical p-to-p divisions}) \times (\text{VOLTS/DIV})} \qquad (9\text{-}4)$$

With the *VOLTS/DIV* control at 100 mV, as illustrated, the peak-to-peak voltages of each wave are:

Wave *A*, $\qquad\qquad V = (4.6\ divisions) \times 100\ \text{mV} = 460\ \text{mV}$

Wave *B*, $\qquad\qquad V = (2\ divisions) \times 100\ \text{mV} = 200\ \text{mV}$

   If the waveforms shown in Figure 9-18 were outputs from an amplifier, for example, they might have dc components as well as the ac components illustrated. Suppose that the dc level of wave *A* were 10 V. The dc level would produce a deflection of

$$10\ \text{V}/100\ \text{mV} = 100\ \text{divisions}$$

Clearly, the wave would be deflected right off the screen if the oscilloscope were dc coupled. With ac coupling the wave is on screen.

Frequency Determination

The time period of a sine wave is determined by measuring the time for one cycle in horizontal divisions and multiplying by the setting of the TIME/DIV control:

$$\boxed{T = (\text{horizontal divisions/cycle}) \times (\text{TIME/DIV})} \qquad (9\text{-}5)$$

The frequency is then calculated as the inverse of the time period. Here again, before measuring the time period of the wave, *it is necessary to check that the central vernier knob on the TIME/DIV control is set in its calibrated (CAL) position.* With the TIME/DIV control set to 0.5 ms, as illustrated, the time period and frequency of each wave in Figure 9-18 are:

Wave *A*, $\qquad\qquad T = \dfrac{(8.8\ \text{divisions}) \times 0.5\ \text{ms}}{2\ \text{cycles}} = 2.2\ \text{ms}$

$$f = \dfrac{1}{2.2\ \text{ms}} \simeq 455\ \text{Hz}$$

Wave *B*, $\qquad\qquad T = \dfrac{(8.8\ \text{divisions}) \times 0.5\ \text{ms}}{6\ \text{cycles}} = 0.73\ \text{ms}$

$$f = \dfrac{1}{0.73\ \text{ms}} \simeq 1.36\ \text{kHz}$$

Phase Measurement

The phase difference between two waveforms is measured by the method illustrated in Figure 9-19. Each wave has a time period of 8 horizontal divisions, and the time between commencement of each cycle is 1.4 divisions. One cycle = 360°. Therefore, 8 div = 360°, and

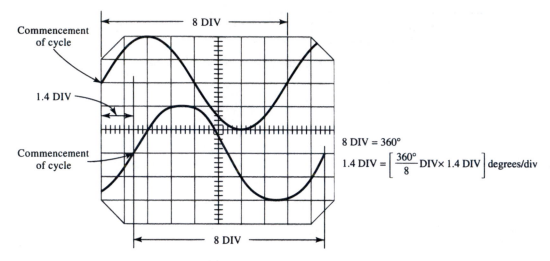

8 DIV = 360°

$$1.4 \text{ DIV} = \left[\frac{360°}{8} \text{DIV} \times 1.4 \text{ DIV} \right] \text{degrees/div}$$

Figure 9-19 The phase difference between two sine waves may be determined by first calculating the horizontal degrees/division for one cycle. This factor is then multiplied by the horizontal divisions between commencement of the cycles.

$$1 \text{ div} = \frac{360°}{8} = 45°$$

Thus, the phase difference is

$$\phi = 1.4 \text{ div} \times (45°/\text{div})$$
$$= 63°$$

$$\boxed{\phi = (\text{phase difference in divisions}) \times (\text{degrees/div})} \qquad (9\text{-}6)$$

Example 9-3

Determine the amplitude, frequency and phase difference between the two waveforms illustrated in Figure 9-20.

Solution

$$V_A \text{ peak-to-peak} = (6 \text{ vertical div}) \times (200 \text{ mV/div})$$

$$= 1.2 \text{ V}$$

$$T_A = (6 \text{ horizontal div}) \times (0.1 \text{ ms/div})$$

$$= 0.6 \text{ ms}$$

$$f_A = \frac{1}{T} = \frac{1}{0.6 \text{ ms}}$$

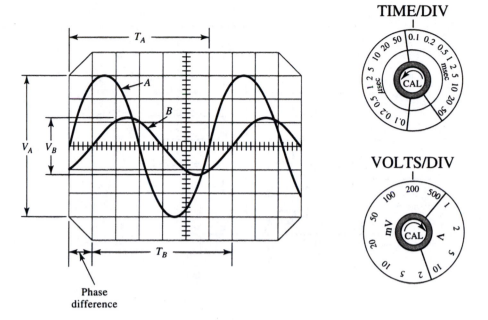

TIME/DIV

VOLTS/DIV

Figure 9-20 Frequency and phase difference determination for Example 9-3.

$$\simeq 1670 \text{ Hz}$$

$$V_B \text{ peak-to-peak} \simeq (2.4 \text{ vertical div}) \times (200 \text{ mV/div})$$

$$= 480 \text{ mV}$$

$$T_B = 0.6 \text{ ms}$$

$$f_B \simeq 1670 \text{ Hz}$$

$$\text{one cycle} = (6 \text{ horizontal div}) = 360°$$

$$\text{phase difference} \simeq 1 \text{ div}$$

$$= \frac{360°}{6} = 60°$$

9-8 PULSE MEASUREMENTS

Pulse Amplitude, Pulse Width, and Space Width

Two pulse waveforms are displayed in Figure 9-21. The pulse amplitude (PA) of each wave is easily measured in terms of the sensitivity (V/DIV) setting. If the sensitivity is 0.1 V/div, the upper wave amplitude is

$$PA = 2 \times 0.1 \text{ V/div} = 0.2 \text{ V}$$

and for the lower wave,

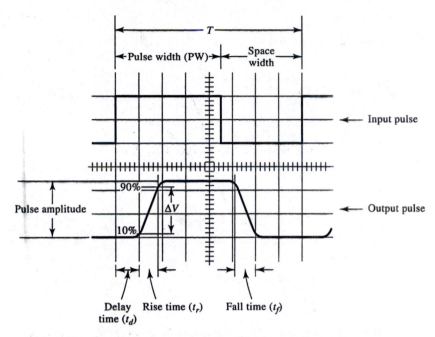

Figure 9-21 Pulse width is determined by multiplying the TIME/DIV setting by the horizontal divisions occupied by the pulse. Pulse amplitude is VOLTS/DIV × (waveform vertical divisions). Rise time is TIME/DIV × (horizontal divisions from 10% to 90% of pulse amplitude). Fall time equals TIME/DIV × (horizontal divisions from 90% to 10% of pulse amplitude).

$$PA = 2.4 \times 0.1 \text{ V/div} = 0.24 \text{ V}$$

The pulse width (PW), space width (SW), and time period (T) can be determined from the time base (TIME/DIV) selection. With the TIME/DIV control at 1 μs, the upper wave quantities in Figure 9-21 are

$$PW = 4.5 \text{ μs}$$
$$SW = 3.5 \text{ μs}$$

and
$$T = 8 \text{ μs}$$

Rise Time, Fall Time, and Delay Time

The upper waveform in Figure 9-21 has steep leading and lagging edges, while the lower wave has measurable *rise time* (t_r) and *fall time* (t_f). If the upper wave is the input to a circuit and the lower is an output, a delay time (t_d) might be measured. As illustrated, the rise time is the time required for the leading edge of the pulse to go from 10% to 90% of the pulse amplitude. Similarly, the fall time is the time taken for the trailing edge to fall from 90% to 10% of the pulse amplitude. The delay time is the time elapsed from commencement of the input pulse until the output pulse reaches 10% of the pulse amplitude. If the TIME/DIV control is set at 1 μs, the rise, fall, and delay times for the lower wave in Figure 9-21 are

$$t_r = 0.7 \ \mu s$$
$$t_f = 0.9 \ \mu s$$
$$t_d = 1 \ \mu s$$

Example 9-4

Determine the pulse amplitude, frequency, rise time, and fall time of the waveform in Figure 9-22.

Solution

$$PA \approx (4 \text{ vertical div}) \times (2 \text{ V/div})$$

$$\approx 8 \text{ V}$$

$$T \approx (5.6 \text{ horizontal div}) \times (5 \ \mu s/\text{div})$$

$$\approx 28 \ \mu s$$

$$f = \frac{1}{T} \approx \frac{1}{28 \ \mu s}$$

$$\approx 35.7 \text{ kHz}$$

$$t_r \approx (0.5 \text{ div}) \times (5 \ \mu s/\text{div})$$

$$\approx 2.5 \ \mu s$$

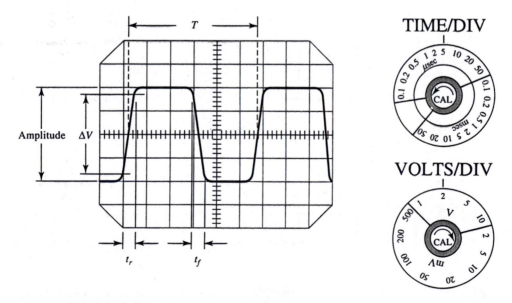

Figure 9-22 Pulse rise time, fall time, and frequency determination for Example 9-4.

$$t_f \approx (0.6 \text{ div}) \times (5 \text{ μs/div})$$

$$\approx 3 \text{ μs}$$

Pulse Distortion

The pulse waveform in Figure 9-23(a) is directly coupled to the oscilloscope. Because it is directly coupled, there should be no distortion in the displayed waveform. If the wave is ac coupled, as shown in Figure 9-23(b), the coupling capacitor (C_c) imposes *tilt* (also termed *slope,* or *sag*) or low-frequency distortion, at the top and bottom of the waveform. This is the result of the input capacitor charging and discharging during the pulse width and space width. In this situation, the time constant of the oscilloscope input is

$$\boxed{\tau = R_i C_c} \tag{9-7}$$

Using the typical 1 MΩ and 0.1 μF quantities for R_i and C_c,

$$\tau = 1 \text{ MΩ} \times 0.1 \text{ μF}$$

$$= 0.1 \text{ s}$$

The pulse width illustrated (50 ms) is exactly half of the (typical) input circuit time constant. If the pulse width and space width are very much shorter than the R_iC_c, there will be very little charge and discharge of the capacitor, and consequently, the pulse distortion will be insignificant. *To avoid low-frequency distortion, the pulse width and space width should not be greater than one-tenth of R_iC_c.*

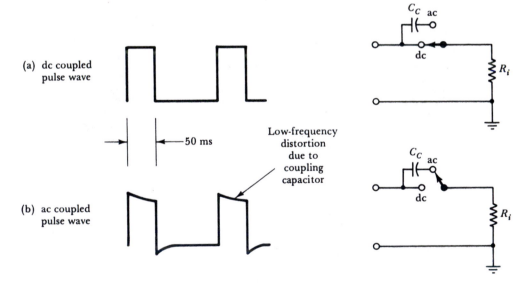

(a) dc coupled pulse wave

50 ms

Low-frequency distortion due to coupling capacitor

(b) ac coupled pulse wave

C_C ac

dc

R_i

C_C ac

dc

R_i

Figure 9-23 A low-frequency pulse waveform will be faithfully reproduced on an oscilloscope if it is dc coupled. When ac coupled, the coupling capacitor may produce low-frequency distortion.

Cathode-Ray Oscilloscopes   Chap. 9

Example 9-5

A pulse wave is ac coupled to an oscilloscope with $R_i = 10$ MΩ and $C_c = 0.1$ μF. Determine the longest pulse width (or space width) that can be displayed without noticeable low-frequency distortion being introduced by the oscilloscope.

Solution

Equation 9-7,
$$\tau = R_i C_c = 10 \text{ MΩ} \times 0.1 \text{ μF}$$

$$= 1 \text{ s}$$

$$PW = \frac{R_i C_c}{10} = \frac{1 \text{ s}}{10}$$

$$= 100 \text{ ms}$$

Figure 9-24 shows high-frequency distortion on a pulse waveform. Very short duration pulses are involved, and the distortion is present with both ac and dc coupled inputs. The dashed-line wave shows that the input pulse has a flat top and vertical leading and lagging edges. The solid-line representation of the displayed pulse shows noticeable rise and fall times as well as rounding of corners.

High-frequency distortion of pulse waveforms can be produced by the combination of signal source resistance R_s and oscilloscope input capacitance C_i (*not* coupling capacitance). In this case the circuit time constant is $\tau = R_s C_i$, and it can be shown that the rise time produced by $R_s C_i$ is*

$$\boxed{t_r = 2.2 \, R_s C_i} \tag{9-8}$$

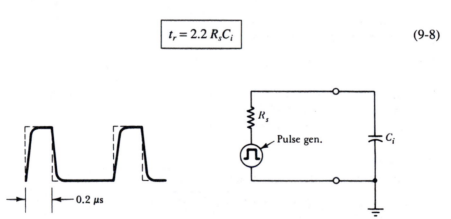

Figure 9-24 High-frequency distortion of a pulse waveform displayed on an oscilloscope occurs when the instrument input capacitance produces rise times and fall times that are not present on the original pulse.

*David A. Bell, *Solid State Pulse Circuits,* 4th ed. (Englewood Cliffs, NJ: Prentice Hall, 1992), p. 42.

This can be minimized by using a low signal source resistance and/or by the use of an attenuator probe to reduce the input capacitance (see Section 9-9). As explained in Section 9-12, pulse rise and fall times can also be affected by the instrument upper cutoff frequency.

The input pulse rise time (t_{ri}) and the rise time imposed by the oscilloscope (t_{ro}) do not add together directly; instead, the displayed rise time (t_{rd}) is

$$t_{rd} = \sqrt{(t_{ri})^2 + (t_{ro})^2}$$ (9-9)

For the rise time due to $R_s C_i$ to be insignificant, the input pulse rise time should be at least three times greater than t_{ro}. If the input pulse rise time is so short that it is not noticeable, then for the displayed wave to have a minimal rise time, the pulse width should be at least $10t_{ro}$.

Example 9-6

A pulse waveform with a 3.3 kΩ source resistance is to be displayed on an oscilloscope with an input capacitance of 15 pF. Determine the shortest pulse width that can be displayed without noticeable distortion being introduced by the combination of R_s and C_i.

Solution

Equation 9-8, $t_{ro} = 2.2 R_s C_i = 2.2 \times 3.3 \text{ k}\Omega \times 15 \text{ pF}$

$$= 109 \text{ ns}$$

$$PW_{min} = 10t_{ro} = 10 \times 109 \text{ ns}$$

$$= 1.09 \text{ μs}$$

Example 9-7

A pulse waveform with a 3.3 kΩ source resistance is to be displayed on an oscilloscope with an input capacitance of 15 pF, as in Example 9-6. Determine the displayed rise time if the input pulse rise time is (a) 109 ns and (b) 327 ns.

Solution

Equation 9-9, $t_{rd} = \sqrt{(t_{ri})^2 + (t_{ro})^2}$

(a) $t_{rd} = \sqrt{(109 \text{ ns})^2 + (109 \text{ ns})^2}$

$$= 154 \text{ ns}$$

(b) $t_{rd} = \sqrt{(109 \text{ ns})^2 + (327 \text{ ns})^2}$

$$= 345 \text{ ns}$$

1:1 Probes

The input signals to an oscilloscope are usually connected via coaxial cables with *probes* on one end [see Figure 9-25(a)]. These are normally just convenient-to-use insulated connecting clips. As illustrated, each probe has two connections, an *input* and a *ground*. The coaxial cable consists of an insulated central conductor surrounded by a braided circular conductor which is covered by an outer layer of insulation. The central conductor carries the input signal, and the circular conductor is grounded so that it acts as a screen to help prevent unwanted signals being *picked up* by the oscilloscope input. This type of probe is usually referred to as a *1:1 (one-to-one) probe,* because it does not contain resistors to attenuate the input signal.

The coaxial cable connecting the probe to the oscilloscope has a capacitance (C_{cc}) which can overload a high-frequency signal source. The input impedance of the oscilloscope *at the front panel* is typically 1 MΩ in parallel with 30 pF. The coaxial cable can add another 100 pF to the total input capacitance. The circuit of a signal source, probe, and oscilloscope input is illustrated in Figure 9-25(b). The total impedance offered by the coaxial cable and the oscilloscope input should always be much larger than the signal

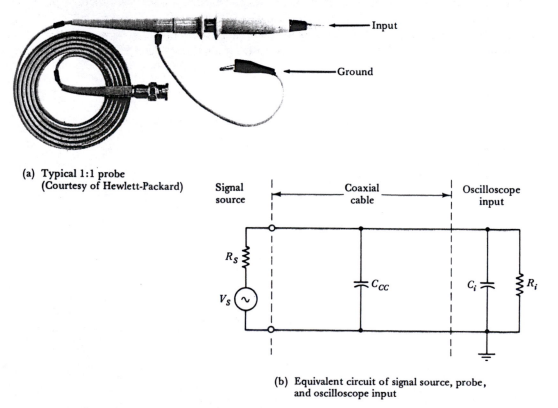

(a) Typical 1:1 probe
(Courtesy of Hewlett-Packard)

(b) Equivalent circuit of signal source, probe, and oscilloscope input

Figure 9-25 A 1:1 probe for an oscilloscope is simply a coaxial cable with convenient connecting clips at its end. The coaxial cable can typically add 100 pF to the oscilloscope input capacitance.

source impedance. Where this is not the case, the signal is attenuated and phase shifted when connected to the oscilloscope.

At frequencies where the reactance of $(C_{cc} + C_i)$ is very much larger than R_s and R_i, the capacitances have a negligible effect and the oscilloscope terminal voltage is

$$V_i = V_s \frac{R_i}{R_s + R_i}$$

The capacitive reactance becomes progressively smaller as the signal frequency increases, so that the signal may be significantly attenuated at the oscilloscope terminals. An attenuation of 3 dB will occur at the frequency at which the capacitive reactance equals R_s. There is also a 45° phase shift at this frequency. *To avoid significant attenuation and phase shift, the signal frequency should not exceed one-tenth of the 3 dB attenuation frequency.*

Example 9-8

A 1 V signal with a source resistance of $R_s = 600 \ \Omega$ is connected to an oscilloscope which has an input impedance of $R_i = 1 \ M\Omega$ in parallel with $C_i = 30$ pF. The coaxial cable has a capacitance of $C_{cc} = 100$ pF. Calculate the oscilloscope terminal voltage (V_i) when the signal frequency is 100 Hz. Also determine the frequency at which V_i is 3 dB below V_s.

Solution

$$\text{Total capacitance} = C_i + C_{cc} = 30 \text{ pF} + 100 \text{ pF}$$

$$= 130 \text{ pF}$$

$$X_c = \frac{1}{2\pi f C}$$

At 100 Hz,

$$X_c = \frac{1}{2\pi \times 100 \text{ Hz} \times 130 \text{ pF}}$$

$$= 12.2 \text{ M}\Omega$$

$$R_s = 600 \ \Omega$$

$$X_c \gg R_s \quad \text{and} \quad R_i$$

$$V_i = V_s \frac{R_i}{R_s + R_i}$$

$$= 1 \text{ V} \times \frac{1 \text{ M}\Omega}{600 \ \Omega + 1 \text{ M}\Omega}$$

$$= 0.9994 \text{ V}$$

When $V_i = (V_s - 3\ dB)$,

$$X_c = R_s = 600\ \Omega$$

$$= \frac{1}{2\pi f C}$$

or

$$f = \frac{1}{2\pi C R_s} = \frac{1}{2\pi \times 130\ pF \times 600\ \Omega}$$

$$= 2.04\ MHz$$

Attenuator Probes

Attenuator probes attenuate the input signal, usually by a factor of 10. They also normally offer a much larger input impedance than a 1:1 probe, thereby minimizing loading effects on the circuit under test. Compensation is included for oscilloscope input capacitance and coaxial cable capacitance. Because of the 10-fold attenuation, these probes are usually referred to as *10:1 probes*; however, other probes are available with different attenuation factors.

Figure 9-26(a) and (b) show the circuit and equivalent circuit of a typical 10:1 probe. A 9 MΩ resistor is included in series with the input terminal, and an adjustable capacitor (C_1) is connected in parallel with the 9 MΩ resistor, as illustrated. C_2 is the sum of oscilloscope input capacitance C_i and coaxial cable capacitance C_{cc}. At low and medium frequencies, the capacitive impedances are too large to be effective, and the oscilloscope input voltage is

$$V_i = V_s\ \frac{R_i}{R_1 + R_s + R_i} \qquad \text{[see Figure 9-26(b)]}$$

When $R_s \ll R_1$,

$$V_i \simeq V_s \frac{R_i}{R_1 + R_i}$$

With $R_1 = 9$ MΩ and $R_i = 1$ MΩ,

$$V_i = V_s \frac{1 M\Omega}{9\ M\Omega + 1\ M\Omega}$$

$$= \frac{V_s}{10}$$

The attenuation of the signal due to the capacitors acting alone can be calculated from

$$V_i = V_s\ \frac{X_{C2}}{X_{C1} + X_{C2}}$$

$$= V_s \frac{1/\omega C_2}{1/\omega C_1 + 1/\omega C_2}$$

$$= V_s \frac{1}{(C_2/C_1) + 1}$$

$$V_i = V_s \frac{C_1}{C_2 + C_1}$$

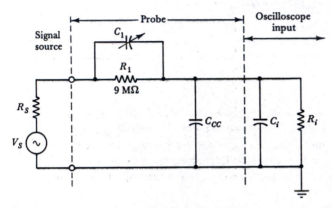

(a) Circuit of 10:1 probe

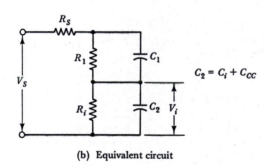

$$C_2 = C_i + C_{cc}$$

(b) Equivalent circuit

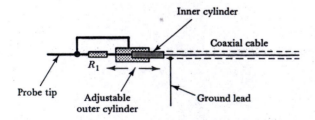

(c) Inner construction of 10:1 probe

Figure 9-26 An oscilloscope 10:1 probe contains a 9 MΩ resistor to attenuate the signal, and an adjustable capacitor (C_1) to compensate for the cable capacitance (C_{cc}) and the input capacitance (C_i).

Cathode-Ray Oscilloscopes   Chap. 9

When the capacitive network attenuates the signal in the same proportion as the resistive network attenuation, V_i across R_i equals V_i across C_2 and

$$\frac{R_i}{R_1 + R_i} = \frac{C_1}{C_2 + C_1}$$

(9-10)

A phasor diagram can be drawn to show that the voltages across C_2 and R_i are in phase, as well as being equal in amplitude. Therefore, C_1 completely compensates for the presence of C_2. The additional attenuation and phase shift at high frequencies due to the presence of C_2 (without C_1) is now eliminated, and the probe-oscilloscope combination is compensated for *all* frequencies.

The value of C_1 required to compensate for C_2 is determined from Equation 9-10 as

$$C_1 = C_2 \frac{R_i}{R_1}$$

(9-11)

Figure 9-26(c) shows the typical construction of an attenuator probe. C_1 is the capacitance between concentric metal cylinders that are connected to opposite ends of the 9 MΩ resistor (R_1). Screw threads facilitate adjustment of the cylinders for variation of capacitance C_1.

Example 9-9

Calculate the value of C_1 required to compensate a 10:1 probe when the oscilloscope input capacitance is 30 pF and the coaxial cable capacitance is 100 pF. Also calculate the probe input capacitance *seen* from the source.

Solution

$$C_2 = C_{cc} + C_i$$

$$= 130 \text{ pF}$$

Equation 9-11,
$$C_1 = C_2 \frac{R_i}{R_1}$$

$$= 130 \text{ pF} \times \frac{1 \text{ M}\Omega}{9 \text{ M}\Omega}$$

$$= 14.4 \text{ pF}$$

The probe input capacitance C_T is C_1 in series with C_2.

$$\frac{1}{C_T} = \frac{1}{C_1} + \frac{1}{C_2}$$

$$= \frac{1}{14.4 \text{ pF}} + \frac{1}{130 \text{ pF}}$$

$$C_T \simeq 13 \text{ pF}$$

Example 9-10

Determine the signal frequency at which the probe in Example 9-9 causes a 3 dB reduction in the signal from a 600 Ω source.

Solution

$$X_c = R_s = \frac{1}{2\pi f C}$$

or
$$f = \frac{1}{2\pi C R_s} = \frac{1}{2\pi \times 13 \text{ pF} \times 600 \text{ } \Omega}$$

$$= 20.4 \text{ MHz}$$

Another 10:1 probe and its equivalent circuit are shown in Figure 9-27. In this case capacitor C_1 is a fixed quantity, and an additional variable capacitor (C_3) is included in parallel with C_i and C_{cc}. C_2 in Equation 9-10 is now the total capacitance of C_i, C_{cc}, and C_3 in parallel. C_2 can be calculated from Equation 9-11.

The input capacitance and input resistance vary from one oscilloscope to another, even for otherwise identical instruments. *It is important that every probe be correctly adjusted when it is first connected for use with a particular oscilloscope.*

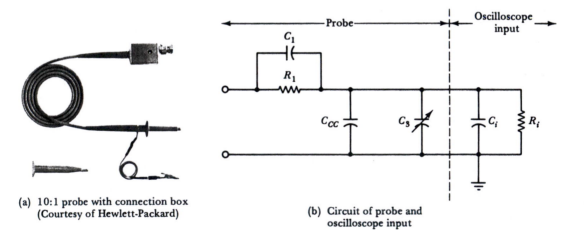

(a) 10:1 probe with connection box
(Courtesy of Hewlett-Packard)

(b) Circuit of probe and oscilloscope input

Figure 9-27 An oscilloscope 10:1 probe with a correction box has the compensating capacitor (C_3) in parallel with the instrument input terminals.

Cathode-Ray Oscilloscopes   Chap. 9

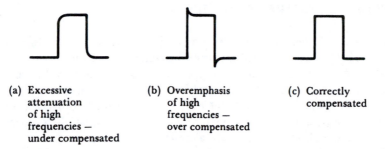

(a) Excessive
attenuation
of high
frequencies –
under compensated

(b) Overemphasis
of high
frequencies –
over compensated

(c) Correctly
compensated

Figure 9-28 Oscilloscope 10:1 probes are calibrated by connecting a square wave (usually provided at a terminal on the oscilloscope front panel), and adjusting the capacitor to obtain an undistorted waveform display.

As mentioned in Section 9-6, a square-wave calibration voltage is generated within the oscilloscope and connected to a terminal on the front panel. The probe is connected to this terminal to display the square wave on the screen. The variable capacitor is now adjusted until the displayed wave is perfectly square. The shape of waveforms with correctly compensated, undercompensated, and overcompensated probes are illustrated in Figure 9-28.

Active Probes

Active probes contain electronic amplifiers that increase the probe input resistance and minimize its input capacitance. Typical active probes use FET input stages, or FET input operational amplifiers. The circuit is connected to function as a voltage follower (see Section 4-2). The amplifier has a gain of 1 and a typical input impedance of 1 $M\Omega \| 3.5$ pF. Input impedances of 10 $M\Omega$ or greater are also possible with FET input stages, and the input capacitance effect can be further reduced by resistive attenuation. Power must be supplied to operate the amplifier. This may be derived from the oscilloscope, or the probe may contain its own regulated power supply with a line cord.

Example 9-11

Determine the signal frequency at which a probe with $Z_i = 10$ $M\Omega \| 3.5$ pF reduces the signal from a 600 Ω source by 3 dB.

Solution

$$X_c = R_s$$

$$\frac{1}{2\pi f \times 3.5 \text{ pF}} = 600 \ \Omega$$

$$f = \frac{1}{2\pi \times 3.5 \text{ pF} \times 600 \ \Omega}$$

$$= 75.8 \text{ MHz}$$

9-10 DISPLAY OF DEVICE CHARACTERISTICS

The characteristics of electronic devices can be displayed on an oscilloscope screen when the time base is disconnected from the horizontal input. Pushing the A-B SWP button on the oscilloscope in Figure 9-17 disconnects the time base and connects channel A inputs to the horizontal deflection amplifier. Channel A input should now be made proportional to the horizontal component of the device characteristics, and channel B input should be proportional to the vertical component. Figure 9-29 illustrates the method of displaying the characteristics of a semiconductor diode.

In Figure 9-29(a), a 1 kΩ resistor (R_1) is connected in series with the device, and a sawtooth voltage wave is applied, as illustrated. The frequency of the sawtooth waveform could be anywhere between about 100 Hz and 1 kHz. The amplitude of the sawtooth should be equal to the maximum voltage drop that is going to occur across

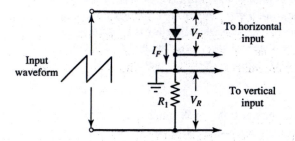

(a) Circuit for displaying diode forward characteristics

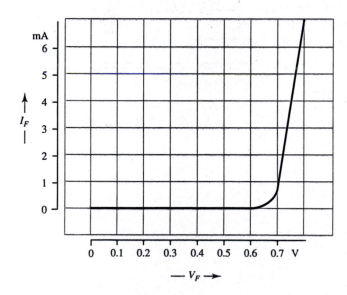

(b) Oscilloscope display of diode forward
characteristics

Figure 9-29 Method of displaying the forward characteristics of a semiconductor diode on an oscilloscope. The device forward voltage is applied to the horizontal input (the time base is disconnected), and a voltage V_R proportional to the forward current is applied to the vertical input.

Cathode-Ray Oscilloscopes   Chap. 9

R_1 plus the voltage drop across the device. For a maximum forward current of 7 mA, as shown in Figure 9-29(b), the sawtooth waveform amplitude should exceed 7.7 V. The voltage across R_1 is applied to the oscilloscope vertical input channel, and the invert button is pressed to give an up-going deflection. The diode voltage is applied to the horizontal input. Note that the common terminal is the grounded terminal for both channels.

With $R_1 = 1$ kΩ, the resistor voltage for each 1 mA of diode current is 1 mA $\times$ 1 kΩ = 1 V. Therefore, with the oscilloscope vertical deflection control set to 1 V/DIV, each 1 V of vertical deflection represents 1 mA of diode forward current. For the display shown in Figure 9-29(b), the horizontal deflection control is set to 0.1 V/DIV. Thus, each screen division represents 0.1 V of diode forward voltage.

During each cycle of the sawtooth, the applied voltage increases from zero, causing I_F and V_F to grow from zero and trace the device characteristics on the screen. When the sawtooth goes to ground level again, I_F and V_F are returned to zero. The characteristic of the diode is repeatedly traced at the frequency of the sawtooth waveform. A sine wave will produce the characteristic display just as satisfactorily as a sawtooth.

A permanent record of the device characteristics can be obtained by using an oscilloscope camera. The scales for I_F and V_F should be drawn along the edges of the photograph, as shown in Figure 9-29(b). It should be noted that *curve tracers* designed solely for tracing device characteristics are commercially available.

9-11 *X–Y* AND *Z* DISPLAYS

When the oscilloscope time base is disconnected and sine waves are applied to both horizontal and vertical inputs, the resulting display depends on the relationship between the two sine waves. Very simple displays occur when the waveforms are equal in frequency. Quite complex figures may be produced with sine waves having different frequencies. In all cases these are known as *Lissajou figures*.

When only one input is applied, either a vertical line or a horizontal line results [Figure 9-30(a) and (b)]. Perfectly in-phase sine waves with equal amplitudes produce a straight line at an angle of 45° from the horizontal, as shown in Figure 9-30(c). This is explained as follows. Both waveforms are at zero at the same instant. Therefore, there is neither vertical nor horizontal deflection, and the electron beam strikes the center of the screen [points 1 and 3 in Figure 9-30(c)]. When both waveforms are at maximum positive amplitude, maximum vertical and horizontal deflection is produced. The positive peak vertical input produces maximum positive (up-going) vertical deflection, and the positive peak horizontal input produces maximum right-hand-side deflection. At this instant the electron beam is at point 2 in Figure 9-30(c). When the sine waves are at their negative peaks, maximum negative (down-going) vertical deflection and maximum left-hand-side horizontal deflection occur. These conditions cause the electron beam to strike the screen at point 4 in Figure 9-30(c). The simultaneously changing voltage levels cause the beam to trace out the straight line between points 2 and 4. Figure 9-30(d) shows that when the waveforms are in antiphase a line is traced at an angle of 135° with respect to horizontal.

A circular display is produced when a 90° phase difference exists between vertical and horizontal inputs [Figure 9-30(e)]. When the vertical input is zero and the horizontal

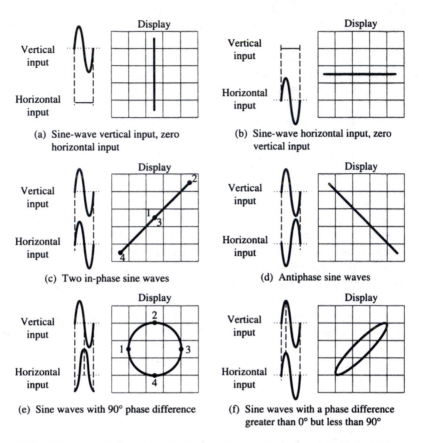

(a) Sine-wave vertical input, zero horizontal input

(b) Sine-wave horizontal input, zero vertical input

(c) Two in-phase sine waves

(d) Antiphase sine waves

(e) Sine waves with 90° phase difference

(f) Sine waves with a phase difference greater than 0° but less than 90°

Figure 9-30 The display produced when equal-frequency sine waves are applied to the horizontal and vertical inputs of an oscilloscope depends on the phase relationship between the waveforms.

is at its negative peak, zero vertical deflection and maximum left-hand horizontal deflection occurs (point 1). When the horizontal input is zero and the vertical is at its positive peak, maximum positive vertical deflection and zero horizontal deflection results (point 2). At the instant of zero vertical input and maximum positive horizontal input, the beam experiences maximum right-hand-side horizontal deflection and zero vertical deflection (point 3). Point 4 on the display is the result of zero horizontal input and maximum negative vertical input. The point at which the beam strikes the screen rotates continuously, tracing the circular display.

When the phase difference between vertical and horizontal inputs is greater than zero but less than 90°, an elliptical display is traced, as illustrated in Figure 9-30(f). This is a thick (almost circular) ellipse when the phase difference is close to 90°. A near-zero phase difference produces a narrow ellipse. An ellipse tilted in the opposite direction is produced when the vertical and horizontal inputs have a phase difference greater than 90° but less than 180°. Once again, the ellipse may be thin or thick depending upon whether the phase difference is close to 180° or close to 90°. The actual phase difference between the two waves may be determined from measurements made on elliptical displays. However, it is much more convenient to use a horizontal time base for measuring phase differences (see Section 9-7).

The oscilloscope displays in Figure 9-31 occur when the vertical and horizontal inputs have different frequencies. In Figure 9-31(a), the vertical input frequency is twice the horizontal input frequency. One cycle of horizontal input causes the electron beam to travel from the center of the screen to the right-hand side, then back through the center to the left-hand side, and finally back to center again. During this time the vertical input deflects the beam from center, up, down, back to center; then up, down, and back to center once more. The ratio of vertical frequency (f_1) to horizontal frequency (f_2) can be determined from the display:

$$\frac{f_1}{f_2} = \frac{\text{number of positive peaks}}{\text{number of right-hand side peaks}} \qquad (9\text{-}12)$$

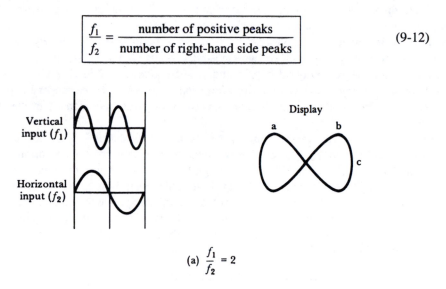

(a) $\dfrac{f_1}{f_2} = 2$

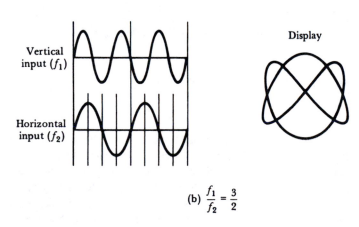

(b) $\dfrac{f_1}{f_2} = \dfrac{3}{2}$

Figure 9-31 Lissajou figures are generated when sine waves having different frequencies are applied to the vertical and horizontal inputs of an oscilloscope.

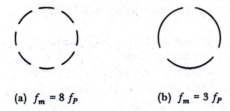

(a) $f_m = 8 f_P$

(b) $f_m = 3 f_P$

Figure 9-32 A circular display [as in Figure 9-30(e)] can be made into a dashed-line display by applying a separate signal to the Z input of the oscilloscope.

The more complex display in Figure 9-31(b) is the result of inputs with a frequency ratio of 3:2. The displayed Lissajou figures can become extremely complex when the frequency ratios are other than the simple 2:1 and 3:2 ratios used in Figure 9-31. Also, for a stationary figure there must be an exact ratio between the two frequencies. When the frequency ratio is not exact, the displayed figure changes continuously.

Two dashed-line circular Lissajou figures are shown in Figure 9-32. As already explained, the circular display is a result of two sinusoidal inputs with a 90° phase difference. The waves must be exactly equal in amplitude and frequency, and this can be best achieved by applying a single sine wave to two *phase shift networks*. The dashed-line effect is achieved by using another, higher-frequency wave to modulate the intensity of the electron beam.

Most oscilloscopes have a (rear) connector for intensity modulation inputs. This is usually termed *Z-axis modulation*. The input wave actually modulates the voltage on the grid of the oscilloscope (see Figure 9-1). Thus, the grid voltage is driven more negative during each negative half-cycle of the modulating wave. This either dims the intensity of the trace or causes it to be blanked completely. Each short line and gap in the circular figure represents one cycle of modulating frequency. Each trace of circular figure is completed by one cycle of the vertical and horizontal inputs. Therefore, the ratio of modulating frequency to deflecting plate frequency is

$$\frac{f_m}{f_p} = \frac{\text{number of gaps in circle}}{1} \qquad (9\text{-}13)$$

where the ratio of f_m/f_p is not an exact quantity, the gaps in the circle rotate.

9-12 OSCILLOSCOPE SPECIFICATIONS AND PERFORMANCE

Sensitivity

The *sensitivity* defines the amplitude of the input voltage that may be displayed on the screen. Sensitivity typically ranges from 2 mV/div to 10 V/div. This means, for example, that a sine wave with a 2 mV peak-to-peak amplitude would occupy only one vertical division on the screen when the sensitivity is set to 2 mV/div. A higher V/div setting gives an even smaller display. For careful examination, a displayed wave should typically fill half the oscilloscope screen (four vertical divisions). Thus, the smallest wave that can be usefully displayed must have a peak-to-peak amplitude of 8 mV when the sensitivity is

Cathode-Ray Oscilloscopes   Chap. 9

2 mV/div. If a 10:1 probe is used, the signal amplitude must be 80 mV (p-to-p) to give a four-division display with a sensitivity of 2 mV/div. As noted in Section 9-6, the vernier knob at the center of the V/div control permits continuous uncalibrated amplitude decrease of the displayed waveform by a factor of approximately 2.5. Using a 10 V/div sensitivity, the largest wave amplitude that can be displayed is 80 V (occupying the entire eight vertical screen divisions). With a 10:1 probe and 10 V/div sensitivity, a maximum of 800 V can be displayed.

Voltage Measurement Accuracy

The accuracy of the V/div sensitivity is typically specified as ±3%. The waveform measurement accuracy depends on the displayed amplitude and the specified accuracy. Because the thickness of the trace is approximately 5% of one vertical division, the best reading accuracy is ±5% of one division. For a waveform with a peak-to-peak amplitude of 5 screen divisions, this gives a reading accuracy of ±5%/5, or ±1%. Adding the ±3% V/div accuracy produces an overall measuring accuracy of ±4%. For a wave that occupies only one vertical division, the measurement accuracy becomes ±(5% + 3%) (see Figure 9-33).

Frequency Response

The *frequency response* of an oscilloscope defines the highest and lowest waveform frequency that may be displayed with no more than 3 dB of attenuation. This is usually specified as *DC to 20 MHz* (or perhaps to *60MHz or 100 MHz*). For ac coupling, a low 3 dB frequency of 2 Hz is often listed. The upper cutoff frequency (f_H) is largely dependent on the frequency response of the vertical deflection amplifiers, while the lower (ac) cutoff frequency (f_L) results from the coupling capacitor (C_1) being switched into the input circuit (see Figure 9-4).

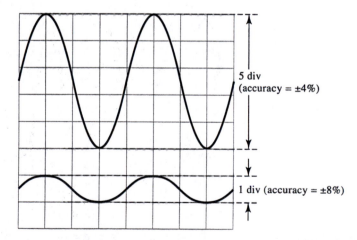

5 div
(accuracy = ±4%)

1 div (accuracy = ±8%)

Figure 9-33 The accuracy of oscilloscope measurement of voltage depends on the amplitude of the displayed waveform.

At the high end of the oscilloscope bandwidth, consider the cases of a sine-wave output from a signal generator that has an upper cutoff frequency of 20 MHz. At 20 MHz, the waveform amplitude should be 3 dB below its normal (midfrequency) amplitude. If this waveform is applied to an oscilloscope that has a 20 MHz cutoff frequency, the displayed wave will be 6 dB below its normal level: −3 dB due to the amplifier and −3 dB due to the oscilloscope. Thus, an oscilloscope with a 20 MHz cutoff frequency would give a misleading result in this situation (Figure 9-34). *For the oscilloscope upper cutoff frequency (f_H) to have a negligible effect on the displayed waveform, the signal frequency should not exceed $f_H/10$.*

As well as attenuation, the oscilloscope normally produces a −45° phase shift at the upper cutoff frequency. However, if the input and output waveforms of an amplifier are being displayed, for example, they will be attenuated and phase shifted by equal amounts.

As explained in Section 9-9, the instrument input capacitance (C_i) combines with the signal source resistance (R_s) to attenuate and phase-shift high-frequency waveforms. So, although the f_H value of the oscilloscope might be much higher than the signal frequency, *substantial attenuation and phase shift can be produced by R_s and C_i at signal frequencies lower than the oscilloscope upper cutoff frequency.*

The oscilloscope lower 3 dB frequency occurs when the impedance of the coupling capacitor equals the input resistance. Using typical quantities of 0.1 μF and 1 MΩ, f_L = 1.59 Hz. It should also be noted that as well as being attenuated, the waveform developed across the 1 MΩ input resistance is phase shifted by +45° with respect to the input waveform. At frequencies slightly higher than f_L, there is still some attenuation and phase shift. *For the oscilloscope lower cutoff frequency (f_L) to have a negligible effect on the displayed waveform, the signal frequency should be at least $10f_L$.*

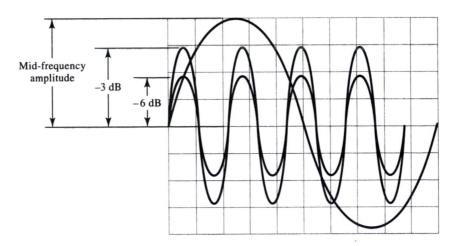

Figure 9-34 The upper cutoff frequency of an oscilloscope attenuates the displayed waveform by 3 dB. If the wave amplitude is already down by 3 dB (due to a circuit cutoff frequency, for example), the total attenuation is 6 dB.

Time Base Measurements

Consider the case of an oscilloscope with a bandwidth of 10 MHz. The time period of one cycle of 10 MHz waveform is 100 ns. The time base specification for such an oscilloscope is typically 0.5 s/div to 200 ns/div. The number of horizontal divisions occupied by any waveform can be determined as

$$\text{horizontal divisions} = \frac{T}{\text{time/div}}$$

or

$$\boxed{\text{horizontal divisions} = \frac{1}{f \times \text{time/div}}} \tag{9-14}$$

So at 200 ns/div, one cycle of 10 MHz waveform would occupy 0.5 division.

If it is desired that for careful examination, one cycle of wave should occupy a minimum number of horizontal divisions, the highest frequency that can be displayed is determined from

$$f_{max} = \frac{1}{(\text{time/div}) \times (\text{horizontal divisions})}$$

For four horizontal divisions and a time base setting of 200 ns/div, f_{max} is 1.25 MHz.

The time base magnifier may be used to expand the horizontal divisions occupied by one cycle of displayed waveform by a (typical) factor of 10. Thus, the 200 ns/div setting becomes 20 ns/div, and the maximum displayed frequency calculated above becomes 12.5 MHz (for one cycle occupying four horizontal divisions).

The measurement accuracy of the time base is typically stated as ±5%. This becomes ±7% when the time base magnifier is set to its maximum position. At all other magnifier settings between zero and maximum, the time base is not calibrated.

As discussed for amplitude measurements, the reading accuracy of the oscilloscope is approximately ±5% of one division. So, for one cycle of waveform occupying five horizontal divisions, the reading accuracy of the time period is ±5%/5, or ±1%. Adding the ±5% time/div accuracy, the overall time (and frequency) measuring accuracy is ±6%. For a time period occupying only one horizontal division, the measuring accuracy is ±(5% + 5%) (see Figure 9-35). These accuracy figures also apply to measurements of phase, pulse width, rise time, and so on.

The lowest-frequency waveform that can be displayed at a setting of 0.5 s/div can be determined by first assuming that the time period occupies (the full screen) 10 horizontal divisions. So

$$f = \frac{1}{T} = \frac{1}{10 \times 0.5 \text{ s}}$$

$$= 0.2 \text{ Hz}$$

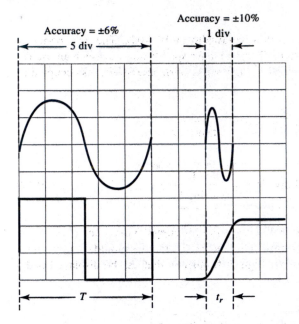

Figure 9-35 The accuracy of oscilloscope measurement of time depends on the horizontal divisions occupied by the displayed waveform.

Example 9-12

Determine the minimum time/division sensitivity for an oscilloscope that is to be used to investigate a 50 MHz waveform. Assume that the time base magnifier expands the horizontal display by a factor of 5.

Solution

$$T = \frac{1}{f} = \frac{1}{50 \text{ MHz}}$$

$$= 20 \text{ ns}$$

For one cycle occupying four horizontal divisions,

$$\text{minimum time/div} = \frac{T}{4} = \frac{20 \text{ ns}}{4}$$

$$= 5 \text{ ns/div}$$

Using the five-times magnifier to give 5 ns/div,

$$\text{minimum time/div setting} = 5 \times 5 \text{ ns/div}$$

$$= 25 \text{ ns/div}$$

Rise Time Measurements

The oscilloscope *rise time* is the rise time imposed by the oscilloscope on an input pulse wave. The rise time is directly related to the upper cutoff frequency by the equation*

$$t_{ro} = \frac{0.35}{f_H} \qquad (9\text{-}15)$$

Therefore, for example, an oscilloscope with a 100 MHz upper cutoff frequency has a rise time of 3.5 ns. For a practical pulse waveform with a finite rise time t_{ri} applied to an oscilloscope that has a rise time t_{ro}, the resultant rise time on the displayed wave is (from Section 9-8)

Equation 9-9, $\qquad t_{rd} = \sqrt{(t_{ri})^2 + (t_{ro})^2}$

Recall that Equation 9-9 can be used to show that *to avoid any significant error due to the oscilloscope rise time (t_{ro}), the input waveform rise time should be at least $3t_{ro}$.*

Example 9-13

Determine the displayed rise time when a pulse waveform with a rise time of 21 ns is applied to an oscilloscope that has an upper cutoff frequency of (a) 20 MHz and (b) 50 MHz.

Solution

Equation 9-15, $\qquad t_{ro} = \dfrac{0.35}{f_H}$

(a) For $f_H = 20\ MHz$, $\qquad t_{ro} = \dfrac{0.35}{20\ \text{MHz}} = 17.5 \text{ ns}$

Equation 9-9, $\qquad t_{rd} = \sqrt{(t_{ri})^2 + (t_{ro})^2}$

$\qquad\qquad t_{rd} = \sqrt{(21\text{ns})^2 + (17.5\text{ns})^2}$

$\qquad\qquad = 27 \text{ ns}$

(b) For $f_H = 50\ MHz$, $\qquad t_r = \dfrac{0.35}{50\ \text{MHz}} = 7 \text{ ns}$

$\qquad\qquad t_{rd} = \sqrt{(21 \text{ ns})^2 + (7 \text{ ns})^2}$

$\qquad\qquad = 22 \text{ ns}$

*David A. Bell, *Solid State Pulse Circuits*, 4th ed.

It is explained in Section 9-8 that the signal source resistance (R_s) combines with the oscilloscope input capacitance (C_i) to impose a rise time on the displayed waveform. So although the rise time of an oscilloscope may be very much shorter than the rise time of the waveform to be displayed, *substantial attenuation rise time can be generated by R_s and C_i*. As noted under "Time Base Measurements," the rise time measurement accuracy depends on the number of horizontal divisions involved. For t_r equal to one division, the accuracy is approximately ±10% (see Figure 9-35).

REVIEW QUESTIONS

9-1 Sketch the basic construction of a cathode ray tube. Identify each section of the tube, and show typical supply voltages at each appropriate point. Carefully explain the operation of the CRT.

9-2 Sketch the focusing section of a CRT. Show the equipotential lines and their effect on electrons traveling from the cathode to the screen. Explain.

9-3 Draw a sketch to illustrate electrostatic deflection. Explain. Define *deflection factor* and *deflection sensitivity* for a CRT.

9-4 Discuss the screen of a CRT and the factors affecting the brightness of the display.

9-5 Sketch the basic circuit of an oscilloscope deflection amplifier together with an input attenuator. Carefully explain the operation of the circuit.

9-6 Sketch the basic circuit and output waveforms of an oscilloscope sweep generator. Carefully explain the circuit operation.

9-7 Sketch the block diagram of an automatic time base for an oscilloscope. Show the waveforms at various points in the system, and carefully explain the operation of the time base.

9-8 Using illustrations, explain the function of a hold-off circuit in an oscilloscope time base.

9-9 Using illustrations, show how a dc output level control on a triggering amplifier can be used to adjust the instant at which a displayed waveform commences on an oscilloscope.

9-10 Explain blanking and unblanking in an oscilloscope, and discuss the need for blanking.

9-11 Sketch the construction of a dual-beam oscilloscope. Briefly explain.

9-12 Sketch the deflection system for **(a)** a dual-beam oscilloscope and **(b)** a single-beam dual-trace oscilloscope. Explain the operation of the single-beam dual-trace system. Also explain the use of this system in chop mode and in alternate mode. Sketch waveforms to illustrate the two modes of operation.

9-13 Briefly discuss each of the following oscilloscope controls: intensity, focus, position, dc shift, volts/div, time/div, trigger level.

9-14 Briefly discuss the procedure for making amplitude and time measurements on an oscilloscope.

9-15 Using illustrations, describe the procedures for oscilloscope measurement of pulse amplitude, pulse width, delay time, rise time, and fall time.

9-16 Using illustrations, explain the kind of pulse-wave distortion that can occur with **(a)** a coupling capacitor and **(b)** oscilloscope input capacitance. Define the circuit time constant in each case.

9-17 Describe a 1:1 probe for use with an oscilloscope. Sketch the circuit diagram for the probe together with the signal source and the oscilloscope input circuit. Explain briefly.

9-18 Sketch the construction of a 10:1 attenuator probe for use with an oscilloscope. Sketch the circuit diagram of the probe together with the signal source and the oscilloscope input. Explain briefly.

9-19 Describe a 10:1 oscilloscope probe that uses a correction box. Sketch its equivalent circuit.

9-20 Discuss the adjustment of oscilloscope probes, and show the various waveforms that can occur when adjusting a probe.

9-21 Discuss active probes for use with oscilloscopes.

9-22 Sketch a circuit diagram for displaying the characteristics of a diode on an oscilloscope. Explain the operation of the circuit, and sketch the type of display that is created.

9-23 Sketch the oscilloscope display that occurs with sine wave vertical and horizontal inputs that **(a)** are in phase, **(b)** are in antiphase, **(c)** have a phase difference of 90°, and **(d)** have a phase difference greater than zero but less than 90°.

9-24 Sketch the oscilloscope display that occurs with two synchronized sine waves when **(a)** the horizontal input has a frequency twice that of the vertical input, and **(b)** the ratio of vertical input frequency f_1 to horizontal input f_2 is $f_1/f_2 = 3:2$.

9-25 Describe how a dashed-line circular display may be produced on an oscilloscope. Explain the frequency ratio involved.

9-26 Discuss oscilloscope sensitivity, and explain how it is affected by the use of a 10:1 probe and by the vernier knob at the center of the V/div control.

9-27 Discuss the accuracy of voltage amplitude and time measurements on an oscilloscope.

9-28 State typical oscilloscope upper and lower cutoff frequencies, discuss their origins, and explain how the cutoff frequencies can affect waveform displays.

9-29 State rules for avoiding oscilloscope-introduced attenuation and phase shift on high- and low-frequency waveforms.

9-30 Explain how oscilloscope rise time can affect the rise time of a displayed waveform. State a rule for avoiding oscilloscope rise-time effect on the rise time of a displayed waveform.

PROBLEMS

9-1 The deflection amplifier in Figure 9-4 has the following components: $R_1 = R_7 = 10$ kΩ, $R_2 = R_8 = 5.6$ kΩ, $R_3 = R_6 = 15$ kΩ, $R_5 = 2.2$ kΩ, $R_9 = R_{11} = 15$ kΩ, and $R_{10} = 2$ kΩ. If the supply is ±15 V, determine the dc voltage levels throughout the circuit when the input level is zero and the moving contact of R_{10} is at its center position.

9-2 A 1 kHz triangular wave with a peak amplitude of 10 V is applied to the vertical deflecting plates of a CRT. A 1 kHz sawtooth wave with a peak amplitude of 20 V is applied to the horizontal deflecting plates. The CRT has a vertical deflection sensitivity of 0.4 cm/V, and a horizontal deflection sensitivity of 0.25 cm/V. Assuming that the two inputs are synchronized, determine the waveform displayed on the screen.

9-3 Repeat Problem 9-2 with the triangular-wave frequency changed to 2 kHz.

9-4 A sweep generator, as shown in Figure 9-7, has the following components: $R_1 = 4.7$ kΩ, $R_2 = 8.2$ kΩ, $R_3 = 3.3$ kΩ, $C_1 = 0.5$ μF, $R_5 = 12$ kΩ, and $R_6 = 4.7$ kΩ. The supply voltages are ±12 V, and the transistors are silicon. Calculate the peak-to-peak amplitude and the time period of the ramp waveform.

9-5 The ramp output from the circuit described in Problem 9-4 is to have its time period doubled. How should C_1 be modified? If the time period is to be adjustable by ±10%, what modifications should be made?

9-6 If the waveforms illustrated in Figure 9-18 occur when the time/div control is set to 0.1 ms, and the volts/div control is at 500 mV, determine the peak amplitude and frequency of each waveform.

9-7 For the waveforms illustrated in Figure 9-19, the time/div control is at 50 ms, and the volts/div control is set to 20 mV. Determine the peak amplitude and frequency of each waveform, and calculate the phase difference.

9-8 In Figure 9-19, how many horizontal divisions would there be between the beginning of each waveform cycle for a phase difference of 25°?

9-9 If the waveforms shown in Figure 9-20 occur when the time/div control is set to 10 ms and the volts/div control is at 5 V, determine the peak amplitude and frequency of each waveform.

9-10 Two waveforms (A and B), each occupying five horizontal divisions for one cycle, are displayed on an oscilloscope. Wave B commences 1.6 divisions after commencement of wave A. Calculate the phase difference between the two.

9-11 If the waveforms shown in Figure 9-21 occur when the volts/div control is set to 0.1 V and the time/div control is at 20 μs, determine each pulse amplitude, the pulse frequency, the delay time, the rise time, and the fall time.

9-12 A signal with an amplitude of $V_s = 500$ mV and a source resistance of 1 kΩ is connected to an oscilloscope with $R_i = 1$ MΩ in parallel with $C_i = 40$ pF. The coaxial cable of the 1:1 probe used has a capacitance of $C_{cc} = 80$ pF. Calculate the signal voltage level (V_i) at the oscilloscope terminals when the signal frequency is 120 Hz. Also calculate the signal frequency at which V_i is 3 dB below V_s.

9-13 A pulse wave from a signal generator with a source resistance of 600 Ω is to be displayed on an oscilloscope with $R_i = 1$ MΩ, $C_i = 30$ pF, and $C_c = 0.1$ μF.

(a) Determine the width of the longest pulse that can be displayed without distortion when the oscilloscope input is set to ac.

(b) Calculate the width of the shortest pulse that may be displayed without distortion when the input is set to dc.

9-14 A 10:1 oscilloscope probe, as shown in Figure 9-26, is used with an oscilloscope with $R_i = 1$ MΩ and $C_i = 40$ pF. If the probe uses a 9 MΩ series resistor and the coaxial cable has a capacitance of 80 pF, determine the value of capacitor C_1 that should be connected in parallel with the 9 MΩ resistor. Also calculate the signal frequency at which this probe will produce a 3 dB reduction in signal from a 1 kΩ source.

9-15 The circuit in Figure 9-29 is to be used to display diode characteristics up to a maximum forward current of 20 mA. Determine a suitable resistance for R_1, select appropriate vertical and horizontal deflection settings, and specify the sawtooth waveform requirements.

Special Oscilloscopes

10

Objectives

You will be able to:

1. Explain the need for a delayed-time-base oscilloscope. Draw the block diagram of a delayed time base, and explain how it operates.
2. Describe analog storage oscilloscopes and analog sampling oscilloscopes, and discuss their applications.
3. Explain the basic operation of a digital storage oscilloscope, and discuss the relationship between sampling rate and bandwidth.
4. Discuss digital storage oscilloscope modes of operation and applications.

Introduction

An oscilloscope with a *delayed time base* has two time base generators: a normal time base, and an additional delayed time base, which is superimposed on the normal time base output. This facility allows any portion of a displayed waveform to be brightened when the oscilloscope is operating on a normal time base. Switching to delayed time base causes the brightened portion to completely fill the screen for detailed investigation.

Many special oscilloscopes use a normal type of cathode ray tube (CRT) together with special control and input circuitry. An exception to this is the *analog storage oscilloscope,* which employs a special type of CRT. Signal waveforms are stored on electrodes within the CRT. This instrument is particularly useful for investigation of nonrepetitive, single-event signals.

A signal with a frequency too high for displaying on a normal oscilloscope can be sampled to create a dot waveform display. A *sampling oscilloscope* samples a repetitive signal once per cycle at different points during the repeating cycle. The samples are used to construct a low-frequency representation of the high-frequency waveform. Single-event signals cannot be investigated in this way.

A *digital storage oscilloscope* might be thought of as a combination of the sampling and storage oscilloscopes. Input signals are sampled, and the samples are stored in a digital memory. No special CRT is involved. The stored samples are recalled from the memory to reconstruct the original input waveform for display. The digital oscilloscope can be used for both single-event and repetitive signal investigations.

10-1 DELAYED-TIME-BASE OSCILLOSCOPES

Need for a Time Delay

Oscilloscopes normally use a *delay line* to delay the input waveform before it is passed to the vertical deflection amplifier (see Figure 10-1). The waveform prior to the delay line is used to trigger the time base, so that all of the leading or lagging edges of a pulse-type waveform can be carefully investigated. However, the time delay is not large enough to permit careful investigation of all parts of the waveform. In a *delayed-time-base oscilloscope*, a variable time delay is introduced in the time base circuit (instead of in the vertical deflection circuit). This permits the sweep time to be triggered after the delay time. As illustrated in Figure 10-2(a), the leading edge of a pulse waveform might be used to trigger the delay time (t_d), so that the lagging edge can be investigated. The portion of the waveform during time t_x can be expanded to fill the oscilloscope screen for careful study.

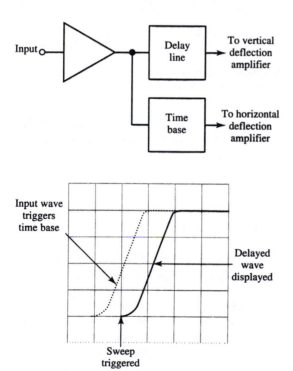

Figure 10-1 Oscilloscopes normally use a delay line to allow the time-base to be triggered right at the start of a waveform that is to be displayed.

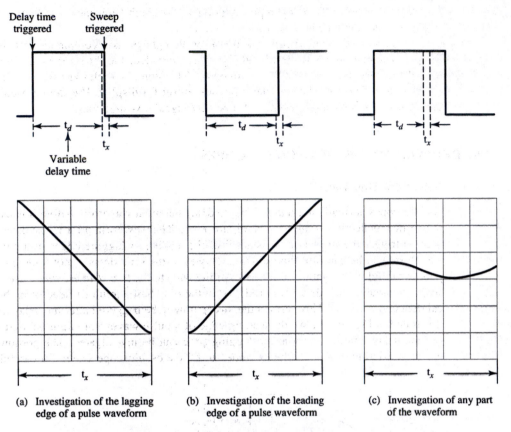

(a) Investigation of the lagging edge of a pulse waveform

(b) Investigation of the leading edge of a pulse waveform

(c) Investigation of any part of the waveform

Figure 10-2 When a time-delay circuit is triggered at the leading edge of a pulse, and the oscilloscope time base is triggered after a delay time t_d, the portion of the wave during t_x can be studied in detail.

Triggering the delay time on the lagging edge of the wave permits investigation of the leading edge [Figure 10-2(b)], and adjusting the delay time facilitates study of any portion of the wave [Figure 10-2(c)].

Delayed-Time-Base System

The block diagram and waveforms for a delayed-time-base system are illustrated in Figure 10-3. The *main time base* (MTB) is triggered as explained in Section 9-4. Also, as already explained, the MTB blanking circuit generates an unblanking pulse to turn on the electron beam in the CRT during the display sweep time. The MTB ramp output is fed via switch S (in Figure 10-3) to the horizontal deflection amplifier. The MTB ramp also goes to one input of a *voltage comparator*. The voltage comparator *compares* the levels of two input voltages and produces a negative (or positive) output spike at the instant that the two voltage levels become exactly equal. The voltage level at the other input of the comparator is controlled by a potentiometer, as illustrated. When the instantaneous level of the MTB ramp becomes equal to the potentiometer voltage, the comparator generates an output spike to trigger the *delayed time base* (DTB). The delay time t_d for the MTB

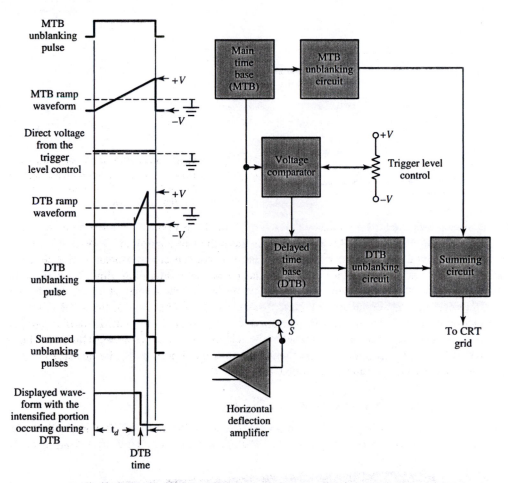

Figure 10-3 Oscilloscope delayed-time-base system. The delayed time base (DTB) is triggered when the main time base (MTB) voltage applied to one input terminal of the voltage comparator becomes equal to the trigger level voltage at the other input of the comparator. The MTB and DTB unblanking pulses are summed to brighten the display.

ramp to become equal to the potentiometer voltage can obviously be altered by adjusting the potentiometer.

Like the MTB, the DTB also has a blanking circuit that generates an unblanking pulse during the DTB ramp time. The two unblanking pulses are added together in a *summing circuit* before being applied to the grid of the CRT. The MTB unblanking pulse acting alone gives a displayed trace of uniform intensity. The summed unblanking pulses, as illustrated, apply a voltage to the CRT grid that is approximately doubled during the DTB ramp time. This increases the density of the electron beam in the CRT and thus intensifies (or brightens) the displayed waveform for the duration of the DTB ramp. Since delay time t_d can be adjusted by the potentiometer, the brightened portion of the displayed waveform can be moved around as desired.

When the part of the wave that is to be further investigated has been identified and brightened as described above, the DTB ramp is switched to the input of the horizontal

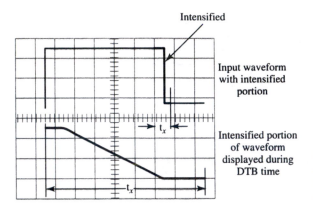

Intensified

Input waveform
with intensified
portion

t_x

Intensified portion
of waveform
displayed during
DTB time

t_x

Figure 10-4 Both the complete waveform
and the delayed portion can be displayed on
the oscilloscope screen by use of the alternate time base mode.

amplifier via switch S in Figure 10-3. Although the DTB ramp usually has a much shorter
time duration than the MTB ramp, it does have the same amplitude ($-V$ to $+V$) as the
MTB ramp. Consequently, it can cause the oscilloscope electron beam to be deflected
from one side of the screen to the other during the shortened ramp time. Now, only the
formerly bright portion of the waveform is displayed on the screen. Horizontal deflection
of the electron beam commences after delay time t_d from the start of the MTB sweep. A
very small section of the waveform under investigation can now be made to fill the screen
by adjusting the DTB time/div control. Alternate mode selection of the MTB and DTB is
possible, so that the entire waveform and the magnified portion can be displayed as two
separate waveforms. This is illustrated in Figure 10-4.

Representative Delay-Time-Base Oscilloscope

The front panel of a representative delay-time-base oscilloscope (the Tektronix TAS 455)
is shown in Figure 10-5. This particular analog oscilloscope has the same kind of facili-

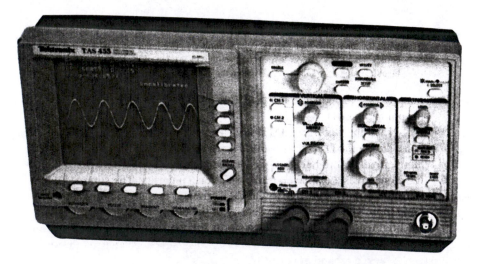

Figure 10-5 Front panel of a Tektronix TAS455 oscilloscope. (Courtesy of Tektronix,
Inc.)

ties for measuring and printing voltage, frequency, and time on the screen as found in a digital storage oscilloscope (see Section 10-4). This is a two-channel instrument with a 60 MHz bandwidth and a sensitivity of 2 mV/div to 5 V/div. The range of sweep speed is 0.5 s/div to 2 ns/div for the main time base, and 5 ms/div to 2 ns/div for the delayed time base.

10-2 ANALOG STORAGE OSCILLOSCOPE

Need for Trace Storage

The phosphor materials used on the screen of an oscilloscope normally glow for a period of only milliseconds. This is referred to as *persistence.* A short persistence display is satisfactory for most waveforms studied. However, when very low frequency signals are displayed, a short persistence screen shows a dot tracing out the wave, rather than the actual waveform. If the screen can be made to continue glowing along the path of the traced-out wave (long persistence), the actual waveform of the signal can be viewed more easily. Another situation in which ordinary oscilloscopes are unsatisfactory is the case of a *transient.* This is a nonrepetitive waveform that appears only once (e.g., when a power supply is switched *on*). In a *storage oscilloscope,* the waveform is retained and can be displayed continuously for an hour or more. A special type of cathode-ray tube (CRT) is required to achieve this effect.

Bistable Storage CRT

Figure 10-6 shows the basic construction of a *bistable storage CRT.* The term *bistable* is applied to a CRT that can display a stored waveform at only one level of brightness. The waveform is either displayed or not displayed. No variation in display intensity is possible. The screen has a *storage layer* of phosphor material which is capable of *secondary emission* and which has a very high insulation resistance between particles. (Secondary emission occurs when high-energy electrons strike a surface and cause other electrons to be emitted from the surface.) A transparent *metal film* is deposited between the glass of the CRT screen and the storage layer. The *collimator* is another metal film deposited around the neck of the tube.

The *write gun* in Figure 10-6 is made up of the accelerating and deflecting electrodes, as discussed in Section 9-1. The two *flood guns* are simply cathodes heated to generate low-energy electrons. The cathodes are at ground potential, and the collimator may also be at ground level or slightly positive. The metal film is at a potential of +1 V to +3 V with respect to ground. The clouds of electrons emitted by the flood guns are attracted by the positive potential on the metal film so that they flood the screen with electrons, as illustrated.

If the electron beam from the write gun has not been activated, the phosphor layer is not affected by the low-energy electrons from the flood guns. These electrons do not penetrate the phosphor and are collected by the collimator. No display occurs in this circumstance.

Now consider what happens when the write gun is energized and a waveform is applied to the oscilloscope input for a very brief time period. The beam of high-velocity

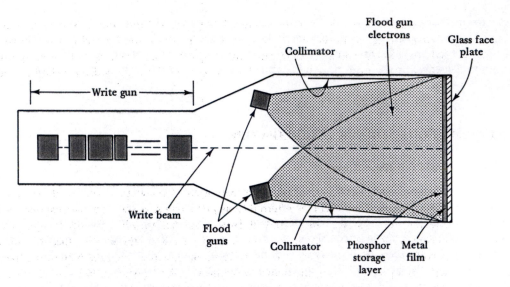

Figure 10-6 In a bistable storage cathode-ray tube, low-energy electrons from the flood guns cause the screen to continue glowing where a transient waveform has been briefly traced.

electrons from the write gun is deflected across the CRT screen. These electrons strike the phosphor layer with sufficient energy to produce secondary emission. The emitted secondary electrons are collected by the collimator. Every point on the screen at which secondary emission occurs becomes positively charged because electrons have been lost from that point. Thus, a positive charge path is traced on the storage layer in the shape of the input waveform. Because of the high insulation properties of the storage layer, there is very little charge carrier leakage, and the positive charge path can remain for hours.

The low-energy electrons from the flood guns are attracted to the positive charge path, and they pass through it to the (more positive) metal film. In passing through the storage layer, the electrons cause the phosphor to continue glowing. In this way, the one-time-occurring transient is displayed continuously. Erasure of the display is effected by making the metal film negative. This repels the flood gun electrons back into the storage layer, where they accumulate and return the written area to the same potential as the surrounding material.

Variable-Persistence Storage CRT

A storage CRT capable of *variable-persistence* operation is shown in Figure 10-7. The major difference between this tube and the CRT in Figure 10-6 is that a fine wire *storage mesh* and a *collector mesh* are now included. The *storage layer* is deposited on the inside surface of the storage mesh, and the screen is just the normal type of CRT screen with a low-persistence phosphor layer backed by an aluminum film. One additional change is that control grids are included in front of each flood gun.

The collector mesh has a positive potential around 100 V, and the voltage on the storage mesh is typically between 0 V and −10 V. The high-energy electron beam from

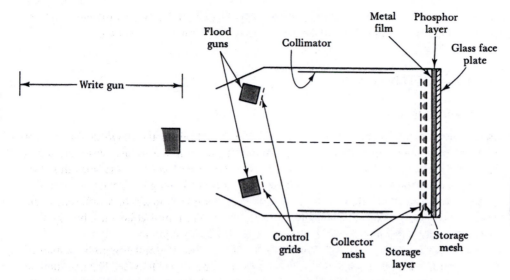

Figure 10-7 A variable-persistence storage cathode-ray tube is similar to a bistable storage tube, except that additional electrodes are included to give greater control over the display persistence.

the write gun produces secondary emission at the storage layer and creates a positive charge path in the shape of the input waveform. Flood gun electrons (low energy) are repelled from all parts of the storage layer except where the write beam has produced the positive charge path. Here, the flood gun electrons pass through and produce a trace on the screen. The secondary emitted electrons and the flood gun electrons that do not pass through the storage layer are attracted back to the highly positive collector mesh. Although the screen phosphor has low persistence in this CRT, the flood gun electrons passing through the positive trace on the storage layer continue to maintain a display on the screen.

Erasure of the stored waveform is effected by connecting the storage mesh to the same high positive voltage as the collector mesh for a brief time period. In this case, the flood gun electrons produce secondary emission all over the storage layer so that the written waveform is wiped out. The storage mesh is then returned to its normal voltage level. Erasure can also be performed by adjusting the display persistence control.

Variation of the display persistence is obtained by slowly erasing the stored waveform. A repetitive negative pulse (−4 V to −11 V) is applied to the storage mesh to discharge the positively charged areas. The pulse has a constant frequency of 1200 Hz typically, and the pulse width is continuously variable. The narrowest pulses give the longest display persistence, and the wider pulses shorten the persistence by discharging the storage layer more rapidly. Alternatively, the pulse width might be maintained constant and the frequency increased or decreased to shorten or lengthen the persistence.

The brightness of the displayed waveform depends on the number and energy of the flood gun electrons that strike the screen. The voltage applied to the grids of the flood guns controls these electrons. A high negative voltage completely cuts off the flood gun electrons. A zero voltage level on the grids permits maximum electron flow.

10-3 SAMPLING OSCILLOSCOPES

Waveform Sampling

Most ordinary oscilloscopes have an upper frequency limit in the range 20 MHz to 50 MHz. Higher input frequencies cause the electron beam to move so fast across the screen that only a very faint trace is produced. The *sampling oscilloscope* overcomes this difficulty by producing a low-frequency dot representation of the signal. Each dot represents an amplitude sample of the input signal, and each sample is taken in a different cycle. The sampling circuits must be capable of operating at very high frequencies, but the CRT and its associated circuitry may be relatively low-frequency equipment.

Figure 10-8 shows 10 cycles of a high-frequency waveform that is to be displayed. One amplitude sample is taken at successively later times in each cycle. The resultant series of samples reproduces the original waveform at a lower frequency. Suppose that the input signal has a time period of $T = 0.01$ μs. Its frequency is $f_1 = 1/0.01$ μs = 100 MHz. The dot display has a time period of 10×0.01 μs or a frequency of $f_2 = 100$ MHz/10 = 10 MHz. Thus, a 100 MHz signal is converted into a 10 MHz waveform for display. If one cycle of dot display is created by sampling 100 cycles of signal, the frequency of the displayed wave is 1/100 of the signal frequency. One disadvantage of this system is that only purely repetitive waveforms can be investigated. If the waveform changes over several

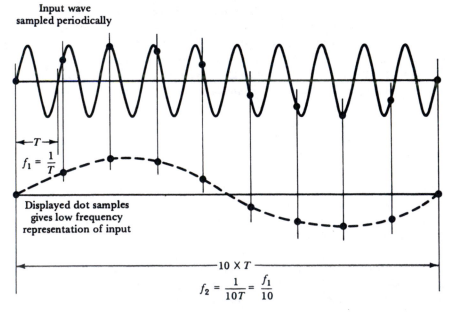

Figure 10-8 A high-frequency repetitive waveform can be investigated by sampling successive cycles to create a lower-frequency dot display on an oscilloscope.

cycles, the dot display will be in error. A transient waveform cannot be investigated by this method of sampling.

System Operation

A basic block diagram and waveforms for a sampling scope are shown in Figure 10-9. Instead of the ramp generator employed in a regular oscilloscope, a *staircase generator* is used for horizontally deflecting the electron beam in the CRT. As the waveform shows, the output of the staircase generator is a series of increasing voltage steps (or staircase), which rise from a negative level (−V) to a positive level (+V). At the end of the sweep time period, the staircase waveform returns to its starting level. The staircase voltage causes the electron beam to be moved from one side of the CRT screen to the other in a series of steps.

As well as being applied to the CRT horizontal deflection amplifier, the staircase generator output is connected to one input of a *voltage comparator*. This compares the staircase voltage to the output of a *ramp generator*. The ramp output commences each time the staircase voltage changes level. This is triggered by means of a pulse signal from the staircase generator. When the instantaneous level of the ramp voltage becomes equal

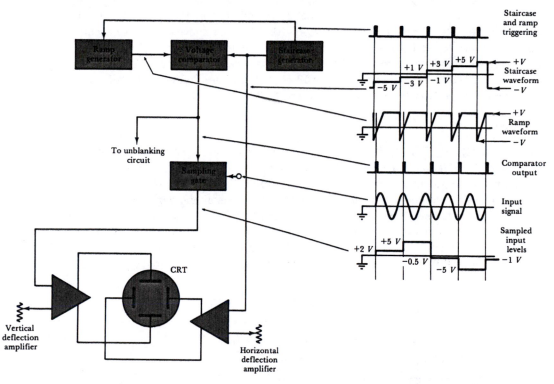

Figure 10-9 Block diagram and waveforms for a sampling oscilloscope. A staircase generator (instead of a ramp generator) is used for horizontal deflection. The waveform is sampled at the beginning of each step of the staircase generator, and the sampled levels are used to produce vertical deflection. When combined with unblanking pulses, the resultant display is a dot waveform.

to the staircase voltage, the comparator produces a short-duration pulse, which turns *on* the *sampling gate*. The sampling gate samples the instantaneous amplitude of the high-frequency input waveform. After switching on for only a brief instant, the sampling gate holds its output level constant until the next sample is taken. The pulse output from the comparator is also applied as an *unblanking pulse* to the CRT grid. This turns *on* the electron beam for a short time to create a dot on the screen.

The waveforms in Figure 10-9 illustrate the process of creating the dot display. The input wave is sampled, as already explained, to create a low-frequency step representation of the original. This step wave is applied to the CRT vertical deflection amplifier, while the staircase wave from the staircase generator is applied to the horizontal deflection amplifier. No waveform display occurs while the CRT grid is biased negatively. Each time an unblanking pulse occurs, a bright dot is created on the screen. The position of each dot depends on the voltages that appear at the horizontal and vertical deflection plates and on the deflection sensitivity of the CRT.

Consider the deflecting voltages when the first (left-hand) unblanking pulse (from the comparator) occurs in Figure 10-9. The vertical deflecting voltage (sampled level) is +2 V, and the horizontal deflecting voltage (staircase wave) is −3 V. Assume that the vertical deflection sensitivity is 0.5 cm/V and the horizontal deflection sensitivity is 1 cm/V. The electron beam is deflected vertically up from the center line on the screen by 2 V × 0.5 cm/V = 1 cm. The beam is also deflected horizontally to the left of center by 3 V × 1 cm/V = 3 cm. This gives point 1 in Figure 10-10. When the unblanking pulse occurs, a bright dot is created at this point. At the next unblanking pulse, the vertical and horizontal deflecting voltages have become +5 V and −1 V, respectively. Therefore, the next dot occurs at 2.5 cm up from the screen center and 1 cm left of center (point 2 in Figure 10-10). Continuing this process, the dot waveform is created as illustrated.

Circuits

Figures 10-11, 10-12, and 10-13, respectively, show the basic circuits of a staircase generator, a voltage comparator, and a sampling gate. The staircase generator circuit is similar to the sweep generator in Figure 9-7, except that the constant current transistor (Q_1) is controlled by a pulse generator. Each negative-going pulse from the pulse generator turns Q_1 on for a brief time period. Each time Q_1 is switched *on*, it provides a current pulse to capacitor C_1. Every current pulse charges C_1 to a new voltage level, which remains constant until the next current pulse arrives. When the capacitor voltage exceeds the intended upper level, the Schmitt trigger circuit is

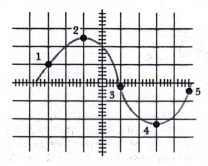

Figure 10-10   Dot waveform produced by staircase time base, sampled waveform, and unblanking pulses.

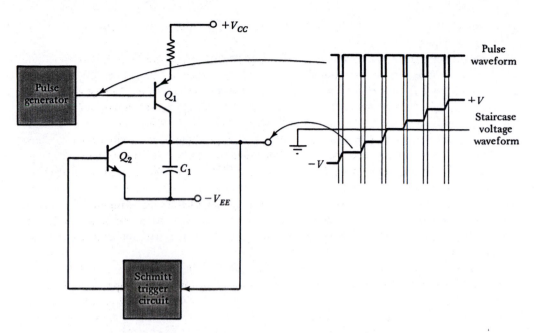

Figure 10-11 Basic circuit and waveforms for a staircase generator. Transistor Q_1 is repeatedly switched *on* and *off* by the pulse generator. Q_1 passes current pulses to capacitor C_1, causing its voltage to increase in steps.

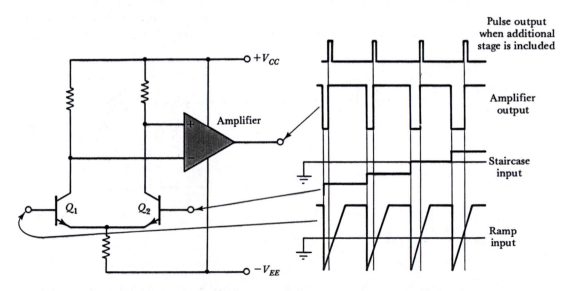

Figure 10-12 Basic circuit and waveforms for a voltage comparator. The output voltage switches rapidly from one extreme to the other when Q_1 base voltage becomes equal to the voltage at Q_2 base.

Sec. 10-3   Sampling Oscilloscope   **295**

activated to turn Q_2 *on* to discharge C_1 to its starting level (as in the linear sweep generator circuit of Figure 9-7).

In the voltage comparator circuit in Figure 10-12, transistors Q_1 and Q_2 constitute a *differential amplifier* (see Section 9-2). Another amplifier stage takes the collector output voltages from Q_1 and Q_2 and amplifies their difference up to the supply voltage saturation levels. With the instantaneous level of the ramp input to Q_1 lower than the staircase voltage, Q_1 is biased *off* and Q_2 is *on*. The collector voltage of Q_1 is *high* (close to $+V_{CC}$), and the collector of Q_2 is low. With the output amplifier terminal polarity illustrated, the output voltage is low at this time. When Q_1 input becomes equal to the input level of Q_2, its collector voltage falls while the collector of Q_2 rises. This produces a positive-going output from the amplifier stage. Therefore, the comparator output is normally low while the ramp voltage is below the staircase level, and it rapidly switches to a high level when the two inputs become equal. A further stage can be added to the circuit in Figure 10-12 to give a spike or short pulse output at the instant that the inputs reach equality.

The sampling gate in Figure 10-13 consists of FET Q_1 and capacitor C_1. A voltage follower (see Section 4-2) is included to provide a low-output impedance. With the input to the gate of Q_1 normally biased negatively, Q_1 is *off*. In this condition, it is like an open switch. When a positive pulse occurs at the gate, Q_1 switches *on* into saturation. The voltage drop from the drain to source terminals of the FET is extremely small when the device is biased *on*, so in the *on* condition it is like a closed switch. Capacitor C_1 is charged to the level of the input voltage each time the FET is biased *on*. When the FET goes *off*, C_1 retains its voltage constant at the level of the last sample.

Expanded Mode Operation

When a sampling scope is operated in *expanded mode*, the results are similar to those obtained with the delayed time base discussed in Section 10-1. Figure 10-14(a) shows a dot representation of an approximately square-wave input signal. Since there are 16 (ampli-

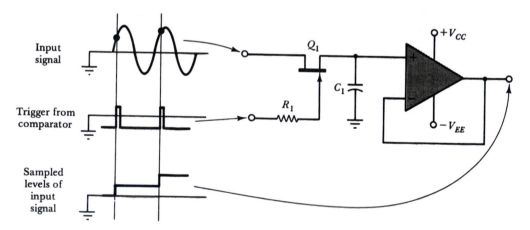

Figure 10-13   Basic sampling gate circuit and waveforms. FET Q_1 is switched *on* and *off* by the trigger pulses to R_1, to sample the instantaneous level of the input signal. The sample voltage is retained on C_1 and passed to the output via the voltage follower.

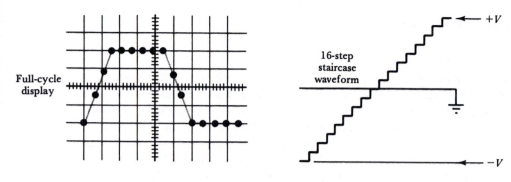

(a) A complete cycle of input signal is sampled and displayed using a normal staircase waveform.

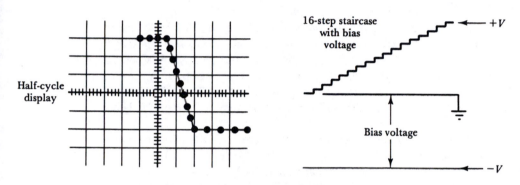

(b) Combining a bias voltage with a more dense staircase allows a portion of the input signal to be sampled and displayed.

Figure 10-14 A sampling oscilloscope can be operated in expanded mode to investigate any portion of the cycle of an input signal.

tude sample) dots shown, the signal was sampled 16 times. To do this, the staircase time base had to have 16 steps, as illustrated. Now, suppose that the time base is adjusted to consist of 16 smaller steps and that this is superimposed upon a dc bias voltage, as illustrated in Figure 10-14(b). This causes all 16 amplitude samples to be taken during the latter half-cycle of the input signal. Thus, only this half of the signal is displayed on the oscilloscope screen. However, this portion is displayed in great detail when the horizontal deflection sensitivity of the oscilloscope is expanded to give a full-screen display of the sampled portion of the input waveform.

By adjusting the bias voltage and the density of the staircase waveform, any portion (including very small portions) of the input signal can be investigated. The section of the waveform to be investigated is usually identified by moving a very bright dot to the desired part of the signal when it is displayed in the normal (full-cycle) mode. This, again, is similar to the method used with a delayed-time-base oscilloscope.

10-4 DIGITAL STORAGE OSCILLOSCOPES

Digital Sampling

As discussed in Section 10-2, analog storage oscilloscopes store the input waveform in a special type of cathode-ray tube. In *digital storage oscilloscopes (DSO)*, the waveform is sampled at regular intervals (see Figure 10-15), and each sample is converted to digital form by means of an analog-to-digital converter (ADC) (Section 5-5). The digitized samples are coded as illustrated in Figure 10-15. The combination of absent and present pulses in each sample represents the amplitude of the analog sample. The 1111 condition represents maximum amplitude with all four pulses present. Because 1111 is the digital equivalent of analog 15, the 0001 condition (only the right-hand pulse present) represents $\frac{1}{15}$ of maximum amplitude. Similarly, 0100 is the digital code for $\frac{4}{15}$ of maximum, and 1101 indicates $\frac{13}{15}$ of maximum waveform amplitude.

Basic DSO Operation

The block diagram of a basic sampling and storage system for a DSO is illustrated in Figure 10-16. The time base generates a pulse waveform at the desired sampling frequency. Each pulse switches the sampling gate (Section 10-3) and the ADC *on* for a brief time period. The sampling gate generates a series of analog samples, as illustrated, and the ADC converts each sample into a coded group of pulses. The pulse groups are passed to a semiconductor (or other type) memory, where they are stored for later recovery.

Figure 10-17 shows a system for recovery of the stored information to recreate the original waveform. The digital-to-analog converter (DAC) (Section 5-6), triggered by pulses from the time base, converts each digital sample back to analog (step) form and passes it to the oscilloscope vertical deflection amplifier. The time base also generates a staircase waveform that is fed to the horizontal deflection amplifier. The vertical and hori-

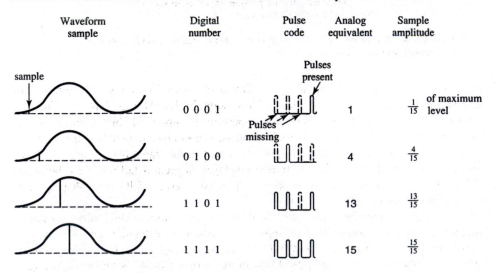

Figure 10-15 A digital storage oscilloscope (DSO) samples a waveform to be displayed, and converts each sample into a code of pulses that represents a digital number.

Special Oscilloscopes   Chap. 10

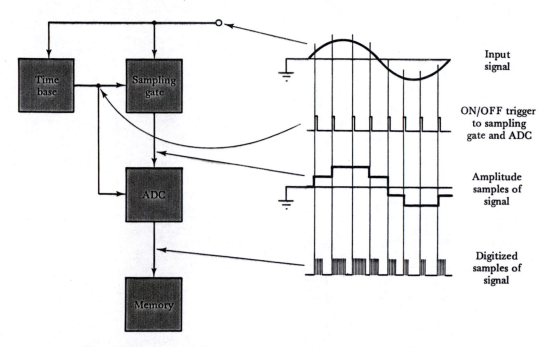

Figure 10-16 Basic sampling and storage system for a DSO. The input waveform is analog sampled, and each sample is converted into digital form by means of an analog-to-digital converter (ADC).

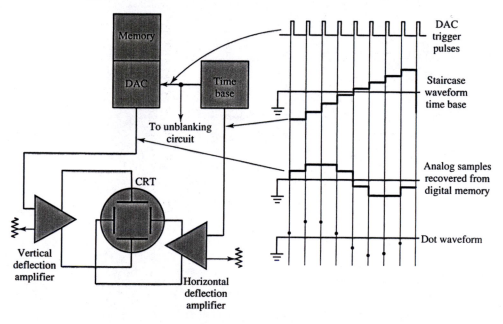

Figure 10-17 Basic DSO system for displaying a stored waveform. A digital-to-analog converter (DAC) converts the digital samples in the memory into analog form for application to the vertical deflection plates. A dot waveform is produced when the samples are combined with the staircase waveform time base and the unblanking pulses.

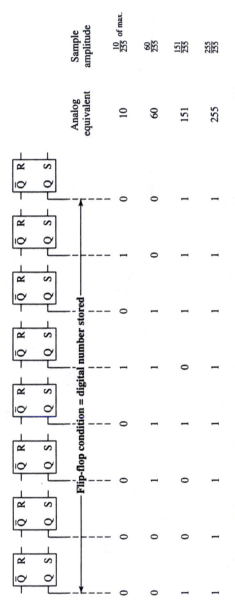

								Analog equivalent	Sample amplitude of max.
0	0	0	0	0	0	1	0	10	$\frac{10}{255}$
0	0	1	1	1	1	0	0	60	$\frac{60}{255}$
1	0	1	1	0	1	1	1	151	$\frac{151}{255}$
1	1	1	1	1	1	1	1	255	$\frac{255}{255}$

Flip-flop condition = digital number stored

Figure 10-18 Flip-flop memory. Eight flip-flops can store a digital number representing up to 255 levels of an analog sample.

zontal inputs can be used to produce either a dot waveform or a step waveform on the oscilloscope screen, as illustrated.

Digital Memory and Resolution

One type of memory for storing digital samples simply consists of a number of set-reset flip-flops (see Section 5-2). Figure 10-18 shows eight such flip-flops together with a portion of the truth table giving the digital representations for various states of the Q outputs. The analog equivalent for each of these states is also listed. Because this *8-bit* memory can represent up to 255 levels of an analog sample, the instantaneous voltage levels of the waveform for each sample is recorded with a resolution of 1 in 255, or better than 0.4%. As explained in Section 2-3, resolution is related to measurement precision. Most DSOs sample with an 8-bit resolution. [The four-pulse (*4-bit*) representation of the samples in Figure 10-15 gives a resolution of only 1 in 15.] One set of eight flip-flops can store only one sample; therefore, many such sets of flip-flops are required for storing sufficient samples to represent several complex waves. A DSO that can store 4000 samples is usually referred to as having a *4000-word* memory, or a *4k* memory.

Interpolation

To obtain a more representive display than the dot or step waveform, digital oscilloscopes normally have a facility for drawing lines between the sampled levels. This is termed *interpolation. Linear interpolation* occurs when straight lines are drawn between the levels.

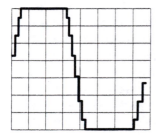

(a) Square waveform without interpolation

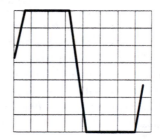

(b) Square waveform with linear interpolation

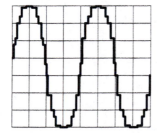

(c) Sinusodial waveform without interpolation

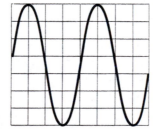

(d) Sinusodial wave with sine-wave interpolation

Figure 10-19   Lines are normally drawn between sample levels on a waveform displayed on a DSO. Linear interpolation is most suitable for pulse waveforms, while sine-wave interpolation gives best results with sinusoidal waveforms.

This is most suitable for pulse waveform displays. *Sine-wave interpolation* introduces a sinusoidal function between the sample levels, for use with sinusoidal waveforms. Figure 10-19 illustrates both types of interpolation. Linear interpolation produces acceptable sine waves when a large number of samples are taken during each cycle. With fewer samples, the straight lines between the sample levels give a distorted display. Sine-wave interpolation can give satisfactory results with only 2.5 samples per cycle for sinusoidal waveforms, but when used with a pulse waveform it can introduce distortion such as overshoots.

Sampling Rate and Bandwidth

As with analog storage oscilloscopes, the quality of the waveform displayed by digital oscilloscopes depends on the number of samples taken during the waveform cycle or transient. The waveform can be reproduced in great detail when a high sampling rate is used [see Figure 10-20(a)]. With a low sampling rate, important details of the waveform may be lost [Figure 10-20(b)]. The type of detail shown in Figure 10-20(a) is easily obtained when a relatively low-frequency signal is being investigated. For example, suppose that a 500 kHz waveform is to be displayed by a DSO that samples at a rate of 50 MS/s (50 megasamples per second). The number of samples taken during one cycle is

$$\frac{T \text{ of signal}}{T \text{ of sampling rate}} = \frac{\text{sampling } f}{\text{signal } f} = \frac{50 \text{ MS/s}}{500 \text{ kHz}}$$

$$= 100$$

Therefore, one cycle of the displayed waveform is made up of 100 samples. If the sampling rate is 5 MS/s, there will be only 10 samples to represent each cycle of the 500 kHz wave. Obviously, the 100 sample/cycle display is likely to be satisfactory, while the 10 sample/cycle representation may not give a clear picture of the waveform.

The sampling rate of the oscilloscope determines the maximum signal frequency that can be displayed satisfactorily. Sampling theory stipulates that to reproduce a repetitive waveform, samples must be taken at a rate greater than two times the highest waveform

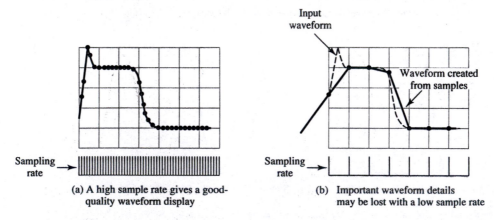

(a) A high sample rate gives a good-quality waveform display

(b) Important waveform details may be lost with a low sample rate

Figure 10-20 A high sampling rate is required for a waveform to be accurately recreated from a series of amplitude samples.

Special Oscilloscopes   Chap. 10

(a) Real-time sampling occurs when a waveform is sampled many times during its cycle.

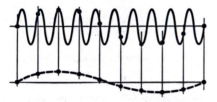

(b) With equivalent-time sampling, the sampling rate is just a little faster than one sample per cycle.

Figure 10-21 With real-time sampling, the waveform is recreated from several samples taken during each cycle. Equivalent-time sampling is used when the waveform frequency is too high for real-time sampling.

frequency (the *Nyquist rate*). In practice a good display requires a minimum of approximately 25 samples per cycle. However, where sinusoidal, square, or other repetitive waveforms are being investigated, a smaller number of samples may give a satisfactory result.

An error known as *aliasing* can occur when a waveform is sampled at a rate lower than the Nyquist rate. A stable waveform is displayed at each of several different time base speeds, giving different frequency readings for a single-frequency input waveform. These frequencies are harmonics of the input frequency. Aliasing can usually be avoided by using the highest possible time base speed.

Some DSO manufacturers specify the instrument's high cutoff frequency as one-fourth of the sampling rate: for example, a dc-to-25 MHz bandwidth for a 100 MS/s sampling rate. This means that only four samples are taken during each cycle of the highest-frequency waveform that can be investigated. Some manufacturers use a two-times ratio, claiming a bandwidth of 50 MHz for 100 MS/s, or two samples per cycle at the high cut-off frequency.

When a waveform is sampled many times during one cycle, the technique is referred to as *real-time sampling,* [Figure 10-21(a)]. The method discussed in Section 10-3 uses a sampling rate just a little faster than one sample per cycle of the waveform investigated. This is termed *equivalent time sampling,* [see Figure 10-21(b), reproduced from Figure 10-8]. Equivalent time sampling is applicable only for repetitive waveforms where the frequency is too high to be real-time sampled.

Signal frequencies are often low enough for the waveform to be displayed directly in analog form on the oscilloscope, and then the DSO advantage is mainly signal storage.

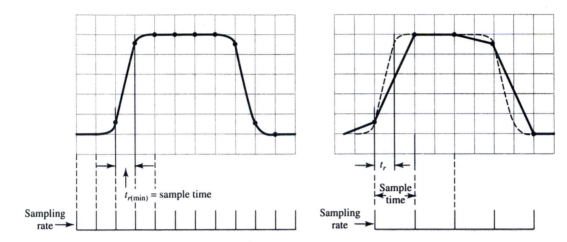

(a) Pulse with a rise time equal to the sampling time interval

(b) Pulse waveform with a rise time less than the sampling time interval

Figure 10-22 The minimum pulse rise time that can be displayed (and measured) accurately by a DSO is limited by the sampling time interval.

Many DSOs are designed for use as both analog and digital instruments; they can display a waveform directly, as well as sample and store it for later recall. These instruments are sometimes referred to as *real-time and storage oscilloscopes (RSO)*.

Pulse Rise Time and Sampling Rate

The pulse rise time (t_r) (see Figure 9-21) is one of the most important quantities involved in the investigation of pulse waveforms. For a detailed study of the pulse leading and lagging edges, perhaps 10 or more samples should be taken during t_r. However, if the rise time is just to be measured accurately, the minimum rise time cannot exceed the time interval between samples. This is illustrated in Figure 10-22. Thus, for a 100 MS/s sampling rate,

$$t_{r(min)} = \frac{1}{\text{sampling rate}} = \frac{1}{100 \text{ MS/s}}$$

$$= 10 \text{ ns}$$

Pulses that have a shorter rise time than the sampling time interval will not be displayed accurately.

10-5 DSO APPLICATIONS

Autoset

Some digital storage oscilloscopes have vertical (volts/div) and horizontal (time/div) controls, just like an analog oscilloscope. Others automatically select the best amplitude and time settings for the displayed waveforms, and print the settings on the

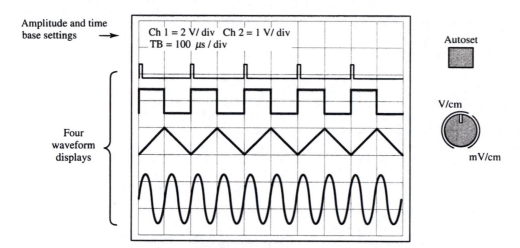

Amplitude and time base settings →

Ch 1 = 2 V/ div   Ch 2 = 1 V/ div
TB = 100 μs / div

Four waveform displays

Autoset

V/cm

mV/cm

Figure 10-23 Most DSOs have an autoset facility that selects the amplitude and time base settings automatically. Many can display four channels of new and stored waveforms.

screen (see Figure 10-23). In this case, manual selection of the quantities is also normally available.

Multichannel Displays

Like analog oscilloscopes, digital storage oscilloscopes are normally capable of displaying waveforms applied to two input channels. However, many DSOs can also display four, or more, channels of new and stored waveforms (Figure 10-23), and some have four separate input channels. As with an analog oscilloscope, input waveforms can be displayed on the screen (sampled and interpolated) as they are received. For comparison purposes, stored waveforms can also be displayed at the same time as new input waves.

Waveform Processing

Digital oscilloscopes normally contain digital voltmeters, digital frequency meters, and time measurement circuits, which print the desired measurements on the oscilloscope screen as the waveform is being displayed. Figure 10-24 illustrates these facilities. Two *cursors* are employed to define the points on the waveform between which the measurement is to be made. The cursors may be crossed lines, arrowheads, or dots, and they can be moved around by controls on the front panel of the oscilloscope, as illustrated. The voltage, time, and frequency are measured much more accurately than could be achieved by means of an analog oscilloscope. Some DSOs automatically process the displayed waveform without reference to cursors, measuring V_{rms}, V_{p-to-p}, f, t_r, T, pulse width, duty cycle, and so on, and printing these quantities on the screen.

Pretriggering and Post-triggering

A DSO can be used to display portions of a waveform that occur before the normal trigger point (*pretriggering*) or after the trigger point (*post-triggering*). Pretriggering is possible because the input waveform is continually being digitized and stored in memory.

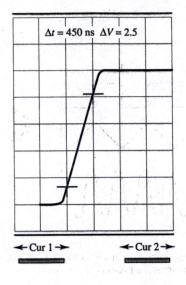

(a) Rise-time measurement

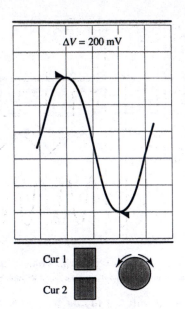

(b) Peak-to-peak voltage measurement

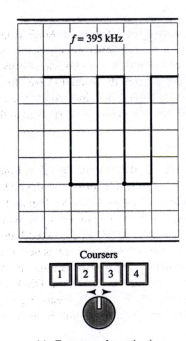

(c) Frequency determination

Figure 10-24 Most DSOs process the displayed waveforms to accurately determine voltage amplitude, rise time, and so on, as defined by movable cursors. The measured quantities are printed on the screen.

Thus, as illustrated in Figure 10-25(a), the stored waveform can be displayed beginning at a point ahead of the normal trigger point. The pretrigger time may be up to 50% or 100% of the waveform time period, depending on the particular oscilloscope. In post-triggering, the commencing point of the display is delayed after the trigger point by a variable *hold-off time,* which may be several times the time period of the displayed waveform [Figure 10-25(b)]. This is similar to the delay time introduced in delayed sweep analog oscilloscopes. Instead of delaying by a specified time, the display may be delayed by a selected number of waveform cycles. The combination of pre- and post-triggering allows any part of the input waveform to be examined.

Zoom and Restart

Zoom and *restart* are essentially additional methods of time-delay selection, and each is available with some (but not all) DSOs. In the first case, the portion of the displayed

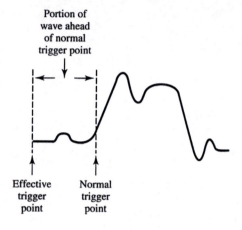

(a) Pretriggering

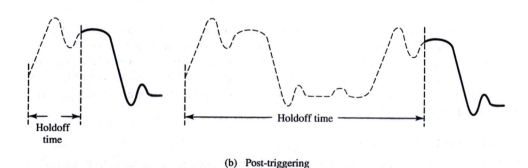

(b) Post-triggering

Figure 10-25 Pretriggering allows the portion of a waveform ahead of the normal triggering point to be displayed on a DSO. Post-triggering introduces a hold-off time (or delay time) after the normal triggering point.

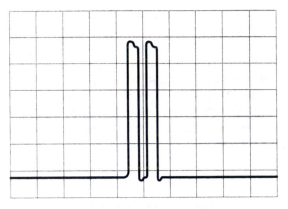

(a) Waveform centered on screen

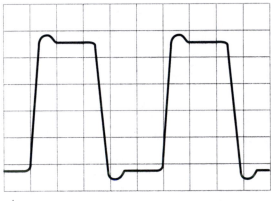

(b) Zoom facility time-expands the waveform

Figure 10-26 Zoom and restart facilities permit a waveform to be time expanded for more detailed study.

waveform to be examined is first centered on the screen or marked by cursors. A *zoom* button is pushed to time-expand the waveform, or *zoom-in* on, the portion of interest. This is illustrated in Figure 10-26. The required time delay and time base changes are introduced automatically. *Restart* selection also makes it possible to view part of a waveform in detail; however, instead of just time expanding, the waveform is resampled at a faster rate for maximum information.

Glitch and Runt Catching

Figure 10-27(a) illustrates the fact that very fast spike-type changes in a displayed waveform (*glitches*) can be missed by a DSO if the sampling rate is not high enough. Glitches occurring during the sampling-time interval are invisible to the oscilloscope, so

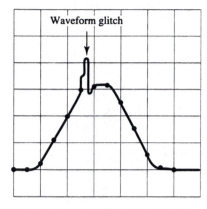

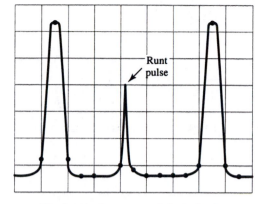

(a) A waveform glitch can be missed by the sampling time interval.

(b) A runt pulse can occur in logic circuits.

Figure 10-27   Glitch and runt pulses could be missed by a DSO if they occur during the sampling time. A DSO with a maximum/minimum level detector catches the waveform anomolies.

where such waveform anomalies are important, an oscilloscope with a very high sampling rate should be used. The fast *on/off* switching of flip-flops in high speed digital logic circuits can produce troublesome waveform glitches; consequently, this is one field in which signal waveforms require careful study.

A *runt pulse* is a special type of glitch that occurs in logic systems [Figure 10-27(b)]. This is a pulse that is not quite large enough to produce circuit triggering. A *maximum/minimum* level detector circuit is employed in some DSOs to detect the presence of a glitch. The oscilloscope may then be made to trigger on the glitch, so that the waveforms before and after the glitch can be displayed.

Baby-Sitting Mode

Some waveform transients in electronics systems may occur only once during a period of hours. A DSO can be set in a *baby-sitting mode* to deal with this situation. The waveform is sampled and recorded continuously, so that at any instant several immediately previous cycles are stored. When an anomaly is detected, the stored samples are retained in the memory for later playback. This is illustrated in Figure 10-28(a).

Roll Mode

Many quantities vary slowly over a period of hours, and traditionally these are monitored by chart recording instruments (see Chapter 13). A DSO time base, which is usually derived by frequency division of the output of a crystal oscillator (see Section 5-4) can be adjusted to give an accurate sampling time interval of minutes or hours. Thus, the amplitude of the slowly changing quantity can be sampled and stored over a long period of time. The stored waveform can be displayed on the oscilloscope screen at a speeded-up

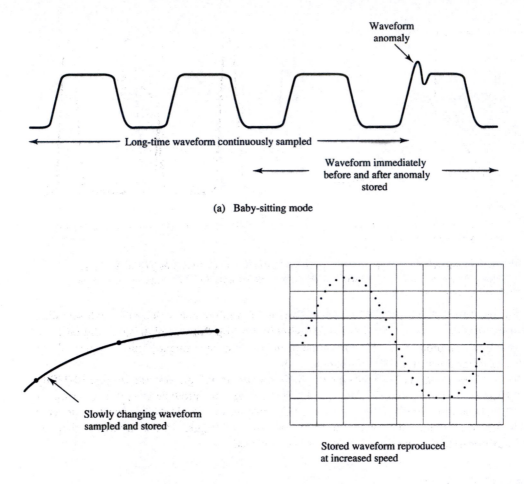

(a) Baby-sitting mode

Long-time waveform continuously sampled

Waveform immediately
before and after anomaly
stored

Waveform anomaly

Slowly changing waveform
sampled and stored

Stored waveform reproduced
at increased speed

(b) Roll mode operation

Figure 10-28 A DSO can be operated in baby-sitting mode to detect infrequent waveform anomolies, and in roll mode to store and display slowly changing quantities.

rate, rolling from right to left, as shown in Figure 10-28(b). When operated in this fashion, the DSO is said to be in *roll mode*.

Documentation and Analysis

Most DSOs can provide outputs to a plotter or computer-type printer to produce a hard copy of any waveform stored in memory. Although the waveform samples can be stored indefinitely, a printed copy is convenient for publication and/or further study. This also allows the DSO memory to be used for further waveform storage. A facility for interfacing with computers is also usually made available, for further analysis of stored waveforms, and for remote control of the oscilloscope by the computer.

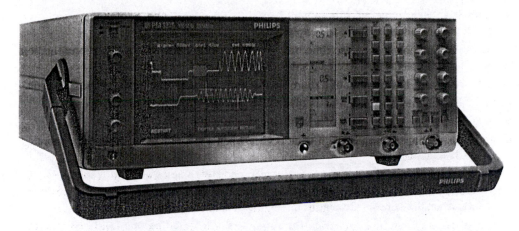

Figure 10-29 Philips 3375 digital storage oscilloscope. (© 1990, John Fluke Mfg. Co., Inc. All rights reserved. Reproduced with permission.)

10-6 REPRESENTATIVE DSO

The Philips PM3375 DSO (Figure 10-29) uses pushbutton controls for manual selection of volts/div and time/div. Automatic setting is also available, and an LCD panel is provided to display amplitude, time base, and other information. Cursor controls are directly below the screen, and all waveform measurements are printed on the screen. The bandwidth is specified as 100 MHz with a sampling rate of 250 MS/s, but the instrument can also be operated as an analog oscilloscope with a 100 MHz bandwidth. A 16 k memory allows up to eight traces to be stored for recall and display.

REVIEW QUESTIONS

10-1 Discuss the need for a time-delay system in an oscilloscope. Using illustrations, explain how a delayed time base can improve waveform investigations.

10-2 Sketch a basic block diagram of a delayed-time-base (DTB) system. Sketch the waveforms throughout the system, and explain the system operation.

10-3 Explain how the portion of the waveform to be investigated is intensified by the DTB system.

10-4 Sketch a diagram to show the construction of a bistable storage CRT. Explain its operation.

10-5 Sketch a diagram to show the construction of a variable-persistence storage CRT. Explain its operation.

10-6 Show how a high-frequency waveform can be sampled to create a low-frequency dot representation of the waveform. Explain the relationship between the signal frequency, the dot waveform frequency, and the number of samples per cycle in the low-frequency wave.

10-7 Sketch the basic block diagram of a sampling oscilloscope. Sketch the waveforms throughout the system, and explain its operation.

10-8 Explain how a staircase time base waveform combines with a step representation of a sampled signal to create a dot representation of the signal.

10-9 Draw the basic circuit of a staircase waveform generator and explain its operation.

10-10 Sketch the basic circuit of a voltage comparator. Explain the circuit operation.

10-11 Sketch the basic circuit of a sampling gate. Explain how it operates.

10-12 Using illustrations, explain what occurs when a sampling oscilloscope is operated in expanded mode.

10-13 Explain what occurs when a waveform is sampled and the samples are digitized. Prepare a chart showing several amplitude samples of a waveform, the resultant pulse codes when the samples are digitized, and the analog equivalents.

10-14 Sketch the basic system block diagram for a digital storage oscilloscope (DSO). Sketch the system waveforms, and explain its operation.

10-15 Sketch the system block diagram, and the associated waveforms for displaying a dot representation of a sampled waveform on an oscilloscope screen. Explain the system operation.

10-16 Show how several set-reset flip-flops can be employed as a digital memory. Prepare a truth table showing the stored digital number for the various flip-flop conditions. Explain how the number of flip-flops affects the resolution of the sample amplitude.

10-17 Discuss linear interpolation and sine-wave interpolation, and sketch DSO waveforms with and without interpolation.

10-18 Discuss the relationship between the sampling rate and the bandwidth of a DSO. Show how waveform detail may be missed by a sampling rate that is too low. Explain aliasing.

10-19 Define real-time sampling and equivalent-time sampling.

10-20 Discuss the relationship between sampling rate and pulse waveform rise time.

10-21 For a DSO, discuss autoset, multichannel displays, and waveform processing.

10-22 Explain pretriggering, post-triggering, zoom, and restart facilities for a DSO.

10-23 Discuss DSO glitch and runt catching, and explain baby-sitting mode and roll mode operation.

PROBLEMS

10-1 The waveform shown in Figure 10-2(a) is the positive half-cycle of a 1 kHz square wave. If t_x is approximately 5% of T, determine the delay time required for investigation of the falling edge of the wave.

10-2 A delayed-time-base oscilloscope is used to investigate the leading edge of a square waveform which has a frequency that varies from 100 kHz to 120 kHz. Estimate the range of adjustment for the required delay time.

10-3 A 200 MHz repetitive waveform is to be sampled and recreated in pulse form, as illustrated in Figure 10-8. If the displayed frequency is to be 100 kHz, estimate the sampling frequency and the number of samples in each displayed cycle.

10-4 The staircase generator circuit in Figure 10-11 uses a 0.01 µF capacitor for C_1. The staircase voltage waveform is to increase from −5 V to +5 V in 40 steps during a 100 µs time period. Determine the required charging current supplied by C_1 if the charging time is to be 50% of the total step time.

10-5 The voltage comparator circuit in Figure 10-12 has a first stage (Q_1 and Q_2) voltage gain of 50 and an output stage gain of 5000. If the supply voltage is ±12 V, determine the minimum input voltage difference at the bases of Q_1 and Q_2 for the output to be saturated at +12 V or − 12 V.

10-6 The sampling gate in Figure 10-13 has a 500 µV drop across FET Q_1 when it is switched *on*. Determine the minimum signal sample amplitude if the maximum error due to the FET voltage drop is not to exceed 5%.

10-7 The 16-step staircase waveform in Figure 10-14(a) goes from −5 V to +5 V in a time of 1 ms. The square wave shown has a time period of 1 ms, and rise and fall times that are approximately 10% of the time period. Calculate the necessary bias voltage in Figure 10-14(b) for the fall time to be displayed as illustrated.

10-8 A digital storage oscilloscope (DSO) has a sampling rate of 100 MS/s. Determine the number of samples taken during one cycle of a 3 MHz sine wave, and during a 15 µs pulse. Also, estimate the maximum time period of a glitch that might be missed by the sampling process.

10-9 A DSO with a 100 MS/s sampling rate is used to investigate the waveform illustrated in Figure 10-20(a). The waveform frequency is 1 MHz, the rise time is 100 ns, the fall time is 250 ns, and the spike at the top of the leading edge has a time period of 150 ns. Estimate the total samples per cycle, and the samples taken during the rise time, the fall time, and the spike time.

Signal Generators

11

Objectives

You will be able to:

1. Sketch a circuit for a low-frequency sine-wave generator, and explain its operation. Show how the oscillator amplitude is stabilized, and how the output frequency is varied.

2. Sketch the basic circuit for a square/triangular waveform function generator. Explain how it operates, and how the output frequency is adjusted.

3. Show how a sine wave can be converted into a square waveform, and how a triangular waveform may be converted into a sine wave.

4. Draw a block diagram for a pulse generator, sketch the system waveforms, and explain how the generator functions.

5. Sketch basic square-wave generator and monostable multivibrator circuits for use in a pulse generator. Explain how they operate.

6. Draw a circuit diagram of an attenuator with an output range control and a dc level shift, for use with pulse and function generators. Explain how the circuit operates.

7. Sketch a block diagram for an RF signal generator. Explain its operation, and show how the system is screened and decoupled.

8. Sketch appropriate oscillator circuits for use with an RF signal generator, and show how amplitude and frequency modulation of the RF wave may be produced.

9. Using diagrams, show how an RF signal generator should be correctly loaded. Discuss the consequences of incorrect loading.

10. Draw a complete block diagram for a sweep frequency generator, show the system waveforms, and explain the generator operation.

11. Sketch the block diagram of a frequency synthesizer, show the system waveforms, and explain how the synthesizer operates.

314

Introduction

The signal generators usually found in electronics laboratories may be classified as *low-frequency (LF) sine-wave generators, radio-frequency (RF) sine-wave generators, function generators, pulse generators, and sweep frequency generators.*

LF signal generators usually have a maximum output frequency of 100 kHz and an output voltage adjustable from 0 to 10 V. Function generators are also usually LF instruments which provide three types of output waveforms: sine, square, and triangular.

The circuit techniques employed for RF signal generation are substantially different from those used in LF instruments. RF screening is necessary. Also, RF generators are normally equipped with an output level meter and a calibrated attenuator. The *frequency synthesizer* is another high-frequency instrument, but in this case the output frequency is stabilized by a piezoelectric crystal. A phase-locked-loop technique is employed to facilitate adjustment of the frequency.

Pulse generators produce pulse waveforms, and controls are provided for adjustment of pulse amplitude, pulse repetition frequency, and pulse width. Some pulse generators have facilities for adjustment of rise time, fall time, delay time, and dc bias level.

The output of a sweep frequency generator is a sine wave that increases gradually from a minimum frequency to a maximum frequency over a selected time period. A ramp voltage with an amplitude proportional to the instantaneous frequency is also generated. Investigation of circuit frequency response is a major application of this instrument.

11-1 LOW-FREQUENCY SIGNAL GENERATORS

Wein Bridge Oscillator

There are several types of sine-wave oscillator circuits that can be used for signal generation. The *Wein bridge oscillator* is one circuit that gives an output with good frequency and amplitude stability, and a low distortion waveform. The Wein bridge is an ac bridge in which balance is obtained only at a particular supply frequency dependent on the values of the bridge components. When used in an oscillator, the Wein bridge forms a feedback network between the output and input terminals of an amplifier.

In the Wein bridge oscillator circuit in Figure 11-1, the bridge components are R_1, R_2, R_3, R_4, C_1, and C_2. The operational amplifier together with R_3 and R_4 forms a noninverting amplifier (see Section 4-2), and R_1, R_2, C_1, and C_2 constitute the feedback network. Analysis of the bridge shows that balance is obtained when

$$\frac{R_3}{R_4} = \frac{R_1}{R_2} + \frac{C_2}{C_1} \tag{11-1}$$

and

$$f = \frac{1}{2\pi\sqrt{R_1 C_1 R_2 C_2}} \tag{11-2}$$

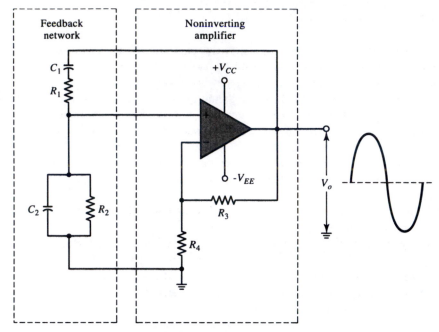

(a)   A Wein bridge oscillator circuit consists of a
      noninverting amplifier and a feedback network

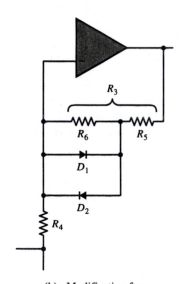

(b)   Modification for
      output amplitude
      stabilization

Figure 11-1   The Wein bridge oscillator circuit produces a sine-wave output with a
frequency of $f = 1/(2\pi CR)$, where $R_1 = R_2 = R$ and $C_1 = C_2 = C$. The diode circuit
stabilizes the output amplitude by reducing the gain of the amplifier at high output
amplitudes.

If $R_1 = R_2 = R$, and $C_1 = C_2 = C$, then from Equation 11-1,

$$R_3 = 2 R_4 \qquad (11\text{-}3)$$

and from Equation 11-2,

$$f = \frac{1}{2\pi CR} \qquad (11\text{-}4)$$

At the balance frequency of the bridge, the amplifier input voltage (developed across R_2 and C_2) is in phase with the output voltage. At all other frequencies the bridge is off balance, and the fed-back voltage does not have the correct phase relationship to the output to sustain oscillations.

The voltage gain of the noninverting amplifier is

$$A_v = \frac{R_3 + R_4}{R_4}$$

So, with $R_3 = 2R_4$ (from Equation 11-3), $A_v = 3$. In fact, the amplifier gain must be slightly greater than this to sustain oscillations. However, the amplitude of the output tends to approach supply voltages $+V_{CC}$ and $-V_{EE}$, and these voltage limits can introduce distortion. To avoid the problem, R_3 is split into two components, R_5 and R_6, and diodes D_1 and D_2 are connected in parallel with R_6, as illustrated in Figure 11-1(b). When the output amplitude is small, the voltage drop across R_6 is not large enough to forward bias the diodes. In this case,

$$A_v = \frac{R_4 + R_5 + R_6}{R_4}$$

When the output amplitude is large enough to forward bias the diodes, R_6 is short-circuited, and the gain is reduced to

$$A_v = \frac{R_4 + R_5}{R_4}$$

If $(R_4 + R_5)/R_4$ is arranged to be less than 3, large output oscillations will not be sustained, but small-amplitude oscillations continue. The arrangement in Figure 11-1(b) is only one of several methods that can be employed to stabilize the output amplitude of a Wein bridge oscillator.

Example 11-1

A Wein bridge oscillator, as in Figure 11-1, has the following components: R_1 and R_2 variable from 500 Ω to 5 kΩ and $C_1 = C_2 = 300$ nF. Calculate the maximum and minimum output frequencies.

Solution

Equation 11-4,
$$f = \frac{1}{2\pi CR}$$

$$f_{(min)} = \frac{1}{2\pi \times 300 \text{ nF} \times 5 \text{ k}\Omega}$$

$$= 106 \text{ Hz}$$

$$f_{(max)} = \frac{1}{2\pi \times 300 \text{ nF} \times 500 \text{ }\Omega}$$

$$= 1.06 \text{ kHz}$$

Frequency Range Changing

Referring to Equation 11-4, it is seen that the frequency of oscillations can be altered by adjustment of either R or C. Going back to Equation 11-2, R_1 and R_2 must be adjusted simultaneously to alter the value of R. Similarly, C_1 and C_2 must be adjusted simultaneously in order to alter C. The most convenient way to accommodate these requirements is to change C_1 and C_2 by switching to various standard values, and use continuously variable resistors for R_1 and R_2. Switching the capacitor values provides for frequency range changing, and variation of R_1 and R_2 facilitates continuous frequency adjustment over each range. Figure 11-2 illustrates how C and R should be adjusted.

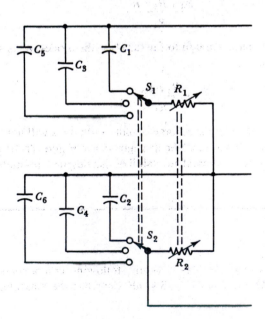

Figure 11-2 The frequency range of a Wein bridge oscillator can be changed by switching capacitor values and maintaining $C_1 = C_2 = C$. Frequency adjustment within each range is made by adjusting R_1 and R_2 simultaneously while maintaining $R_1 = R_2 = R$.

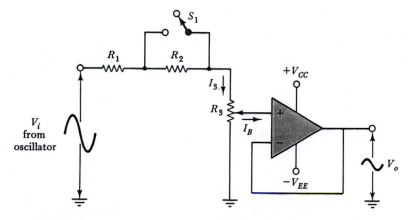

Figure 11-3 The output amplitude of an oscillator may be adjusted by a simple potential divider and potentiometer attenuator circuit. Shorting R_2 by switch S_1 changes the range of output voltage. The op-amp voltage-follower circuit provides low output impedance.

Output Controls

A sine-wave generator for laboratory use must have its output amplitude adjustable, as well as its output frequency. In Figure 11-3 R_1, R_2, and R_3 form a potential divider that attenuates the oscillator output. The operational amplifier is connected as a voltage follower (see Section 4-2) to provide a low output impedance from the signal generator. R_3 is a potentiometer for adjustment of the output amplitude, and switch S_1 allows the output to be switched between two amplitude ranges.

Example 11-2

A 5 V sine wave is fed from a Wein bridge oscillator to the attenuator circuit illustrated in Figure 11-3. Calculate the values of R_1, R_2, and R_3 to give output voltage ranges of 0–0.1 V and 0–1 V. The input bias current to the operational amplifier is $I_B = 500$ nA.

Solution

With R_1 and R_2 in the circuit:

$$V_{R3} = 0.1 \text{ V}$$

and
$$V_{R1} + V_{R2} = V_i - V_{R3}$$

$$= 5 \text{ V} - 0.1 \text{ V}$$

$$= 4.9 \text{ V}$$

$$I_3 \gg I_B$$

Select
$$I_3 = 100 \ \mu A$$

Then,
$$R_3 = \frac{0.1 \ V}{100 \ \mu A} = 1 \ k\Omega \text{ (potentiometer)}$$

and
$$R_1 + R_2 = \frac{4.9 \ V}{100 \ \mu A} = 49 \ k\Omega$$

With R_2 switched out of the circuit:

$$V_{R3} = 1 \ V$$

and
$$I_3 = \frac{V_{R3}}{R_3} = \frac{1 \ V}{1 \ k\Omega}$$

$$= 1 \ mA$$

$$V_{R1} = 5 \ V - 1 \ V$$

$$= 4 \ V$$

$$R_1 = \frac{4 \ V}{1 \ mA} = 4 \ k\Omega$$

$$R_2 = 49 \ k\Omega - 4 \ k\Omega$$

$$= 45 \ k\Omega$$

Square-Wave Conversion

Figure 11-4 shows one method of converting the sine-wave output of an oscillator into a square wave. The operational amplifier is connected to function as a noninverting amplifier (see Section 4-2). The amplifier has a very high gain, so that the amplified output tends to be very large (see the waveforms in Figure 11-4). Diodes D_1 and D_2 together with zener diodes D_3 and D_4 and the associated resistors form a *clipping circuit*. This has the effect of clipping off the positive and negative half-cycles of the amplifier output at a certain voltage level. As illustrated, the amplified and clipped sine wave becomes (approximately) a square wave.

Suppose that in Figure 11-4 $V_{Z3} = V_{Z4} = 6.3 \ V$. Also assume that D_1 and D_2 are silicon diodes with a forward voltage drop of 0.7 V. When the amplifier output goes positive, D_1 becomes forward biased and prevents the output from exceeding $(V_{Z3} + V_{D1})$. Similarly, when the output goes negative, D_2 is forward biased and the output cannot fall below $-(V_{Z4} + V_{D2})$. Thus, the square-wave output amplitude is

$$V_o = \pm(V_Z + V_D)$$

$$= \pm(6.3 \ V + 0.7 \ V)$$

$$= \pm 7 \ V$$

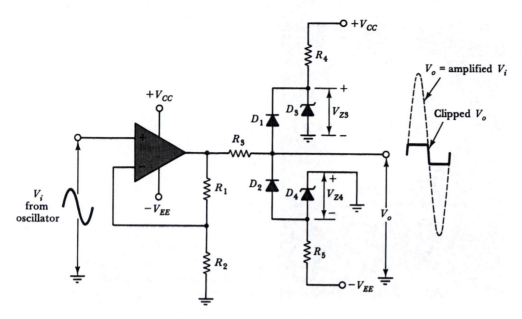

Figure 11-4 A square wave can be produced by amplifying and clipping a sinusoidal waveform. Amplification is produced by the noninverting amplifier (the op-amp together with R_1 and R_2), and the wave is clipped at $+(V_{Z3} + V_{D1})$ and $-(V_{Z4} + V_{D2})$.

Block Diagram

A low-frequency signal generator normally consists of a sinusoidal oscillator, a sine-to-square wave converter, and an attenuator output stage (see Figure 11-5). When required, the square-wave shaping circuit is switched into the system between the oscillator output and the attenuator input.

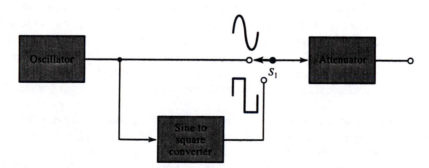

Figure 11-5 Block diagram of a sine/square-wave generator. The sinusoidal or square-wave output is selected at the attenuator input.

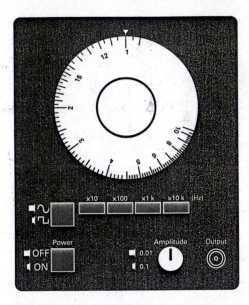

Figure 11-6 Low-frequency sine/square-wave generator.

Representative Signal Generator

The front panel of a signal generator that performs the functions discussed above is shown in Figure 11-6. Square- or sine-wave output is selected by depressing or releasing the left-hand pushbutton. Four frequency ranges are available from a minimum of 10 Hz–100 Hz to a maximum of 10 kHz–100 kHz. Continuous frequency adjustment on each range is provided by the large control knob and scale. The output amplitude is also continuously adjustable on two ranges: 0–0.1 V and 0–10 V.

Application

A typical audio signal generator application is illustrated in Figure 11-7. A low-frequency signal generator is used as a signal source for an audio amplifier. A dual-trace oscilloscope is connected to monitor the input and output waveforms of the amplifier under test, and a frequency meter is included to give an accurate indication of signal frequency. Normally, the amplifier is tested for frequency response and phase shift. The phase shift between input and output is easily measured by comparing the two waveforms on the oscilloscope (see Section 9-7). Unless the signal generator is equipped with an output level meter which can be observed, the signal voltage cannot be assumed to remain constant. Before each measurement is made of amplifier output amplitude, the input level should be checked on the oscilloscope, and the signal generator amplitude control adjusted as necessary.

The signal frequency is first set approximately at the middle of the amplifier frequency response. From this point, the frequency is decreased in convenient steps, and the amplifier gain and phase shift are recorded at each step. The procedure is continued until the amplifier lower 3 dB frequency is found. The generator frequency is then increased in steps from the midfrequency until the upper 3 dB frequency of the amplifier is found. At each step the amplifier gain and phase shift are again noted. Using the recorded data, the gain/frequency and phase/frequency response graphs are plotted for the amplifier. The

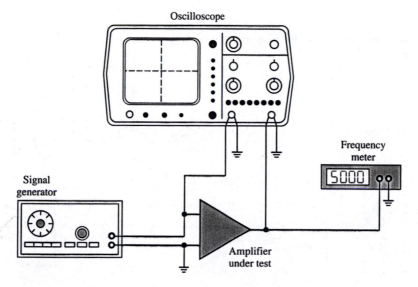

Figure 11-7 Signal generator used to test an amplifier. An oscilloscope monitors the input and output waveforms of the amplifier, and a frequency meter displays the waveform frequency.

process described above can be performed very much faster by using a sweep frequency generator (see Section 11-5).

11-2 FUNCTION GENERATORS

Basic Circuit

A *function generator* produces sine, square, and triangular waveform outputs. Sometimes a ramp waveform is also generated. Output frequency and amplitude are variable, and a dc offset adjustment may be included. The sine/square-wave generator described in Section 11-1 could be used as a function generator if a circuit for converting the square wave into a triangular wave is employed. However, the usual method of generating a triangular wave is to use an *integrator* and a *Schmitt trigger circuit.* The arrangement is shown in Figure 11-8.

The Schmitt trigger circuit is a noninverting type. (Schmitt trigger circuits are also discussed in Section 9-4.) When the input voltage increases to the *upper trigger point* (UTP), the output suddenly rises from its most negative level to its most positive level. Similarly, when the input goes to the *lower trigger point* (LTP), the op-amp output voltage rapidly drops to its most negative level. Note that the inverting input terminal of the Schmitt op-amp is grounded. Also recall that the operational amplifier has a voltage gain of about 200 000. Only a very small voltage difference is required between inverting and noninverting terminals to drive the op-amp output to saturation in either a positive or negative direction. If V_{CC} and V_{EE} are ±15 V, the output is typically ±14 V. The minimum voltage difference between the inverting and noninverting input terminals to produce output saturation is

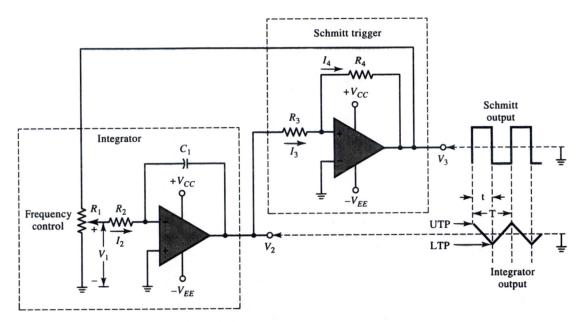

Figure 11-8 A basic function generator circuit consists of an integrator and a Schmitt trigger circuit. The integrator output is a negative-going ramp voltage when the Schmitt output is positive, and vice versa. The Schmitt output changes state when the integrator output ramp reaches the Schmitt upper or lower trigger point.

$$V_i = \frac{14 \text{ V}}{200\ 000} = 70\ \mu\text{V}$$

Refer to the integrator section of Figure 11-8. In this circuit the op-amp noninverting terminal is grounded, and the inverting terminal is connected via capacitor C_1 to the amplifier output. If C_1 were replaced by a short circuit, the op-amp would behave as a voltage follower. Both the output and inverting input terminals would be at ground level because the noninverting input is grounded. With C_1 in circuit and with zero charge on C_1, the circuit again behaves as a voltage follower. All three terminals of the op-amp are at ground level. Now suppose that C_1 becomes charged with a terminal voltage of 1 V, + on the right side and − on the left side. The inverting input terminal still remains at ground level, and the output is +1 V with respect to ground. The circuit voltages remain stable at these levels while C_1 terminal voltage remains constant. Similarly, if C_1 is charged to 1 V with the opposite polarity (+ on the left, − on the right), the inverting terminal remains at ground level and the output voltage becomes −1 V with respect to ground.

Now assume that a positive input voltage ($+V_1$) is applied to R_2 as shown in Figure 11-8. The left-hand terminal of R_2 is at $+V_1$, while the right-hand terminal is at ground level (because the noninverting input remains at ground). Therefore, all of V_1 appears across R_2, and a *constant current* I_2 flows through R_2:

$$I_2 = \frac{V_1}{R_2}$$

I_2 is selected very much larger than the input bias current to the operational amplifier. Consequently, virtually all of I_2 flows into C_1, charging it with a polarity: + on the left, − on the right. As C_1 charges, its voltage increases linearly, and because its left-hand (+) terminal is at ground level, the op-amp output voltage decreases linearly. When the polarity of V_1 is inverted, I_2 is reversed, and C_1 commences to charge with the opposite polarity. This causes the integrator output voltage to reverse direction.

Returning again to Figure 11-8, it is seen that the integrator input voltage is derived from the Schmitt trigger output. Also, the integrator output is applied as an input to the Schmitt circuit. To understand the combined operation of the two circuits, assume that the Schmitt output (V_3) is +14 V and that the integrator output is at ground level. V_1 is positive because the Schmitt output is +14 V, and consequently, I_2 is charging C_1: + on the left, − on the right. Thus, the integrator output voltage (V_2) is decreasing linearly from ground level. When V_2 arrives at the LTP of the Schmitt, the output voltage of the Schmitt switches rapidly to $V_3 = -14$ V. This causes V_1 to reverse polarity and results in I_2 reversing direction. Now C_1 commences to charge in the opposite direction, and V_2 increases linearly from the LTP (see the waveforms in Figure 11-8). C_1 continues to charge in this direction until the integrator output becomes equal to the Schmitt UTP. When V_2 arrives at the UTP, the Schmitt output immediately reverses polarity once again to $V_3 = +14$ V. V_1 is now positive once more, and I_2 charges C_1 with a polarity that makes V_2 go in a negative direction once again.

The process described above is repetitive, and the integrator output produces a triangular waveform, as illustrated. The Schmitt output is a square wave, which is positive while the integrator output is negative going, and negative during the time that V_2 is positive going.

The frequency of the output waveforms is determined by the time for C_1 to charge from the UTP to the LTP, and vice versa. The equation for a capacitor charging linearly is

$$C = \frac{It}{\Delta V}$$

or

$$t = \frac{C \, \Delta V}{I} \qquad (11\text{-}5)$$

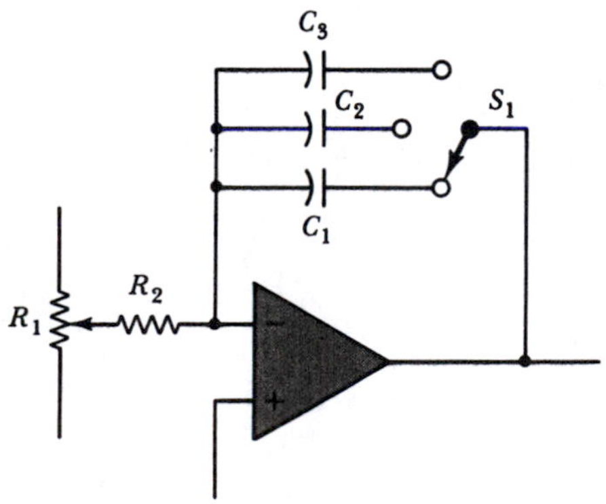

Figure 11-9 The output frequency of the basic function generator in Figure 11-8 is adjusted continuously by means of potentiometer R_1, and the frequency range is changed by capacitor selection.

In this case, C is C_1, $\Delta V = \text{UTP} - \text{LTP}$, I is I_2, and t is the time for V_2 to go between UTP and LTP. Time t is also half the time period of the output waveform. V_1 is adjustable by means of potentiometer R_1, and because $I_2 = V_1/R_2$, I_2 is also controllable by R_1. Adjustment of I_2 results in a change in t; consequently, R_1 is a frequency control. Frequency range changing is produced by switching different capacitor values into the circuit, as illustrated in Figure 11-9.

Example 11-3

The integrator circuit in Figure 11-8 has $C_1 = 0.1\ \mu\text{F}$, $R_1 = 1\ \text{k}\Omega$, and $R_2 = 10\ \text{k}\Omega$. If the Schmitt trigger circuit has ± 3 V trigger points, calculate the output frequency when the moving contact of R_1 is at the top of the potentiometer, and when it is at 10% of R_1 from the bottom. The supply voltage is $V_{CC} = \pm 15$ V.

Solution

$$V_3 \simeq \pm(V_{CC} - 1\ \text{V}) = \pm(15\ \text{V} - 1\ \text{V})$$

$$= \pm 14\ \text{V}$$

For contact at top of R_1:

$$V_1 = V_3 = 14\ \text{V}$$

$$I_2 = \frac{V_1}{R_2} = \frac{14\ \text{V}}{10\ \text{k}\Omega}$$

$$= 1.4\ \text{mA}$$

$$\Delta V = \text{UTP} - \text{LTP} = 3\ \text{V} - (-3\ \text{V})$$

$$= 6\ \text{V}$$

Equation 11-5,
$$t = \frac{C\,\Delta V}{I_2} = \frac{0.1\ \mu\text{F} \times 6\ \text{V}}{1.4\ \text{mA}}$$

$$\simeq 0.43\ \text{ms}$$

$$f = \frac{1}{2t} = \frac{1}{2 \times 0.43\ \text{ms}}$$

$$= 1.17\ \text{kHz}$$

For R_1 contact at 10% from bottom:

$$V_1 = 10\% \text{ of } V_3 = 10\% \text{ of } 14\ \text{V}$$

$$= 1.4\ \text{V}$$

$$I_2 = \frac{1.4\ \text{V}}{10\ \text{k}\Omega} = 0.14\ \text{mA}$$

$$t = \frac{0.1 \, \mu F \times 6 \, V}{0.14 \, mA}$$

$$\simeq 4.3 \, ms$$

$$f = \frac{1}{2 \times 4.3 \, ms}$$

$$\simeq 117 \, Hz$$

Sine-Wave Conversion

A widely used method for converting a triangular wave into an approximate sinusoidal waveform is illustrated in Figure 11-10. If diodes D_1 and D_2 and resistors R_3 and R_4 were not present in the circuit of Figure 11-10(a), R_1 and R_2 would simply behave as a voltage divider. In this case, the output from the circuit would be an attenuated version of the triangular input wave:

$$V_o = V_i \frac{R_2}{R_1 + R_2}$$

With D_1 and R_3 in the circuit, R_1 and R_2 still behave as a simple voltage divider until V_{R2} exceeds $+V_1$. At this point D_1 becomes forward biased, and R_3 is effectively in parallel with R_2.

Now
$$V_o \simeq V_i \frac{R_2 \| R_3}{R_1 + R_2 \| R_3}$$

Output voltage levels above $+V_1$ are attenuated to a greater extent than levels below $+V_1$. Consequently, the output voltage rises less steeply than without D_1 and R_3 in the circuit [see Figure 11-10(a)]. When the output falls below $+V_1$, diode D_1 is reverse biased, R_3 is no longer in parallel with R_2, and the attenuation is once again $R_2/(R_1 + R_2)$. Similarly, during the negative half-cycle of the input, the output is $V_o = V_i[R_2/(R_1 + R_2)]$ until V_o goes below $-V_1$. Then, D_2 becomes forward biased, putting R_4 in parallel with R_2 and making

$$V_o \simeq V_i \frac{R_2 \| R_4}{R_1 + R_2 \| R_4}$$

With $R_3 = R_4$, the negative half-cycle of the output is similar in shape to the positive half-cycle.

When six or more diodes are employed, all connected via resistors to different bias voltage levels [see Figure 11-10(b)], a good sine-wave approximation can be achieved. With six diodes, three positive bias voltage levels, and three negative bias levels, the slope of the output wave changes three times during each quarter cycle. Assuming correctly selected bias voltages and resistor values, the output wave shape is as shown in Figure 11-10(b).

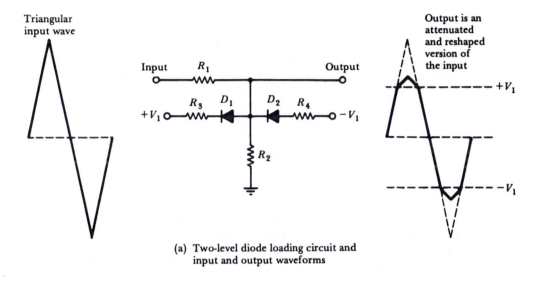

(a) Two-level diode loading circuit and
input and output waveforms

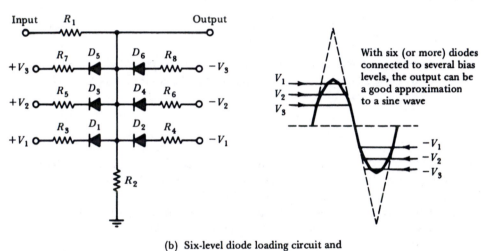

(b) Six-level diode loading circuit and
its effect on input waveform

Figure 11-10 A triangular waveform can be shaped into a good approximation of a sine wave by diode/resistor loading. In (a), low-amplitude inputs are potentially divided across resistors R_1 and R_2. At higher amplitudes, diodes D_1 and D_2 become forward biased, causing either R_3 or R_4 to parallel R_2 and produce further attenuation.

Function Generator Block Diagram

The block diagram of a basic function generator is shown in Figure 11-11. The integrator output is fed into the Schmitt trigger and the sine wave converter. As explained earlier, the integrator must have the square wave from the Schmitt as an input. A switch is provided for selection of sine, triangular, or square waves. The output stage can be the type of attenuator shown in Figure 11-3. This circuit provides low output impedance and out-

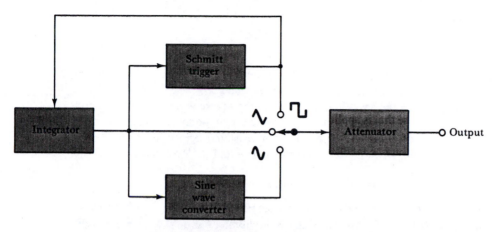

Figure 11-11 Basic function generator block diagram.

put amplitude control. A means of synchronizing the output frequency to an external source is sometimes included in a function generator. In Figure 11-8 a sync input can be arranged by connecting a resistor between ground and the Schmitt op-amp inverting input terminal. The sync input pulses are then capacitor coupled to the inverting input terminal. The method is explained in Section 9-4.

Commercially available function generators typically have the output frequency in decade ranges from a minimum of 0.2 Hz to a maximum of 2 MHz. Output amplitude is usually 0–20 V p-to-p and 0–2 V p-to-p. The amplitude control is often marked as 0–20 V, with a 20 dB attenuation pushbutton which changes the output to 0–2 V. Output impedances of function generators are typically 50 Ω. It should be remembered that the signal voltage is potentially divided across the output impedance and the load impedance. Thus, a 20 V output on open-circuit may become 10 V when a 50 Ω load impedance is connected. Accuracy of frequency selection is usually around ±2% of full scale on any particular range, and distortion content in sinusoidal outputs is normally less than 1%. Figure 11-12 shows the controls of a typical function generator.

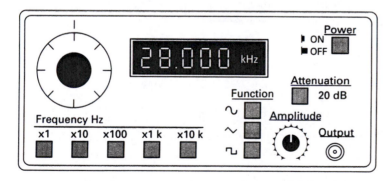

Figure 11-12 Typical function generator. Frequency range and function are selected by pushbutton switches, and the frequency is controlled continuously by the rotatable knob. The frequency is displayed digitally.

11-3 PULSE GENERATORS

Block Diagram

A basic pulse generator is made up of a square-wave generator, a monostable multivibrator, and an attenuator output stage. In the block diagram shown in Figure 11-13, the monostable multivibrator is triggered by the negative-going edge of the square wave, to produce a constant-width pulse that is applied to the output stage. Variation of the square-wave frequency varies the pulse frequency, and adjustment of the monostable adjusts the output pulse width. The attenuator facilitates output amplitude control and dc level shifting.

Astable Multivibrator as Square-Wave Generator

A very simple square-wave generator circuit is shown in Figure 11-14. The circuit is an operational amplifier *astable multivibrator*. An astable multivibrator has no stable state; it oscillates continuously between the conditions of *output high* and *output low*. Capacitor C_1 is charged from the op-amp output via resistor R_1. The op-amp, together with resistors R_2 and R_3, constitutes an inverting-type Schmitt trigger circuit. When the capacitor voltage reaches the upper trigger point (UTP) of the Schmitt, the op-amp output switches to *low*, as illustrated by the waveforms. With the op-amp output *low*, C_1 charging current is reversed, so that the

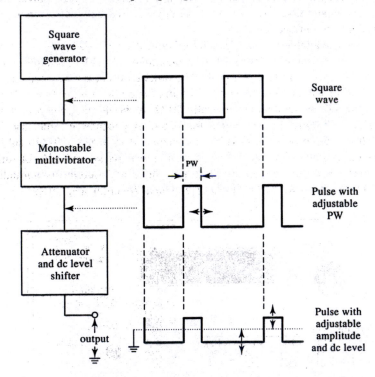

Figure 11-13 Basic pulse generator block diagram. A square-wave generator triggers the monostable multivibrator to produce the pulse waveform. The pulse frequency is determined by the frequency of the square wave, the pulse width can be set by adjusting the monostable, and the pulse amplitude is controlled by the output attenuator.

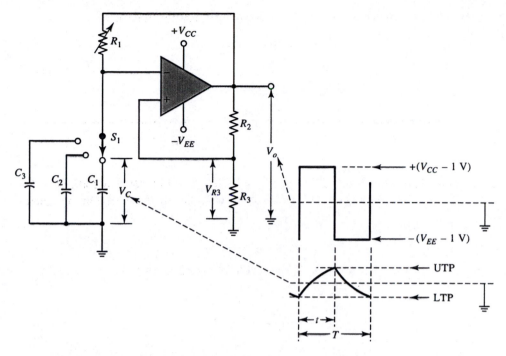

Figure 11-14 An op-amp astable multivibrator is the simplest form of square-wave generator. The op-amp together with resistors R_2 and R_3 constitute a Schmitt trigger circuit. C_1 is charged via R_1 from the Schmitt output.

capacitor discharges, and then recharges with negative polarity. When C_1 voltage arrives at the Schmitt LTP, the op-amp output switches back to *high,* and the cycle begins again.

The process repeats continuously as described, producing a square-wave output from the Schmitt circuit, as well as generating an exponential wave across C_1. The frequency of the square wave depends on the time (t) for the capacitor to charge between LTP and UTP:

$$f = \frac{1}{T} = \frac{1}{2t}$$

Time t can be adjusted by variable resistor R_1. So R_1 is a continuous frequency control. Frequency range is changed by selecting various capacitor values by means of switch S_1, as illustrated. The time t is calculated from the basic equation for a capacitor charged via a resistor:

$$e_c = E - (E - E_o)\varepsilon^{-t/CR}$$

which can be rearranged

$$t = CR \ln\left(\frac{E - \text{LTP}}{E - \text{UTP}}\right)$$

(11-6)

Sec. 11-3   Pulse Generators

331

The Schmitt upper and lower trigger voltage levels are determined from

$$|UTP| = |LTP| = V_o \frac{R_3}{R_2 + R_3} \qquad (11\text{-}7)$$

Example 11-4

The astable multivibrator circuit in Figure 11-14 has $R_1 = 20$ kΩ, $R_2 = 6.2$ kΩ, $R_3 = 5.6$ kΩ, and $C_1 = 0.2$ μF. The supply voltage is ±12 V. Calculate the frequency of the square-wave output.

Solution

$$V_o \simeq \pm(V_{CC} - 1\ \text{V}) = \pm(12\ \text{V} - 1\ \text{V})$$

$$= \pm11\ \text{V}$$

Equation 11-7,
$$|UTP| = |LTP| = V_o \frac{R_3}{R_2 + R_3}$$

$$= \pm11\ \text{V} \times \frac{5.6\ \text{k}\Omega}{6.2\ \text{k}\Omega + 5.6\ \text{k}\Omega}$$

$$= \pm5.22\ \text{V}$$

Equation 11-6,
$$t = C_1 R_1 \ln\left(\frac{E - \text{LTP}}{E - \text{UTP}}\right)$$

$$= 0.2\ \mu\text{F} \times 20\ \text{k}\Omega \times \ln\left[\frac{11\ \text{V} - (-5.22\ \text{V})}{11\ \text{V} - 5.22\ \text{V}}\right]$$

$$= 4.13\ \text{ms}$$

$$f = \frac{1}{2 \times 4.13\ \text{ms}}$$

$$= 121\ \text{Hz}$$

Monostable Multivibrator

A *monostable multivibrator* has one stable state. When a triggering input is applied, the output changes state for a fixed period of time, and then reverts back to its initial condition. This produces an output pulse with a constant width each time the monostable multivibrator is triggered.

An operational amplifier connected to function as a monostable multivibrator is shown in Figure 11-15. The inverting input terminal has a positive bias voltage (V_B), where V_B is typically 1 V. The op-amp noninverting terminal is grounded via resistor R_2. Thus, the dc condition of the circuit is (inverting input) $= +V_B$ and (noninverting input) $=$ 0 V. This causes the op-amp output to be saturated in a negative direction, $V_{o-} \approx -(V_{EE} -$ 1 V). Capacitor C_2 is grounded via R_2 on its left side, while its right terminal is at V_{o-}. Therefore, C_2 is charged (+ on the left, − on the right) to a level of $V_{C2} = V_{o-}$. These dc voltages are maintained until a triggering input signal arrives via capacitor C_1.

When a square-wave input is applied to the monostable circuit, C_1 charges rapidly via R_1 at each positive-going and negative-going edge of the input waveform. The pulses of charging current through R_1 generate spikes at the op-amp inverting input terminal. Positive spikes occur at the positive-going edges of the square wave, and negative-going spikes coincide with the negative-going edges (see Figure 11-15). C_1 and R_1, in fact, behave as a differentiating circuit, and diode D_1 clips off the positive spikes.

Since the positive spikes are clipped off, they have no effect on the monostable circuit. However, the negative-going spikes drive the op-amp inverting input terminal below the

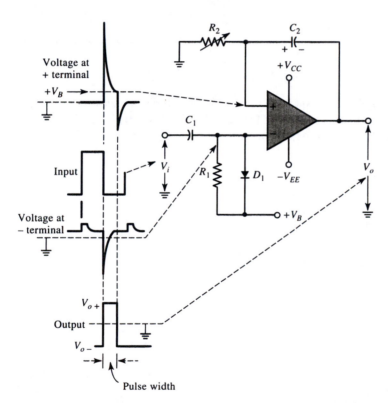

Figure 11-15 Op-amp monostable multivibrator. With $+V_B$ at the op-amp inverting input, the output is normally *low* (at V_{o-}). A negative-going trigger input via C_1 causes the output to switch *high* (to V_{o+}), where it remains until C_2 discharge causes the op-amp noninverting input to fall to $+V_B$.

level of the noninverting input. When this occurs, the op-amp output rapidly switches to its positive saturation level; $V_{o+} \approx V_{CC} - 1$ V. Because of the charge on the capacitor, there is now a high positive voltage at the op-amp noninverting terminal, and this holds the op-amp output at its positive saturation level, even when the negative triggering spike has ended.

Capacitor C_2 now begins to discharge via R_2, and to recharge with opposite polarity. As C_2 charges, the noninverting terminal voltage falls towards ground level. As soon as this voltage (at the noninverting input) falls slightly below the bias $(+V_B)$ at the inverting input, the op-amp output rapidly switches to V_{o-} once again. The time duration for which the output voltage is high depends on the resistance of R_2 and the capacitance of C_2. This time is the width of the positive output pulse which is generated every time the monostable circuit is triggered. Since R_2 is variable, it is a pulse width (PW) control. The range of the PW can be changed by switching different capacitor values into the circuit in place of C_2. The equation for determining the output pulse width can be shown to be

$$PW = CR \ln\left[\frac{(V_{o+}) - (V_{o-})}{V_B}\right]$$ (11-8)

Example 11-5

The monostable multivibrator circuit in Figure 11-15 has a supply voltage of ±10 V, and V_B = +1 V. The circuit components are $R_1 = 22$ kΩ, $R_2 = 10$ kΩ, $C_1 = 100$ pF, and $C_2 = 0.01$ µF. Calculate the output PW. Also determine the new capacitance of C_2 to give PW = 6 ms.

Solution

$$V_{o+} = +(V_{CC} - 1 \text{ V}) = +(10 \text{ V} - 1 \text{ V})$$

$$= +9 \text{ V}$$

$$V_{o-} = -(V_{EE} - 1 \text{ V})$$

$$= -9 \text{ V}$$

Equation 11-8, $$PW = C_2 R_2 \ln\left[\frac{(V_{o+}) - (V_{o-})}{V_B}\right]$$

$$= 0.01 \text{ µF} \times 10 \text{ k}\Omega \times \ln\left[\frac{9 \text{ V} - (-9 \text{ V})}{1 \text{ V}}\right]$$

$$= 289 \text{ µs}$$

for PW = 6 ms:
From Equation 11-8,

$$C_2 = \frac{PW}{R_2 \ln\left[\frac{(V_{o+}) - (V_{o-})}{V_B}\right]}$$

$$C_2 = \frac{6\ \text{ms}}{10\ \text{k}\Omega \times \ln\left[\dfrac{9\ \text{V} - (-9\ \text{V})}{9\ \text{V} - 8\ \text{V}}\right]}$$

$$= 0.2\ \mu\text{F}$$

Output Attenuator

In Figure 11-16, resistors R_1, R_2, and R_3 together with operational amplifier A_1 constitute an output attenuator, as shown in Figure 11-3. This allows the output amplitude of the pulse generator to be adjusted and gives a low output impedance. Operational amplifier A_2 and resistors R_4, R_5, and R_6 provide dc level shifting. Capacitor C_1 passes pulses from the pulse generator into the attenuator circuit while allowing the dc output level of the attenuator to be set to any desired level. A_2 is a voltage follower, and its dc output voltage is set by potentiometer R_5. When the moving contact of R_5 is at ground level, A_2 output is also at ground. This gives an output pulse from A_1 that is symmetrical above and below ground level. When the moving contact of R_5 is +5 V, the output pulse is symmetrical above and below the +5 V level, and when the potentiometer voltage is −5 V, the pulse output is symmetrical above and below the −5 V level.

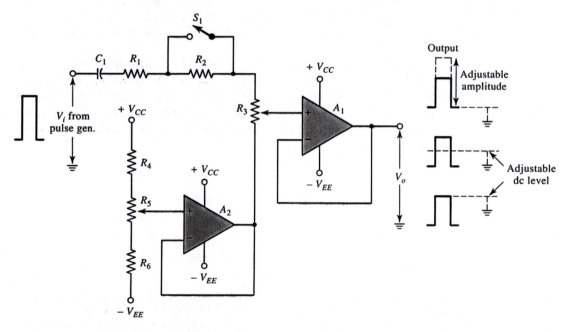

Figure 11-16 Attenuator and dc offset control for use with a pulse generator. R_3 provides continuous output amplitude control, and S_1 changes the attenuation range. R_5 is the dc offset control.

Pulse Shaping

The output pulses from a pulse generator usually have leading and lagging edges which seem to be perfectly vertical when displayed on an oscilloscope. In fact, the leading edges have a finite *rise time,* and the lagging edges have a *fall time* (see Figure 11-17). The rise time (t_r) is defined as the time for the output to go from 10% to 90% of its amplitude, and the fall time (t_f) is the time for the output to go from 90% to 10%. Where t_r and t_f are very much smaller than the PW, the pulse does indeed appear to have perfectly vertical sides. Some pulse generators include facilities for adjusting the rise and fall time of the output pulses. When an external triggering facility is included, an adjustable time delay may also be provided. The *delay time* (t_d in Figure 11-17) is the time between the trigger input and commencement of the output pulse.

Representative Pulse Generator

The pulse/function generator shown in Figure 11-18 can produce sine, square, and triangular waveforms as well as normal or inverted pulse waveforms. The square-wave output is symmetrical about ground level; however, positive and negative square waves are also available, as is a dc voltage level. All of the above are taken from the FUNCTION OUT terminal and selected by the FUNCTION switch. A SYNC OUT terminal is included, and four logic outputs (for ECL—*emitter-coupled logic*—and TTL—*transistor-transistor logic*) are also provided.

The output frequency ranges from 0.0001 Hz to 20 MHz. This is switched through decade steps and is continuously variable at each step. The amplitude of output waveforms is a maximum of 30 V peak-to-peak and by attenuator selection can be reduced in steps to 3 mV. A dc offset up to +15 V can be applied to all outputs from the FUNCTION OUT terminal.

When used as a pulse generator, the output PW can be varied from 25 ns to 1 ms. Here again, the range can be switched in decade steps, and PW is continuously variable for each range. The pulse output can be continuous triggered, manually triggered, or gated by an external triggering source (applied to the TRIG IN terminal). A variable pulse delay can also be introduced, and repetitive double pulses with a variable intervening space can be generated.

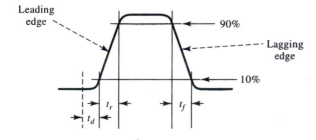

Figure 11-17   Many pulse generators have facilities for adjustment of pulse rise time and fall time, t_r and t_f. A delay time (t_d) from the instant of triggering may also be adjustable.

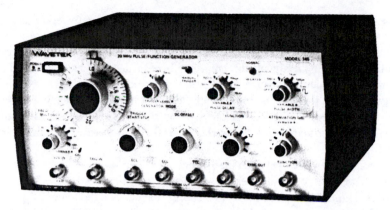

Figure 11-18 Pulse/function generator capable of producing pulse waveforms over a frequency range of 0.00001 Hz to 20 MHz. The pulse width is adjustable from 25 ns to 1 ms, and the output amplitude is variable from 3 mV to 30 V. (Courtesy of WAVETEK.)

11-4 RF SIGNAL GENERATORS

Basic Block Diagram

A *radio-frequency* (RF) signal generator has a sinusoidal output with a frequency range somewhere in the 100 kHz to 40 GHz region. Basically, the instrument consists of an *RF oscillator,* an *amplifier,* a *calibrated attenuator,* and an *output level meter,* as illustrated by the signal generator block diagram in Figure 11-19. The RF oscillator has a continuous frequency control and a frequency range switch, to set the output to any desired frequency. The amplifier includes an output amplitude control. This allows the voltage applied to the attenuator to be set to a *calibration point* on the output level meter. The output level must always be reset to this calibration point every time the frequency is changed. This is necessary to ensure that the output voltage levels are correct, as indicated on the calibrated attenuator.

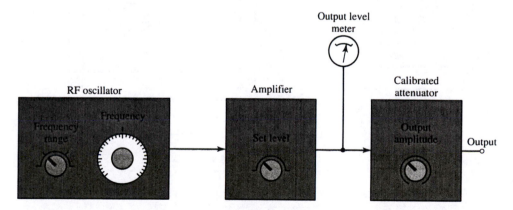

Figure 11-19 Basic block diagram of an RF signal generator. The RF oscillator output is applied to an amplifier that has a *set level* control. This is used in conjunction with the output level meter to ensure a constant input voltage to the attenuator.

The oscillator circuit used in an RF signal generator is usually either a *Hartley oscillator* or a *Colpitts oscillator*. The basic circuits of both types are shown in Figure 11-20. Both circuits consist of an *amplifier* and a phase-shifting *feedback network*. As well as amplifying the input signal, the amplifier inverts it, or phase shifts it through 180°. The amplified output is attenuated and phase shifted through a further 180° by the feedback network. Then it is applied to the amplifier input terminals. The gain of the amplifier equals the reciprocal of the feedback network attenuation. Thus, each oscillator circuit has a *loop gain* of 1 and a *loop phase shift* of 360°, which are the requirements for sustained oscillation.

The essential differences between the Hartley and Colpitts circuits are in the phase shift networks. The Hartley circuit uses two inductors and a capacitor: L_1, L_2, and C in Figure 11-20(a). The Colpitts circuit uses two capacitors and one inductor: C_1, C_2, and L in Figure 11-20(b). Capacitors C_C in both circuits are coupling capacitors. The frequency of oscillations for both circuits is the resonant frequency of the phase shift network:

$$f = \frac{1}{2\pi\sqrt{C_T L_T}}$$

(11-9)

For the Hartley circuit, $C_T = C$, and L_T is the total inductance of L_1 and L_2, including the mutual inductance. For the Colpitts circuit, $L_T = L$, and C_T is the total capacitance of C_1

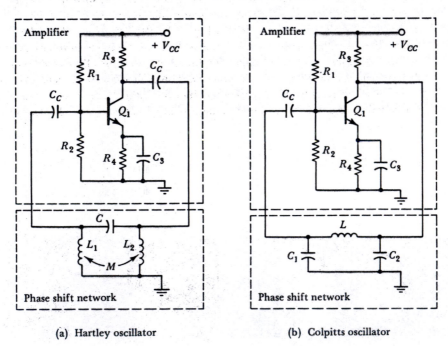

(a) Hartley oscillator   (b) Colpitts oscillator

Figure 11-20   The Hartley and Colpitts oscillators are suitable circuits for use in an RF signal generator. Both circuits consist of an inverting amplifier and a phase-shift feedback network.

and C_2 in series. The oscillating frequency of each circuit can be altered by changing the component values in the phase shift network.

Modulation

Most RF signal generators include facilities for amplitude modulation and frequency modulation of the output. Amplitude modulation is easily accomplished at the amplification stage. Figure 11-21 shows a basic circuit for this purpose. If field-effect transistor Q_2 were not present in the circuit, the amplifier gain would be $A_v = R_3/R_4$. Q_2 is capacitor coupled to R_4 via C_2, so that it has no effect on the dc bias condition in the circuit of Q_1. With Q_2 in the circuit the amplifier gain becomes

$$A_v = \frac{R_3}{R_4 \| R_D} \tag{11-10}$$

where R_D is the drain resistance of the FET. The low-frequency signal applied to the gate of the FET varies the drain resistance of Q_2, and consequently, varies the gain of the amplifier. In this way the amplitude of the RF output is increased and decreased in phase with the low-frequency input (see the waveforms in Figure 11-21).

Frequency modulation is usually performed at the oscillator stage of an RF signal generator. One method of frequency modulating the oscillator output uses a *variable-voltage capacitor diode* (VVC diode). This is a specially constructed semiconductor

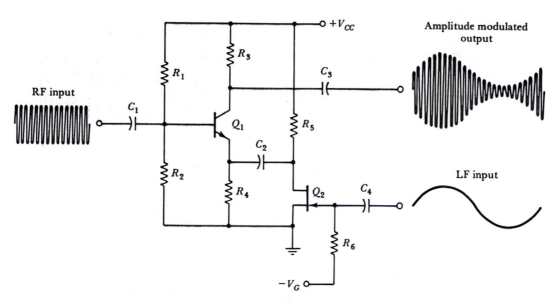

Figure 11-21 Basic circuit for amplitude modulating an RF waveform. The LF input alters the channel resistance (R_D) of FET Q_2, which is in parallel with R_4. Thus, the voltage gain of the circuit $[A_v = R_3/(R_4\|R_D)]$ is changed, and the output amplitude is modulated.

diode operated in reverse bias. Varying the reverse bias on a VVC diode alters its capacitance. In Figure 11-22, Q_1 and its associated components can be employed to vary the voltage across the VVC diode (D_1) when a low-frequency input signal is applied. Capacitor C_3 couples D_1 to the LC tank circuit of the oscillator. The tank circuit capacitance is the diode capacitance C_D in parallel with C_4, and the resonance frequency is

$$f = \frac{1}{2\pi\sqrt{LC_D\|C_4}} \qquad (11\text{-}11)$$

As the capacitance of D_1 is varied, the resonant frequency of the tank circuit varies. Thus, the oscillator output frequency is modulated by the low-frequency input signal.

Complete Block Diagram

A complete block diagram of an RF signal generator is illustrated in Figure 11-23. The basic components from Figure 11-19 (oscillator, amplifier, meter, and attenuator) are reproduced in Figure 11-23, together with FM and AM internal modulating sources. Switches S_1 and S_2 allow selection of *no modulation,* as well as internal or external FM or AM modulation. Note that each section of the system is shielded by enclosing it in a metal box. The whole system is then completely shielded. The purpose of this is to prevent RF interference between the components, and to prevent the emission of RF energy from any point except the output terminals. The power line is also decoupled by means of RF chokes and capacitors (see Figure 11-23) to prevent RF emissions on the power line.

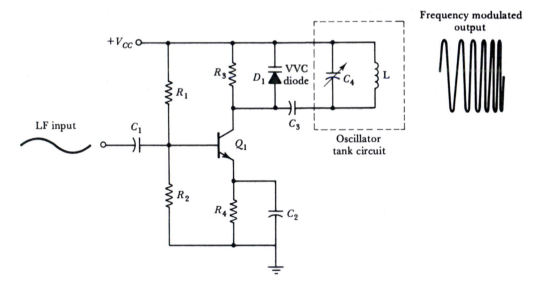

Figure 11-22 Basic circuit for frequency modulation. The VVC diode capacitance (C_D) is in parallel with the oscillator tank circuit capacitor (C_4). The LF input varies C_D by varying the diode reverse bias voltage. Thus, the oscillator output frequency is modulated.

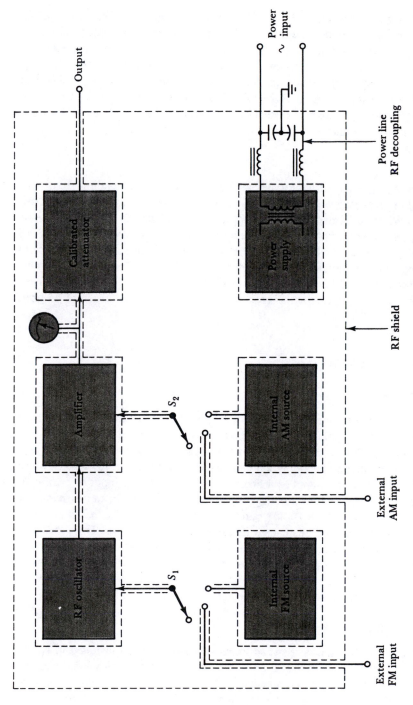

Figure 11-23 Complete block diagram for an RF signal generator. Internal FM and AM sources are included. Each section of the system is shielded, and the entire system is completely shielded.

Application

Although the RF signal generator in Figure 11-23 has an output level meter, the actual output voltage from the calibrated attenuator is the indicated level only when the instrument is correctly loaded. If a 75 Ω load is specified, a 75 Ω load must be connected for the attenuator output to be correct [Figure 11-24(a)]. Where the load is other than the specified load for the signal generator, parallel or series resistors must be included to modify the load as necessary [Figure 11-24(b)]. When series-connected resistors are involved, the signal generator output is further attenuated, and the actual signal level applied to the load must be calculated [Figure 11-24(c)].

Specification

A typical laboratory-type RF generator has an output frequency extending from 0.15 MHz to 50 MHz in eight ranges. The frequency error is less than 1%, and the output impedance is 75 Ω. When connected to a 75 Ω load, the output voltage is 50 mV and can be attenuated by up to 80 dB. Amplitude modulation may be performed to a depth of 30% by an internal 1 kHz source, or by an external source with a frequency of 20 Hz to 20 kHz. Internal frequency modulation may be performed by an internal source which is either the power frequency or 1 kHz. The frequency change effected depends on the selected output frequency range. An external frequency modulating source can also be used with a frequency of 0 to 5 kHz.

11-5 SWEEP FREQUENCY GENERATORS

Simple Block Diagram

The process of testing the frequency response of amplifiers and filters can be simplified and speeded up by using a signal generator that automatically varies its frequency over a predetermined range. Such an instrument is known as a *sweep frequency generator.*

A very much simplified block diagram of a sweep frequency generator is shown in Figure 11-25. A *ramp generator* applies a linear ramp voltage to the input of a *voltage-tuned oscillator.* The basic circuit of a voltage-tuned oscillator is similar to the frequency modulation circuit in Figure 11-22. As the ramp voltage level increases, the reverse bias on the VVC diode increases, and this causes its capacitance to decrease. Thus, the resonance frequency of the tank circuit (which is the oscillator output frequency) increases as the ramp voltage grows. When the ramp voltage returns to its zero level, the diode capacitance and oscillator frequency return to their starting levels. The range over which the oscillator frequency is swept is determined by selection of L and C_4 in Figure 11-22.

More-Detailed Block Diagram

A more complete block diagram of a sweep frequency generator is illustrated in Figure 11-26 on page 345. It should be noted that this is still to some extent a simplified diagram compared to the block diagrams of commercially available sweep frequency generators. The illustration shows that the ramp waveform is amplified and then applied to the voltage-tuned oscillator (VTO). As well as going to the next amplifier stage, the VTO output

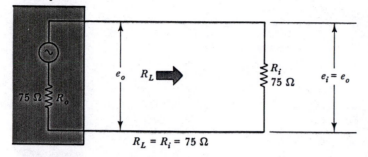

(a) When $R_L = R_o$, the output is
as specified

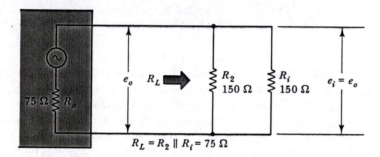

(b) When $R_L > R_o$, a parallel resistor must
be included to make $R_L = R_o$

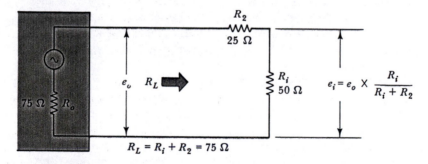

(c) When $R_L < R_o$, a series resistor must
be connected to make $R_L = R_o$

Figure 11-24 For the indicated output level to be correct, an RF generator must be corrected loaded with $R_L = R_o$.

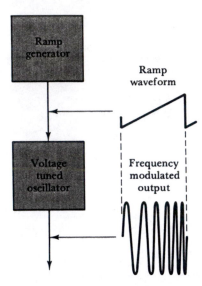

Figure 11-25 In a sweep frequency generator, the output frequency is swept from a low to a high frequency by means of a ramp voltage applied to a voltage tuned oscillator.

is applied to a *discriminator.* The discriminator produces an output voltage which is proportional to its input frequency. Because the discriminator input is the swept frequency from the VTO, its output is a ramp voltage similar to the ramp from the ramp generator. This ramp from the discriminator is applied as an input to the *differential amplifier.* While the VTO output frequency sweeps over the correct range of frequencies, the ramp from the discriminator balances the ramp from the ramp generator. If the VTO output frequency is lower than it should be at any instant, the instantaneous output voltage from the discriminator drops below the instantaneous level of the ramp voltage from the ramp generator. This results in an increased output voltage from the differential amplifier, which causes the VTO output frequency to increase. Similarly, if the VTO output frequency becomes higher than intended, the discriminator output voltage rises above the level of the ramp generator voltage. In this case, the differential amplifier output decreases and produces a lower VTO output frequency. In this way, the output frequency of the VTO is stabilized.

As well as producing an output frequency that sweeps over a desired band of frequencies, the sweep frequency generator must have a stable output voltage level. It is very important that the output voltage remain constant at whatever level it is set, over the entire range of output frequencies. The output voltage level is stabilized by the action of the *automatic level control* (ALC) circuit and the *variable gain amplifier.* The ALC circuit produces a voltage proportional to the power output of the variable gain amplifier. This voltage is compared to an internal reference voltage in the ALC circuit, and the difference between the two is applied to the variable-gain amplifier. Where the power output is lower than required, the amplifier gain is increased. If the power output is too high, the gain is decreased. In this way, the power input to the calibrated attenuator is stabilized. The attenuator offers a constant input resistance, so the constant input power means that the input voltage is constant. Thus, the output voltage from the attenuator remains constant at all frequencies.

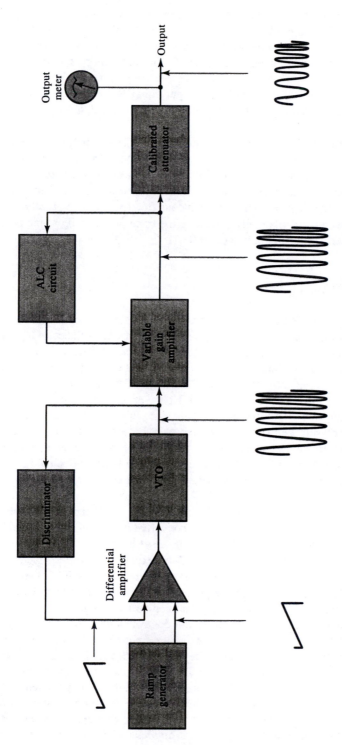

Figure 11-26 Sweep frequency generator block diagram. The output frequency of the voltage-tuned oscillator (VTO) is stabilized by feedback via the discriminator. The amplitude of the waveform applied to the attenuator is stabilized by the interaction of the variable-gain amplifier and the automatic level control (ALC) circuit.

The output amplitude from the sweep frequency generator is adjusted by means of the calibrated attenuator. The range of output frequency may be altered by altering the amplitude of the ramp from the sweep generator. Alternatively, the VTO output frequency might be changed by modifying the parameters of the discriminator. For example, if the discriminator output is potentially divided to half its normal amplitude, the VTO output frequency must double to make the discriminator output equal to the sweep generator ramp voltage. Similarly, if the sweep generator output voltage is doubled, the VTO output frequency must double.

In addition to the controls described above, laboratory-type sweep frequency generators usually have facilities for adjusting the rate of change of the output frequency, and for triggering the sweep from an external source. The sweep can also be arranged to commence at a particular frequency, or to be symmetrical about a selected center frequency. An output voltage directly proportional to the instantaneous frequency is also usually provided. This can be employed for horizontally deflecting the trace on an oscilloscope used for displaying the characteristics of a circuit under test.

11-6 FREQUENCY SYNTHESIZER

The output frequency of the transistor oscillator circuits shown in Figure 11-20 can be made very stable by substituting a piezoelectric crystal in place of one of the coupling capacitors. The crystal offers a high impedance to all frequencies except its own resonance frequency. Thus, the circuit can oscillate only at the resonance frequency of the crystal, which is an extremely stable quantity. The one disadvantage of the crystal oscillator is that its output frequency normally cannot be adjusted. However, the frequency can be altered by using the oscillator as part of a *frequency synthesizer.* In a frequency synthesizer, the output frequency of a crystal oscillator is multiplied by a factor (N) that may be set by a bank of switches. The switches are located on the front panel of the instrument, and are labeled in a way that allows them to be set to indicate the desired output frequency. The system employed to multiply the oscillator frequency is known as a *phase-locked loop* (PLL).

The block diagram of a PLL system as used in a frequency synthesizer is illustrated in Figure 11-27. The output of the crystal oscillator is converted into a square wave and is then fed into one input of a *phase detector.* The other input of the phase detector has another square wave applied to it, as illustrated. As will be explained, the frequencies of these two square waves are identical, and there is a phase difference (ϕ) between them, as illustrated. The output of the phase detector is a pulse waveform with pulse widths controlled by the phase difference. The pulse wave is applied to the *low-pass filter* which converts it into a dc voltage (E). This is used as the control voltage for the *voltage-controlled oscillator (VCO),* and it determines the output frequency of the VCO. The VCO output frequency is directly proportional to the dc voltage level (E) at its input terminal. The VCO produces the instrument output frequency, and this may be passed through another stage to stabilize its amplitude before applying it to the output attenuator. As illustrated, the output of the VCO is fed to a circuit that converts it into a square wave for triggering a *digital frequency divider.* The frequency divider operates in the manner described in Section 5-4, and the ratio (N) by which it divides the VCO frequency is set

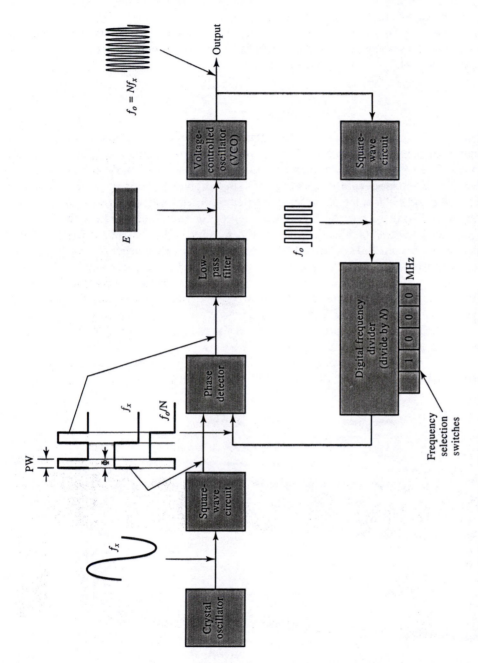

Figure 11-27 The frequency synthesizer uses a phase-locked-loop (PLL) system to multiply the frequency from a crystal oscillator by a selected factor. The multiplying factor is determined by a digital frequency divider.

347

by a bank of switches. These switches may be pushbuttons with digital readouts, or they may be *thumb-wheel* type which indicate their condition numerically. They are connected in such a way that the displayed number is the factor N by which the output frequency is divided before being applied to the phase detector.

Suppose that the crystal oscillator output frequency is $f_x = 1$ MHz, and that the output of the digital frequency divider is also exactly 1 MHz. If the switches are set to divide the VCO output by $N = 1000$, the VCO frequency must be

$$f_o = Nf_x = 1000 \times 1 \text{ MHz}$$

$$= 1000 \text{ MHz}.$$

Thus, the dc input voltage to the VCO must have the appropriate level to produce the 1000 MHz output frequency. Now suppose that the output frequency increases slightly above 1000 MHz. The frequency-divider output is still $f_o/1000$, and this is now greater than $f_x = 1$ MHz. This produces an immediate decrease in the phase difference, and results in a decreased pulse width at the phase detector output. Consequently, the dc voltage output from the filter falls to a lower level. This lower level of E causes the VCO output frequency to drop back toward 1000 MHz. Similarly, if f_o falls below 1000 MHz, the frequency divider output becomes less than 1 MHz, and ϕ and PW are immediately increased, causing an increase in E, which drives f_o back up toward 1000 MHz once again.

It is seen that the VCO output frequency is stabilized at exactly N times the crystal frequency. With $f_x = 1$ MHz and $N = 1000$, f_o is 1000 MHz. If the frequency ratio is now set to 2345, the output frequency is stabilized at $f_o = 2345 \times f_x$. The output frequency of the synthesizer can be set to any multiple of the crystal oscillator frequency simply by selecting the desired frequency-divider ratio.

The PM6160B frequency synthesizer illustrated in Figure 11-28 has two output frequency ranges: 1 MHz to 12 MHz and 10 MHz to 160 MHz. The frequency resolution is

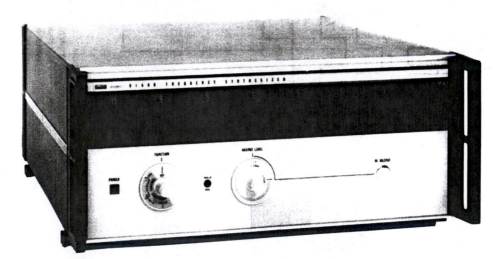

Figure 11-28 The PM 6106B frequency synthesizer produces output frequencies ranging from 1 MHz to 160 MHz with a maximum resolution of 1 Hz. (© 1991, John Fluke Mfg. Co., Inc. All rights reserved. Reproduced with permission.)

Signal Generators   Chap. 11

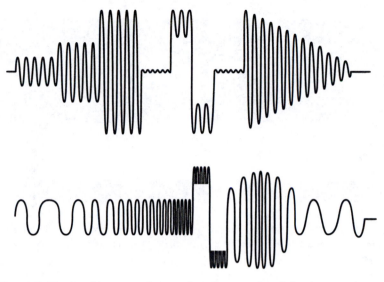

Figure 11-29 An arbitrary waveform consists of a succession of changing waveshapes, typically used for testing electronics equipment.

0.1 Hz on the lower of the two ranges, and 1 Hz on the higher range. The output voltage level is adjustable from 1 mV to 300 mV rms into a 50 Ω load.

11-7 ARBITRARY WAVEFORM GENERATOR

An *arbitrary waveform generator* allows the instrument user to design and generate virtually any desired waveform. For an example of the use of such a waveform, consider the tests necessary on communications equipment. A modulated signal that varies over the entire bandwidth and amplitude range of the equipment could be created for testing purposes (see Figure 11-29). Noise could also be superimposed upon the signal, and gaps might be introduced between waveform bursts, to investigate the response of the equipment. Once such a waveform has been designed, it could be stored and recalled repeatedly for production testing.

The PM5139 signal generator shown in Figure 11-30 is described as a *function generator with arbitrary waveform capability*. This instrument is able to produce virtually all sinusoidal, pulse, ramp, and triangular wave shapes. Amplitude, frequency, and phase modulation of the various waveforms is possible, and dc offset voltages can be superimposed. There are ten standard waveforms, selectable through menu buttons, and precisely set for frequency, amplitude, etc., by means of the control knob. A backlit liquid crystal display reads out the status of the generated waveform.

Arbitrary waveforms may be created by combining and modulating the various standard waveforms. The waveform generating functions include linear and logarithmic amplitude and frequency sweeps. Up to six arbitrary waveforms may be stored and recalled from memory as required. The output voltage range of the generator is 1 mV to 20 V peak-to-peak, and the frequency range is 0.1 mHz to 20 MHz.

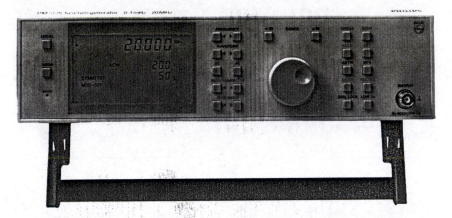

Figure 11-30 The PM 5139 is a function generator which can produce user-designed arbitrary waveforms. (© 1991, John Fluke Mfg. Co., Inc. All rights reserved. Reproduced with permission.)

REVIEW QUESTIONS

11-1 Sketch the circuit of a Wein bridge oscillator. Explain the circuit operation, and write the equations for output frequency and amplifier gain.

11-2 Show how modifications should be made to the Wein bridge circuit in Question 11-1 for **(a)** output amplitude stabilization and **(b)** frequency control. Explain the effect of each modification.

11-3 Sketch the circuit of a two-range adjustable output attenuator for use with a sinusoidal oscillator. Explain the circuit operation.

11-4 Draw the diagram of a sine-to-square-wave conversion circuit using diode clipping. Explain how the circuit operates.

11-5 Sketch the basic block diagram for a sine/square-wave generator. Explain briefly.

11-6 Draw a diagram showing how a sine-wave generator may be used to test the voltage gain and frequency response of an audio amplifier. Explain.

11-7 Sketch a basic function generator circuit for producing sine, triangular, and square waveforms. Explain the circuit operation, and show a modification for frequency range changing.

11-8 Draw a diode/resistor loading circuit for converting a triangular waveform into an approximation of a sine wave. Explain the circuit operation.

11-9 Sketch the block diagram of a basic function generator. Explain briefly.

11-10 List the important items in the specifications for sine/square-wave generators and function generators. Explain each item.

11-11 Sketch a basic block diagram and waveforms for a pulse generator. Explain briefly.

11-12 Sketch the circuit and waveforms for an op-amp astable multivibrator for use as a square-wave generator. Explain its operation.

11-13 Draw an op-amp monostable multivibrator circuit. Show the waveforms at various points in the circuit, and explain its operation.

11-14 Sketch the circuit of an attenuator with a two-range adjustable output and dc output level shifting, for use with pulse and function generators. Explain its operation.

11-15 Draw a pulse waveform showing rise time, fall time, and delay time. Briefly explain. Also, discuss precautions that should be observed when using pulse generators.

11-16 Sketch the basic block diagram of an RF signal generator. Explain briefly.

11-17 Draw circuit diagrams for Hartley and Colpitts oscillators. Write the equation for oscillating frequency, and explain the operation of each circuit.

11-18 Sketch a simple circuit for amplitude modulating the output of an RF oscillator. Show the circuit waveforms, and explain the circuit operation.

11-19 Draw a simple circuit for frequency modulating an RF oscillator. Explain its operation.

11-20 Sketch the complete block diagram for an RF signal generator, and explain its operation. Discuss the need for correct loading of an RF signal generator.

11-21 Draw a basic block diagram and waveforms for a sweep frequency generator. Explain briefly.

11-22 Sketch the complete block diagram for a sweep frequency generator. Show the waveforms at various points in the system, and explain the operation of the instrument.

11-23 Draw a block diagram of a frequency synthesizer, showing the waveforms at various points. Explain carefully how the synthesizer operates.

PROBLEMS

11-1 A Wein bridge oscillator circuit as in Figure 11-1(a) has $C_1 = C_2 = 250$ nF, and $R_1 = R_2 =$ (variable from 200 Ω to 3 kΩ). Calculate the maximum and minimum output frequencies and determine the new capacitor values required to give $f_{max} = 300$ Hz.

11-2 The Wein bridge circuit modification in Figure 11-1(b) has $R_4 = 390$ Ω, $R_5 = 470$ Ω, and $R_6 = 330$ Ω. Calculate the maximum amplitude of the output voltage.

11-3 The attenuator in Figure 11-3 is to have (0 to 0.3) V and (0 to 3) V output ranges. If $V_i = 6$ V, and the op-amp has 300 nA input bias current, determine suitable resistance values for R_1, R_2, and R_3.

11-4 The diode clipper circuit in Figure 11-4 is to produce a ±4 V output square wave. If a ±15 V supply is used, select suitable zener diode voltages.

11-5 The Schmitt trigger circuit in Figure 11-8 has $R_3 = 2.7$ kΩ, $R_4 = 15$ kΩ, and $V_{CC} = $ ±12 V. Calculate the upper and lower trigger points for the circuit.

11-6 Determine the adjustment range of the trigger points for the Schmitt circuit in Problem 11-5, if R_3 is variable from 2.7 kΩ to 3.2 kΩ.

11-7 The integrator in Figure 11-8 has $R_1 = 500\ \Omega$, $R_2 = 4.7\ k\Omega$, and $C_1 = 0.3\ \mu F$. If the Schmitt circuit has ± 1 V trigger points, and the supply voltage is ± 12 V, determine the output frequency when R_1 is at its center point.

11-8 The integrator in Problem 11-7 is connected with the Schmitt circuit in Problem 11-5 to form a function generator. Determine the peak-to-peak amplitude and frequency of the triangular waveform.

11-9 The square-wave generator in Figure 11-13 has $f = 1$ kHz. Determine the space width between the output pulses when the monostable output has **(a)** PW = 100 μs and **(b)** PW = 0.9 ms.

11-10 The op-amp astable multivibrator in Figure 11-14 has $R_1 = 12\ k\Omega$, $R_2 = 4.7\ k\Omega$, $R_3 = 3.3\ k\Omega$, $C_1 = 0.3\ \mu F$, and $V_{CC} = \pm 9$ V. Calculate the frequency of the output square wave.

11-11 The op-amp monostable multivibrator in Figure 11-15 has $R_1 = 15\ k\Omega$, $R_2 = 15\ k\Omega$, $C_1 = 100$ pF, $C_2 = 0.02\ \mu F$, $V_{CC} = \pm 12$ V, and $V_B = +1.5$ V. Calculate the output pulse width.

11-12 The attenuator circuit in Figure 11-16 has $R_1 = 10\ k\Omega$, $R_2 = 135\ k\Omega$, $R_3 = 5\ k\Omega$, $R_4 = 10\ k\Omega$, $R_5 = 5\ k\Omega$, $R_6 = 10\ k\Omega$, $V_{CC} = \pm 12$ V, and $V_i = 10$ V. Determine the range of amplitude adjustment for the output pulse, and the dc offset voltage range.

11-13 Two capacitors ($C_1 = C_2 = 100$ pF), and two inductors ($L_1 = L_2 = 100\ \mu H$) are available for use in the Hartley and Colpitts oscillator circuits in Figure 11-20. Determine the oscillating frequency in each case.

11-14 The amplitude modulation circuit in Figure 11-21 has $R_3 = 5\ k\Omega$, $R_4 = 5\ k\Omega$, and the R_D of Q_2 ranges from 1 $k\Omega$ to 5 $k\Omega$. Calculate maximum and minimum peak-to-peak output voltages if $V_i = 100$ mV p-to-p.

11-15 The crystal oscillator in the frequency synthesizer in Figure 11-27 has a 1 MHz output frequency, and the VCO has a linear output/input ratio of 10 MHz/100 mV. Determine the output frequency and the dc input output level from the low-pass filter when the frequency selection switches are set at 0235. If the crystal oscillator frequency is changed to 0.5 MHz, determine the new output frequency.

Instrument Calibration

12

Objectives

You will be able to:

1. Show how to calibrate digital and analog dc voltmeters and ammeters using standard instruments.
2. Discuss the procedure for calibrating deflection instruments.
3. Show how to calibrate ac voltmeters and ammeters using standard instruments.
4. Discuss electromechanical and electronic ohmmeter calibration.
5. Sketch the circuit for calibrating a wattmeter, and discuss the calibration procedure.
6. Discuss the use of digital multimeters as standard instruments for calibrating other instruments.
7. Sketch the basic circuit of a precision voltage source designed for use as a calibrator. Explain its operation, and calculate the output error for a voltage calibrator with a specified accuracy.
8. Sketch a basic potentiometer circuit, explain its operation, and determine the measured voltage at a given setting.
9. Sketch the circuit of a potentiometer that uses switched resistors. Explain the circuit operation, and determine the measured voltage at a given setting.
10. Show how a potentiometer may be used for calibrating ammeters and voltmeters, and discuss the use of shunt boxes and volt boxes.

Introduction

All voltmeters, ammeters, and wattmeters should be calibrated periodically to check that they are within their prescribed accuracy. The simplest way to do this is to compare the instrument reading with that of a more accurate instrument when they are both measuring the same quantity. Bench-type digital meters are usually sufficiently accurate for calibrating other digital and analog instruments. Highly accurate voltage and current sources, known as *calibrators,* are designed for the sole purpose of instrument calibration. The *potentiometer* can be used for precise measurement of low dc voltage levels. Thus, a potentiometer can be employed in conjunction with precision resistors and potential dividers for ammeter and voltmeter calibration.

12-1 COMPARISON METHODS

DC Voltmeter Calibration

An instrument may be calibrated simply by comparing its reading with that of a more accurate instrument when they are both measuring the same quantity. The more accurate instrument is used as a *standard* for comparison purposes. Figure 12-1(a) illustrates the method for calibrating a dc voltmeter. A convenient dc voltage level is applied from the power supply to the parallel-connected standard instrument and the instrument to be calibrated. The voltage is adjusted in steps over the desired range, and the two instrument readings are noted.

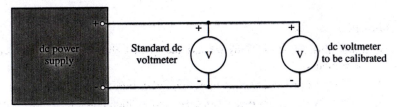

(a) Circuit for dc voltmeter calibration by the use of a standard instrument

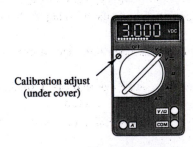

Calibration adjust
(under cover)

(b) A digital voltmeter can
usually be adjusted for
calibration.

Scale reading	Precise voltage	Correction
100	103	+3
90	93	+3
80	82.5	+2.5
70	72.5	+2.5
60	62	+2
50	51.7	+1.7
40	41.5	+1.5
30	31	+1
20	19.7	- 0.3
10	9.5	- 0.5
0	0	0

(c) A calibration chart should be prepared
when calibrating an instrument.

Figure 12-1   A voltmeter may be calibrated by comparing it to a more accurate (standard) voltmeter measuring the same quantity. The accuracy of the standard instrument should be at least four times better than the instrument to be calibrated.

Instrument Calibration   Chap. 12

When a digital instrument is calibrated and found to be inaccurate, a correction may be made by adjusting a variable resistor within the circuit [see Figure 12-1(b)]. For example, the dual-slope-integrator DVM described in Section 6-1 uses a constant-current source that determines the instrument accuracy. This can be adjusted when calibrating, to make the DVM indicate precisely the standard instrument voltage. For both digital and analog instruments, the preparation of a calibration chart [Figure 12-1(c)] can be useful for determining whether or not the instrument is within its specified accuracy.

Example 12-1

For the 10 and 50 readings in the calibration chart in Figure 12-1(c), determine the instrument accuracy as a percentage of the reading and as a percentage of full scale.

Solution

Scale reading = 10:

$$error = \frac{-0.5}{10} \times 100\% \text{ of reading}$$

$$= -5\% \text{ of reading}$$

$$= \frac{-0.5}{100} \times 100\% \text{ of full scale}$$

$$= -0.5\% \text{ of full scale}$$

Scale reading = 50:

$$error = \frac{+1.7}{50} \times 100\% \text{ of reading}$$

$$= +3.4\% \text{ of reading}$$

$$= \frac{+1.7}{100} \times 100\% \text{ of full scale}$$

$$= +1.7\% \text{ of full scale}$$

Deflection Instrument Calibration

Before attempting to calibrate any instrument, the instrument should be cleaned, and all necessary repairs should be made. For example, the pointer on a deflection instrument may need to be straightened. The pointer zero position should be checked, and the mechanical zero control adjusted if necessary while tapping the instrument gently to relieve friction. The supply should be adjusted to increase the voltage or current, progressively setting the pointer at each major scale division. Again, the instrument should be tapped lightly to relieve friction at each setting. At each pointer position of the instrument being calibrated, the precise quantity measured by the standard instrument is noted and recorded on the calibration chart [Figure 12-1(c)]. The *correction,* or difference between the indicated and precise levels, is also charted, as illustrated.

DC Ammeter Calibration

Figure 12-2 shows two dc ammeters connected in series, so that they are both measuring the same current. One ammeter is a standard instrument, known to be accurate, and the other is the ammeter to be calibrated. The current is derived from a dc power supply, and a current-limiting resistor may be included, as illustrated. If the power supply has a current level control, the current-limiting resistor is not required. As for voltmeter calibration, the power supply is adjusted to set the measured quantity in convenient steps over the desired range. The two instrument readings are noted, and a calibration chart is prepared. No adjustment is normally made in the case of a digital multimeter employed as an ammeter, because the DMM current measurement function uses the voltage-measuring facility.

AC Instrument Calibration

The procedure for calibrating ac voltmeters and ammeters is exactly the same as for dc instruments, except that an ac supply and an ac standard instrument must be used. Usually, the ac supply is derived from an autotransformer connected to the 60 Hz, 115 V supply, as shown in Figure 12-3. Where an instrument is to be checked for accuracy at higher frequencies, the supply must be derived from an oscillator with a suitable output voltage level.

Ohmmeter Calibration

An ohmmeter may be calibrated simply by using it to measure known precision resistors. The accuracy of an electromechanical ohmmeter (Sections 3-8 and 3-9) depends on its internal resistors and its deflection instrument. There are mechanical and electrical zero adjust controls, but there is normally no calibration adjustment.

In electronic instruments, both analog and digital, the ohmmeter function utilizes the voltage-measuring circuitry. Consequently, when the instrument is calibrated as a voltmeter, no further calibration adjustment may need to be made. In the case of instru-

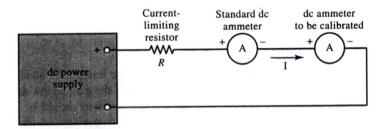

Figure 12-2 An ammeter may be calibrated by comparing it to a more accurate (standard) ammeter measuring the same quantity. The accuracy of the standard instrument should be at least four times better than the instrument to be calibrated.

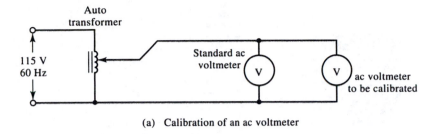

(a) Calibration of an ac voltmeter

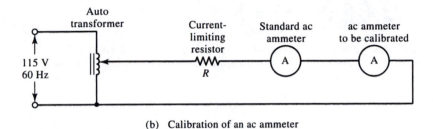

(b) Calibration of an ac ammeter

Figure 12-3 Ac voltmeters and ammeters may be calibrated by comparing them to more accurate standard instruments measuring the same quantity. The accuracy of the standard instrument should be at least four times better than the instrument to be calibrated.

ments that use the type of linear ohmmeter circuit shown in Figure 4-11, the constant-current source may be adjustable for ohmmeter calibration.

Wattmeter Calibration

Wattmeter calibration is simply a combination of the ammeter and voltmeter calibration methods. Figure 12-4 shows a wattmeter with its current coil connected to one dc power supply and its voltage coil connected to another supply. A current-limiting resistor and a standard ammeter are connected in series with the current coil, and a standard voltmeter measures the voltage applied to the voltage coil. Once the precise voltage and current are determined for a given reading on the wattmeter, the exact power level is determined as $P = VI$. To avoid damaging the instrument, care must be taken to avoid the specified maximum voltage and maximum current (see Section 3-11).

Example 12-2

An electrodynamic wattmeter being calibrated as in Figure 12-4 indicates full scale of 120 W on its 120 V, 1 A range (see Figure 3-31). The measured current and voltage are precisely 1 A and 114 V, respectively. Determine the wattmeter error and correction figure.

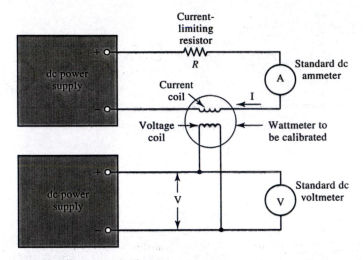

Figure 12-4 Circuit for calibration of a wattmeter. The voltage and current coils are supplied from different sources, and each quantity is measured by a standard instrument. The correct power is then calculated as *VI*.

Solution

$$P = VI = 114 \text{ V} \times 1 \text{ A}$$

$$= 114 \text{ W}$$

$$error = \text{correction figure}$$

$$= 114 \text{ W} - 120 \text{ W}$$

$$= -6 \text{ W}$$

$$error = \frac{-6 \text{ W}}{120} \times 100\%$$

$$= -5\%$$

12-2 DIGITAL MULTIMETERS AS STANDARD INSTRUMENTS

The standard instrument used to calibrate another instrument by the comparison method discussed in Section 12-1 simply has to be any instrument that has an accuracy at least four times better than the instrument to be calibrated. Recall from Sections 3-10 and 4-6 that the best accuracy of analog instrument is usually ±2% of full scale. From Section 6-2, the accuracy for a typical hand-held digital multimeter is ±(0.5% rdg + 1 d). Thus, for example, a $3^1/_2$-digit DMM reading 100.0 V would have a maximum error of ±0.6 V, or ±0.6%. So such a DMM could be suitable for use as a standard instrument when calibrating an analog meter [Figure 12-5(a)].

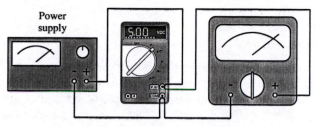

(a) A hand-held digital voltmeter is accurate
enough to calibrate an analog instrument.

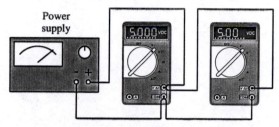

(b) A high-performance hand-held DVM can be
used for calibrating a lower-accuracy DVM.

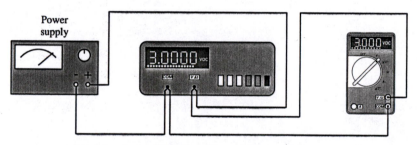

(c) A bench-type DVM is usually accurate enough to calibrate
a high-performance hand-held DVM.

Figure 12-5 Any instrument can be used as a standard to calibrate another instrument, provided that its accuracy is at least four times better than the instrument to be calibrated.

The accuracy of a high-performance hand-held DMM is typically $\pm(0.1\%$ rdg $+$ 1 d), and for a bench-type digital instrument the figure is $\pm(0.01\% + 1$ d) (see Section 6-2). Consequently, a high-performance hand-held DMM could be used as the standard instrument to calibrate either an analog instrument or a lower-accuracy DMM [Figure 12-5(b)]. Similarly, a bench-type digital instrument would be a suitable standard instrument for calibrating any one of the other three lower-accuracy instruments [Figure 12-5(c)].

12-3 CALIBRATION INSTRUMENTS

Precision DC Voltage Source

An instrument designed for the purpose of calibrating dc voltmeters is essentially an adjustable precision dc regulated power supply (see Chapter 16) with a high-resolution display of the output voltage. All dc voltage regulators are based on the use of a zener diode (or breakdown diode) as a reference voltage. The zener diode characteristics illustrated in Figure 12-6(a) show that the device has a very stable anode-to-cathode voltage (V_z) when operated in reverse breakdown. The stability of V_z is improved if the current (I_z) is supplied by a constant-current source [see Figure 12-6(b)] and is fur-

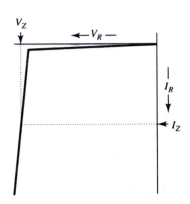

(a) Zener diode characteristics

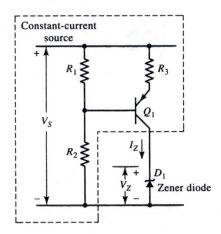

(b) Zener diode with a constant-current source

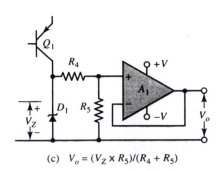

(c) $V_o = (V_Z \times R_5)/(R_4 + R_5)$

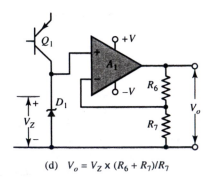

(d) $V_o = V_Z \times (R_6 + R_7)/R_7$

Figure 12-6 A voltage calibrator is essentially a precision voltage reference source. A temperature-controlled zener diode with a constant current is used, and precision resistors are employed to divide or amplify V_z to the desired output level.

Instrument Calibration   Chap. 12

ther enhanced if the device is enclosed in a temperature-controlled container. The level of I_z can be adjusted by the instrument manufacturer to give a precise and reliable level of V_z. By means of amplifiers and precision resistors, V_z can be increased or decreased to give a wide range of precise voltage levels for instrument calibration [Figure 12-6(c) and (d)].

Example 12-3

The voltage reference circuit in Figure 12-6(c) has $R_4 = 1125\ \Omega$, $R_5 = 4017.9\ \Omega$, and $V_z = 6.4$ V. The resistor accuracy is ±100 ppm, and the V_z error is ±0.01%. Determine the output voltage and its accuracy.

Solution

$$V_o = (V_z \pm 0.01\%) \times \frac{R_5 \pm 100\ \text{ppm}}{(R_4 \pm 100\ \text{ppm}) + (R_5 \pm 100\ \text{ppm})}$$

$$= 6.4\ \text{V} \left(1 \pm \frac{0.01}{100}\right) \times \frac{4017.9\ \Omega \pm 0.04\ \Omega}{(1125\ \Omega \pm 0.1\ \Omega) + (4017.9\ \Omega \pm 0.04\ \Omega)}$$

$$= 5\ \text{V} \pm 700\ \mu\text{V}$$

Voltage Calibrator

The DATEL DVC8500 voltage calibrator illustrated in Figure 12-7(a) produces 0 to +19.999 V, or 0 to −19.999 V, selectable in 1 mV steps. The output voltage is set by the front-panel lever switches, and a rotatable vernier control provides ±1.5 mV adjustment. The output voltage accuracy is specified as [±25 ppm (parts per million) of the setting ±1/2 LSB (least significant bit)]. If the output voltage is set to 10.000 V, for example, the maximum possible error would be

$$\text{error} = \pm\left[\left(10.000\ \text{V} \times \frac{25}{1\ 000\ 000}\right) + 0.0005\ \text{V}\right]$$

$$= \pm 750\ \mu\text{V}$$

$$= \pm 0.0075\% \text{ of } 10\ \text{V}$$

The DVC8500 can be used for calibrating voltmeters that have accuracies of ±0.03%, or worse. It is also very suitable for calibrating A/D and D/A converters (see Sections 5-5 and 5-6). A maximum output current of 25 mA can be supplied without affecting the voltage accuracy. For voltmeter calibration, the instrument to be calibrated is simply connected to the calibrator terminals, and the calibrator is adjusted to give the desired voltage levels [see Figure 12-7(b)]. More complex calibrators are available with precision dc and ac reference voltages and currents, and precision resistors.

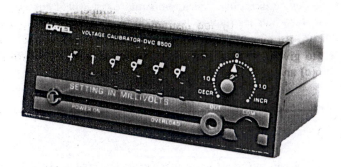

(a) Voltage calibrator (Courtesy of
DATEL, Inc.)

Calibrator

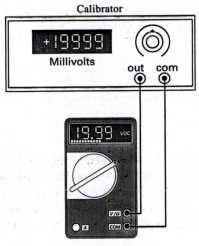

(b) Voltmeter calibration

Figure 12-7   The DATEL DVC 8500 voltage calibrator produces an output of 0 to
± 19 999 mV, adjustable in 1 mV steps by
front-panel lever switches. A ±1.5 mV adjustment is provided by a vernier control.

12-4 POTENTIOMETERS

Basic Potentiometer

Figure 12-8 shows the basic circuit of a potentiometer. A resistance wire (AB) having a
uniform resistance per unit length is placed alongside a calibrated scale such as a meter
stick. A current I supplied from a battery (B_1) flows through the wire. The current level is

controlled by a variable resistance (R_1). A sensitive center-zero galvanometer (G) is connected to the resistance wire via a sliding contact (C). The other terminal of the galvanometer is connected to a switch (S), which facilitates contact to either a standard cell (B_2) or to a voltage to be measured (V_x). Resistor R_2 protects the standard cell against excessive current flow.

The potentiometer must be *calibrated* before it can be used to measure voltage. This is done by setting the switch to connect the standard cell to the galvanometer. The sliding contact is set to the position on the resistance wire (or *slide wire*) which should have a voltage exactly equal to that of the standard cell. Then R_1 is adjusted until the galvanometer indicates zero (or null), that is, until $V_{BC} = V_{B2}$.

Suppose that the resistance wire is exactly 100 cm in length, and assume that the moving contact is set to 50.95 cm from terminal B. When R_1 is adjusted for null on the galvanometer, the voltage across the 50.95 cm is exactly equal to the standard cell voltage. With a standard cell voltage of 1.0190 V, V_{BC} equals 1.0190 V. Thus, the resistance wire now has

$$\text{voltage/unit length} = \frac{1.0190 \text{ V}}{50.95 \text{ cm}} = 20 \text{ mV/cm}$$

When the potentiometer has been calibrated, S can be switched to V_x, then the sliding contact is adjusted to again null the galvanometer. Voltage V_x is now determined by measuring the length of wire from terminal B to the new position of C:

$$V_x = (\text{length } BC) \times 20 \text{ mV/cm}$$

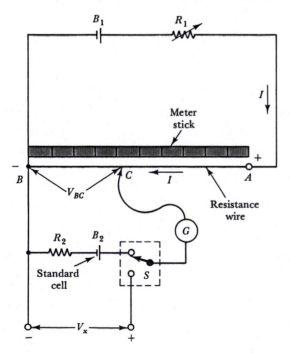

Figure 12-8 A basic potentiometer circuit has a resistance wire AB with a sliding contact C and a current I. A meter stick measures the distance of C from B. The instrument is calibrated to have a given voltage/cm along the resistance wire. When the galvanometer indicates null, the measured voltage is $V_x = V_{BC} = BC$ (in cm) $\times$ V/cm

If the sliding contact can be set to an accuracy of ±1 mm, the resolution of the instrument is the slide wire voltage per mm, that is, (20 mV per cm)/10 = 2 mV. This is another way of saying that V_x can be measured with a precision of ±2 mV. If there are no other sources of error, the measurement accuracy is also ±2 mV.

The potentiometer measures the unknown voltage by comparing it (via the potentiometer) to the standard cell voltage. Because a galvanometer is used to detect null when calibrating, no current is drawn from the standard cell. Also, since null is again detected when measuring the unknown voltage, no current is drawn from V_x, and its open-circuit terminal voltage is accurately measured.

Example 12-4

The resistance of wire AB in a simple potentiometer (as in Figure 12-8) is 100 Ω, and its length is 100 cm. Battery B_1 has a terminal voltage of 3 V and negligible internal resistance. The standard cell voltage is 1.0190 V, and R_1 is adjusted to calibrate the potentiometer when BC is 50.95 cm.

(a) Determine the current through AB and the resistance of R_1 when the potentiometer is calibrated.

(b) Calculate V_x when null is obtained at 94.3 cm.

(c) Determine the resistance of R_2 to limit the standard cell current to a maximum of 20 μA.

Solution

(a) *At calibration:*

$$V_{BC} = V_{B2} = 1.0190 \text{ V}$$

$$\text{volts per unit length} = \frac{1.0190 \text{ V}}{50.95 \text{ cm}}$$

$$= 20 \text{ mV/cm}$$

$$V_{AB} = 100 \text{ cm} \times 20 \text{ mV/cm}$$

$$= 2 \text{ V}$$

$$I = \frac{V_{AB}}{R_{AB}} = \frac{2 \text{ V}}{100 \text{ Ω}}$$

$$= 20 \text{ mA}$$

$$V_{R1} = V_{B1} - V_{AB}$$

$$= 3 \text{ V} - 2 \text{ V} = 1 \text{ V}$$

$$R_1 = \frac{1 \text{ V}}{20 \text{ mA}} = 50 \text{ Ω}$$

(b) $V_x = 94.3$ cm $\times$ 20 mV/cm
$$= 1.886 \text{ V.}$$

(c) *At worst, the terminal voltage of B_2 or B_1 may be reversed, R_1 may be set to zero, and C set to terminal A. The total voltage producing current flow through the standard cell is now*

$$V_{B2} + V_{B1} = 3 \text{ V} + 1.019 \text{ V}$$

$$= 4.019 \text{ V}$$

$$R_2 = \frac{4.019 \text{ V}}{20 \text{ }\mu\text{A}} \approx 200 \text{ k}\Omega$$

Potentiometer with Switched Resistors

The simple potentiometer described above can measure voltage with a maximum accuracy of approximately ± 2 mV. The more complex circuit shown in Figure 12-9 employs resistors in addition to the slide wire to extend the accuracy of the instrument. As illustrated, R_6 through R_{12} are precision resistors connected to a rotary switch. With the switch contact in the position shown when null is obtained on the galvanometer, the measured voltage is $V_x = V_{R11} + V_{R12} + V_{BC}$.

Since the slide wire is arranged in a circle, a circular scale and pointer can be used to indicate the exact position of the sliding contact. Similarly, a pointer and circular scale can be used to indicate the position of switched contact F. Instead of pointers and fixed circular scales, rotatable scales are usually employed, with windows and lines to indicate the scale positions [see Figure 12-9(b)]. Also, rather than giving the positions of the moving contacts, the corresponding voltages (when calibrated) are indicated on the scales.

Two additional precision resistors, R_3 and R_4, are included to permit fast calibration. These resistances are selected so that the instrument is calibrated when V_{R3} is exactly equal to the standard cell voltage. By a pushbutton switching arrangement (not shown) the galvanometer is disconnected from V_x and the slide wire, and reconnected in series with standard cell B_2 and the junction of R_3 and R_4. This position is shown by dashed lines in Figure 12-9(a).

Current to the potentiometer is provided from battery B_1 via variable resistors R_1 and R_2. R_1 may have a low resistance value, so that it acts as a *coarse* control of current I. R_2 should have a resistance many times greater than R_1. Thus, a relatively large adjustment of R_2 has only a small effect on the combined resistance of R_1 and R_2 in parallel. Consequently, R_2 has a correspondingly small effect on I, and it acts as a *fine* control to set I to the exact level required for calibration, after it has been set approximately by R_1.

To calibrate the circuit, the pushbutton switch (discussed above) is depressed, and R_1 and R_2 are adjusted as explained to obtain null on the galvanometer. The pushbutton is then released to connect V_x and the galvanometer into position for measuring V_x. The switched contact F is moved as necessary until the galvanometer approaches null. Then the sliding contact C is adjusted to achieve a precise null. When null is obtained, the

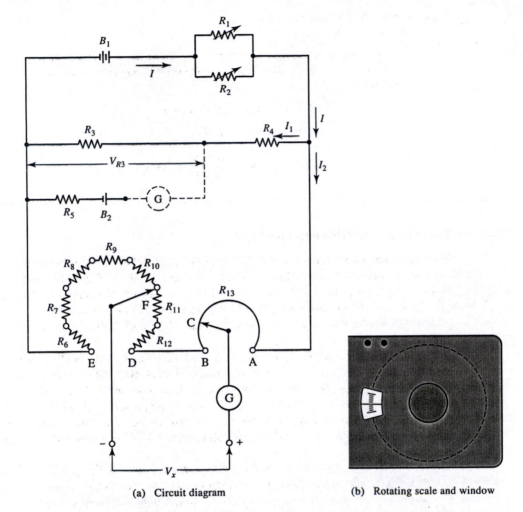

(a) Circuit diagram   (b) Rotating scale and window

Figure 12-9 Potentiometer with switched resistors (R_6 through R_{12}) and slide wire R_{13}. The resistors and slide wire have equal voltage drops. When galvanometer null is obtained at F and C, the measured voltage is $V_x = V_{R11} + V_{R12} + [BC$ (in cm) $\times$ (V/cm for R_{13})]

pushbutton switch is again depressed to check the potentiometer calibration. This is necessary because the battery voltage is always falling slightly, causing the potentiometer to drift away from calibration. Adjustment of R_2 is made as required, to recalibrate. With the instrument recalibrated, the pushbutton is released and contact C is once more adjusted to achieve null. Then, calibration is once again checked and reset if required, before once more checking for galvanometer null with V_x in the circuit. This procedure is repeated until null is obtained both when calibration is checked (pushbutton depressed) and when V_x is being measured (pushbutton released). Finally, the scales of the potentiometer are read to determine V_x.

Example 12-5

The potentiometer shown in Figure 12-9 has a 100 Ω slide wire (R_{13}) and resistors R_6 through R_{12} are each exactly 100 Ω. R_3 is 509.5 Ω and R_4 is 290.5 Ω. If the standard cell voltage is 1.0190 V, determine the maximum voltage that may be measured by the potentiometer. Assuming that the slide wire is 100 cm in length and that the sliding contact position can be read within ± 1 mm, calculate the instrument resolution.

Solution

$$V_{R3} = V_{B2} = 1.0190 \text{ V}$$

$$I_1 = \frac{V_{B2}}{R_3} = \frac{1.0190 \text{ V}}{509.5 \text{ }\Omega}$$

$$= 2 \text{ mA}$$

Maximum voltage measurable:

$$V_{AE} = I_1(R_3 + R_4)$$

$$= 2 \text{ mA}(290.5 \text{ }\Omega + 509.5 \text{ }\Omega)$$

$$= 1.6 \text{ V}$$

Resolution:

$$I_2 = \frac{V_{AE}}{R_6 + R_7 \text{ through } R_{13}}$$

$$= \frac{1.6 \text{ V}}{8 \times 100 \text{ }\Omega}$$

$$= 2 \text{ mA}$$

$$V_{AB} = I_2 R_{13} = 2 \text{ mA} \times 100 \text{ }\Omega$$

$$= 200 \text{ mV}$$

slide wire volts/unit length $= 200 \text{ mV}/100 \text{ cm}$

$$= 2 \text{ mV/cm}$$

$$= 0.2 \text{ mV/mm}$$

instrument resolution $= \pm 0.2 \text{ mV}$

12-5 POTENTIOMETER CALIBRATION METHODS

DC Ammeter Calibration

An ammeter may be calibrated by connecting it in series with a precision resistor, and then accurately measuring the resistor voltage drop. The level of ammeter current is determined by dividing the resistor voltage drop by its resistance value. When the resistor value is known precisely, and the voltage drop is measured by a potentiometer, the level of ammeter current can be determined with great accuracy. Figure 12-10 shows the circuit for using a potentiometer to calibrate an ammeter.

Precision resistors are available in *shunt boxes* solely for the purpose of calibrating ammeters by means of a potentiometer. Figure 12-11 shows the circuit and terminals of a typical shunt box. The shunt box resistors are actually connected in a circular fashion around the terminals of a rotary switch. In Figure 12-11 the resistors and rotary switch are shown in a straight-line formation. The *line* terminals of the shunt box are

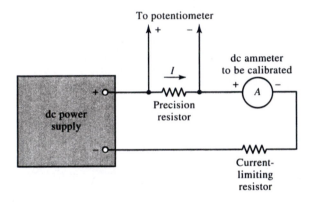

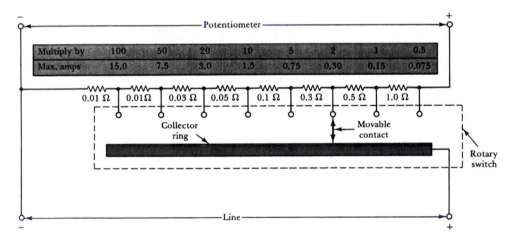

Figure 12-10 An ammeter may be calibrated by using a potentiometer to measure the voltage drop across a precision resistor connected in series with the ammeter.

Figure 12-11 A shunt box is a set of precision series-connected resistors for use in potentiometer calibration of ammeters. A movable contact selects the point at which the resistors are connected in series with the ammeter.

Instrument Calibration   Chap. 12

connected in series with the ammeter, and the shunt box potentiometer terminals go to the *potentiometer.*

When the shunt box is set to the multiply by 1 position, the resistances in series with the line terminals add up to 1 Ω. If the measured voltage is exactly 0.1 V, the ammeter current is 0.1 V/1 Ω = 0.1 A or 0.1 × (multiplier 1) = 0.1 A. When the shunt box is at the multiply by 0.5 position, the total resistance is 2 Ω. The ammeter current is now calculated as (measured voltage) × 0.5 A. Similarly, at all other shunt box positions, the current in amperes is (measured voltage) × multiplier.

With the movable contact in the position shown in Figure 12-11, the total resistance in series with the ammeter is 0.5 Ω (i.e., the total of all resistors to the left of the movable contact). Note also that the resistors to the right of the movable contact are actually in series with the galvanometer of the potentiometer. These total only 1.5 Ω (for the contact position shown), which is much smaller than the galvanometer resistance. Also, at null there is no current flowing in the galvanometer circuit, and there is no voltage drop across the 1.5 Ω resistance. So these resistors have no effect upon the measurement.

For every multiplier position on the shunt box, there is a maximum current level, identified as *MAX AMPS*. In each case, this current multiplied by the resistance in series with the line terminals gives an output (to the potentiometer) of 150 mV from the shunt box. At the multiply by 100 position, the maximum current is 15 A. Thus, the output voltage is 15 A × 0.01 Ω = 150 mV. At the multiply by 5 position, the 0.75 A maximum current gives the output voltage 0.75 A × (0.1 + 0.05 + 0.03 + 0.01 + 0.01)Ω = 150 mV. The maximum current level is selected to minimize resistor power dissipation and the resultant temperature errors.

DC Voltmeter Calibration

The precise voltage applied to a voltmeter can be directly measured by the potentiometer if the voltmeter range is not greater than 1.5 V. For higher ranges, two precision resistors should be used as a potential divider, as illustrated in Figure 12-12. The voltage across R_1 is measured by the potentiometer, and the precise voltmeter potential is determined as $E = V_{R2}(R_1 + R_2)/R_2$.

Like the shunt box for use in ammeter calibration, a *volt box* contains the necessary potential-divider resistors for potentiometer calibration of voltmeters. The circuit and terminals of a volt box are illustrated in Figure 12-13. Once again, the rotary switch and resistors are shown in a straight-line formation. The potentiometer always measures the

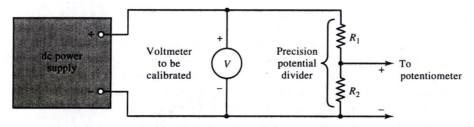

Figure 12-12   A voltmeter may be calibrated by using a potentiometer to measure the voltage from a precision potential divider connected in parallel with the voltmeter.

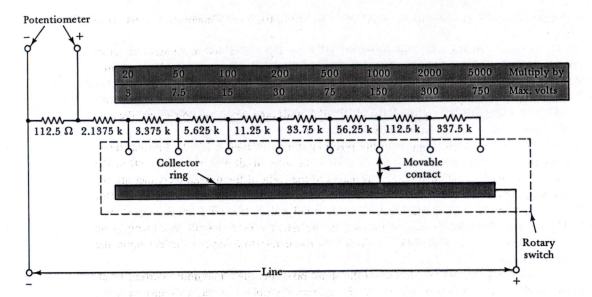

Figure 12-13 A volt box is a precision potential divider for use in potentiometer calibration of voltmeters. A movable contact selects the point at which the volt box is connected in parallel with the voltmeter.

voltage drop across the 112.5 Ω (left side) resistor. As illustrated, the *line voltage* (which is the voltmeter potential) is applied across the 112.5 Ω resistor and at least one other series-connected resistor. The voltage division ratio is altered by switching the moving contact.

When the volt box is set to the *multiply by 20* position, the voltage division ratio is

$$\frac{\text{line voltage}}{\text{potentiometer voltage}} = \frac{112.5\ \Omega + 2.1375\ k\Omega}{112.5\ \Omega}$$

$$= 20$$

Thus, the precise line voltage (or voltmeter potential) is determined by multiplying the potentiometer (measured) voltage by 20. Similarly, at all other volt box positions, the voltmeter potential is (measured voltage) × multiplier.

Every multiplier position on the volt box has a *MAX VOLTS* level stated. This is the maximum line voltage that should be applied to the volt box in each case. When the maximum voltage is applied, the volt box output (to the potentiometer) is always 150 mV. Like the shunt box maximum current, the maximum voltage levels are intended to minimize power dissipation and heating in the volt box resistors. Taking the maximum voltage and total resistance at any setting, the sensitivity of the volt box is calculated as 750 Ω/V, and the maximum current comes out to 1.33 mA. For example, at the *multiply by 20* position, the maximum voltage is 3 V, and the total resistance is

$$R = 112.5 \ \Omega + 2.1375 \ \text{k}\Omega$$

$$= 2.25 \ \text{k}\Omega$$

$$sensitivity = \frac{R}{V} = \frac{2.25 \ \text{k}\Omega}{3 \ \text{V}}$$

$$= 750 \ \Omega/\text{V}$$

and

$$I_{max} = \frac{V_{max}}{R} = \frac{3 \ \text{V}}{2.25 \ \text{k}\Omega}$$

$$= 1.33 \ \text{mA}$$

REVIEW QUESTIONS

12-1 Sketch a circuit to show how a standard voltmeter may be used to calibrate a dc voltmeter. Explain the calibrating procedure, and discuss the use of a calibration chart.

12-2 Discuss the procedure that should be followed before attempting to calibrate a deflection instrument.

12-3 Sketch a circuit to show how a standard ammeter may be used to calibrate a dc ammeter. Explain the calibrating procedure.

12-4 Sketch circuits to show how ac voltmeters and ammeters should be calibrated using standard instruments. Explain.

12-5 Briefly discuss the procedure for calibrating electromechanical and electronic ohmmeters.

12-6 Sketch the circuit for calibrating a wattmeter, and explain the calibration procedure.

12-7 Discuss the required accuracy for standard instruments used in calibrating other instruments. Also, explain why digital instruments may be used as standard instruments for calibrating deflection instruments.

12-8 Sketch the characteristics of a Zener diode, and discuss how it should be used in a precision voltage source. Sketch the basic circuits for a precision voltage source designed for use as a calibrator. Explain the operation of each circuit.

12-9 Sketch the circuit of a basic potentiometer. Explain how the potentiometer should be calibrated, and how it is used for precise measurement of dc voltage.

12-10 Sketch the circuit of a potentiometer that uses switched resistors. Explain how the potentiometer should be calibrated, and how it is used for voltage measurement.

12-11 Sketch a circuit to show how a potentiometer should be used for calibrating a dc ammeter. Briefly explain.

12-12 Sketch the circuit of a shunt box for use with a potentiometer in ammeter calibration. State any precautions that should be observed in using the shunt box.

12-13 Sketch a circuit to show how a potentiometer should be used for calibrating a dc voltmeter. Briefly explain.

12-14 Sketch the circuit of a volt box for use with a potentiometer in voltmeter calibration. State any precautions that should be observed in using the volt box.

PROBLEMS

12-1 Determine the voltmeter accuracy as a percentage of the reading and as a percentage of full scale for the 90, 60, and 30 readings in the calibration chart in Figure 12-1(c).

12-2 An electrodynamic wattmeter calibrated on its 60 V, 0.5 A range indicates 30 W full scale. The measured current and voltage at full scale are precisely 0.5 A and 57 V, respectively. Determine the wattmeter error and correction figure.

12-3 An analog voltmeter which has a ±3% accuracy is to be calibrated. Determine the minimum accuracy for a digital voltmeter with a three-digit display to be used as the standard instrument.

12-4 A DVM with a $3^1/_2$-digit display and an accuracy of ±(0.5% rdg + 1 d) is to be calibrated. Specify a suitable higher-accuracy DVM to be used as the standard instrument.

12-5 The constant-current circuit in Figure 12-6(b) has $V_s = 25$ V, $R_1 = 8.2$ kΩ, $R_2 = 3.9$ kΩ and $R_3 = 820$ Ω. Calculate the Zener diode constant current.

12-6 Determine the output voltage for the voltage reference circuit in Figure 12-6(c) if $R_4 = 1$ kΩ, $R_5 = 1739.1$ Ω, and $V_z = 6.3$ V. Also, calculate the accuracy of the output if the resistor accuracy is ±200 ppm, and the V_z error is ±0.009%.

12-7 Determine the output voltage for the circuit in Figure 12-6(d) if $R_6 = 1292.1$ Ω, $R_7 = 2.2$ kΩ, and $V_z = 6.3$ V. Also, calculate the output voltage accuracy if the resistor accuracy is ±150 ppm, and the V_z error is ±0.012%.

12-8 A basic potentiometer, as shown in Figure 12-8, has the following components: a 3 V battery, a 150 Ω slide wire 1.5 m long, a standard cell with a voltage of 1.0195 V, and variable resistor R_1.
- **(a)** Calculate the slide wire current and the value of R_1 when the potentiometer is calibrated to measure 1.5 V maximum.
- **(b)** Determine the slide wire position BC at which the sliding contact should be set for calibration.
- **(c)** Calculate the measured voltage V_x when null is obtained at 72.5 cm.
- **(d)** Calculate the value of resistor R_2 to be connected in series with the standard cell to limit the maximum current to 20 μA.

12-9 A basic potentiometer has a 200 cm slide wire with a resistance of 100 Ω. A 4 V battery in series with a variable resistor R_1 provides current through the slide wire. The standard cell potential is 1.018 V, and the potentiometer is calibrated when the sliding contact is set to 101.8 cm from the zero voltage end of the slide wire.

(a) Calculate R_1 and the current flowing through R_1.

(b) Determine the value of resistance that must be connected in series with the standard cell to limit its current to 20 μA.

(c) Determine the measured voltage when zero galvanometer deflection is obtained with the slide wire at 94.3 cm from the zero voltage end.

12-10 A potentiometer, as in Figure 12-9, has a 50 Ω slide wire (R_{13}) and seven 50 Ω resistors (R_6 through R_{12}). The standard cell voltage is 1.018 V, and resistors R_3 and R_4 are $R_3 = 470\ \Omega$ and $R_4 = 84.028\ \Omega$.

(a) Determine the maximum voltage that can be measured on the potentiometer.

(b) If the slide wire is 150 cm long, and the sliding contact can be set to within ±1 mm, determine the instrument resolution.

(c) Calculate the value of series resistance ($R_1 \| R_2$ in Figure 12-9) when the supply battery has $V_B = 3$ V.

12-11 A 100 mA deflection-type ammeter is to be calibrated by means of the potentiometer in Figure 12-9. Determine a suitable value of precision resistor and its maximum power dissipation.

12-12 A 30 V voltmeter is to be calibrated by means of the potentiometer in Figure 12-9. Determine the ratio of suitable potential-divider resistors.

Graphic Recording Instruments

13

Objectives

You will be able to:

1. Sketch the mechanical systems and electronic circuitry for galvanometric and potentiometric strip-chart recorders. Discuss the operation and performance of each instrument.

2. Sketch the mechanical systems and electronic circuitry for a potentiometric *X-Y* recorder. Explain the instrument operation, and discuss the typical controls and specification for an *X-Y* recorder.

3. Show how an *X-Y* recorder may be used for tracing the characteristics of various electronic devices.

4. Explain the operation of plotters, and discuss their applications.

Introduction

Permanent (hard copy) records of waveforms displayed on an oscilloscope can be made by use of a camera. However, for some applications, more satisfactory results can be obtained by the use of a *graphic recorder,* or *chart recorder.* The two basic laboratory chart recorders are the *strip-chart recorder* and the *X-Y recorder.* In the strip-chart recorder, a continuously moving strip of paper is passed under a pen or other recording mechanism. The pen is deflected back and forward across the paper in proportion to an input voltage. The resulting trace is a record of input voltage variations over a given period of time. The *X-Y* recorder uses a single sheet of paper and has two inputs; one input deflects the pen horizontally and the other produces vertical deflection. In this case, the resulting trace might represent the characteristics of an electronic device or the frequency response of a circuit. A major disadvantage of chart recorders is that they can operate only at very low

374

frequencies; however, when connected to a digital storage oscilloscope, a chart recorder can produce permanent traces of high-frequency waveforms.

13-1 STRIP CHART RECORDERS

Galvanometric Strip-Chart Recorder

In a *galvanometric* (or *oscillographic*) strip-chart recorder a strip of paper is unrolled and passed under a pen, as illustrated in Figure 13-1. The pen is at the end of a lightweight pointer connected to the coil of a PMMC meter movement (or galvanometer). The pen deflection (or pointer position) is directly proportional to the voltage applied to the moving-coil circuit. When a slowly changing voltage is applied to the coil, the pen is deflected back and forward across the paper. With the paper passing under the pen at a constant velocity, the waveform of the input voltage is traced out on the paper. Because the movement of the paper is proportional to time, a strip-chart recorder is sometimes termed a *YT recorder.*

The pens used in this type of recorder are usually the fiber-tipped type, which are disposable. Instead of a pen, *thermal* writing tips are sometimes employed. These are either tungsten or ceramic tips which are heated by an electric current. The heated tip burns a fine line on the surface of the paper.

Another method of writing on the strip chart is illustrated in Figure 13-2. In this case, the deflection system is a small galvanometer with a mirror instead of a pointer and pen. A finely focused beam of ultraviolet light is reflected from the mirror on to photographically treated paper, producing an instant trace. The one disadvantage of the *light beam* system is that specially treated paper is required. A major advantage is that this type of instrument can record waveforms with frequencies up to 5 kHz, while the galvanometric pen recorder is usually limited to a maximum frequency of 200 Hz.

Apart from the paper-moving mechanism, a galvanometric pen recorder is simply an analog voltmeter. Instead of using a calibrated scale, the pointer (with the pen at its

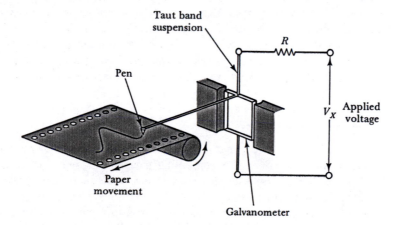

Figure 13-1 A galvanometric strip-chart recorder uses a PMMC (galvanometer) deflection system to move a pen across the chart paper. The system is essentially a deflection voltmeter with a pen and chart instead of a pointer and scale.

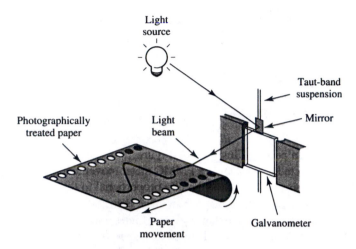

Light
source

Taut-band
suspension

Mirror

Photographically
treated paper

Light
beam

Paper
movement

Galvanometer

Figure 13-2 Galvanometric strip-chart recorder using a light source and a mirror mounted on the deflection coil. The light beam strikes photographically treated paper to produce the waveform trace.

end) is deflected across the recording paper. The circuitry is similar to the analog electronic voltmeters discussed in Section 4-2. An amplifier is used to give a high input resistance and to amplify small voltage levels before applying them to the galvanometer circuit. An attenuator divides high-level input voltages down before they are applied to the amplifier. Thus, the galvanometer current can be set to give (for example) a deflection of 1 cm for each 1 V input (1 cm/V). Alternatively, it might be set to give 2 cm/V, 0.1 cm/V, and so on.

The paper-moving system is usually *traction feed,* in which rotating sprocket wheels drive paper with hole-punched edges. The speed of the drive motor is proportional to the current flowing in its windings. This current can be controlled by means of switched or variable series resistors. The paper velocity might be set (for example) to a high speed of 5 cm/s, or a low of perhaps 5 cm/h. A pen *up/down control* is also usually included, so that the pen can be lifted off the paper while adjustments are made.

Figure 13-3 shows a representative strip chart recorder (the Hioki 8202). This instrument has five chart speeds ranging from 20 cm/min to 2 cm/h, and waveforms are thermally recorded at 32 dots per second. There are nineteen input ranges for ac/dc voltages and current.

Potentiometric Strip-Chart Recorder

The *potentiometric* strip-chart recorder mechanism is essentially that of a *self-balancing potentiometer*. The basic arrangement for controlling the pen position is illustrated in Figure 13-4(a). The pen holder slides along a pen carriage shaft under the control of a *drive cord*. The drive cord passes over an *idler pulley* and a *drive pulley* which is mechanically coupled to a *servomotor*. The pen holder makes electrical contact to a *potentiometer slide wire* [Figure 13-4(b)] which has a dc supply voltage ($\pm E$) applied to its terminals. It also makes contact to a low-resistance slide wire, so that the voltage *picked off* from the potentiometer slide wire is connected to the low-resistance slide wire. This volt-

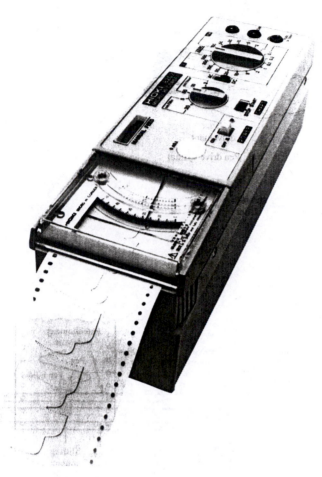

Figure 13-3 Strip chart recorder. Apart from the paper speed controls, the input terminals and selection switch are similar to those of an analog voltmeter/ammeter. (Courtesy of Hioki-RCC, Inc.)

age (V_F) is passed via a voltage follower (A_3) to resistor R_4, which is part of the *summing amplifier.*

To understand the operation of the summing amplifier, first note that the noninverting terminal of op-amp A_2 is connected to the zero voltage level. Because of this, the voltage at the inverting input terminal will always be close to the zero (or ground) level. Therefore, the junction of resistors R_2, R_3, and R_4 is always at zero volts. Consequently, the currents I_2, I_3, and I_4 are: $I_2 = V_2/R_2$, $I_3 = V_i/R_3$, and $I_4 = V_F/R_4$. The voltage drop across R_5 is the output voltage of the amplifier, and for the current directions shown this is a negative quantity:

$$V_o = -R_5(I_2 + I_3 + I_4)$$

$$= -R_5\left(\frac{V_2}{R_2} + \frac{V_i}{R_3} + \frac{V_F}{R_4}\right)$$

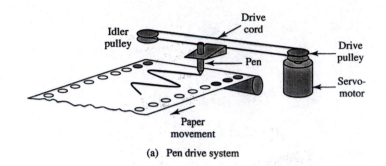

(a) Pen drive system

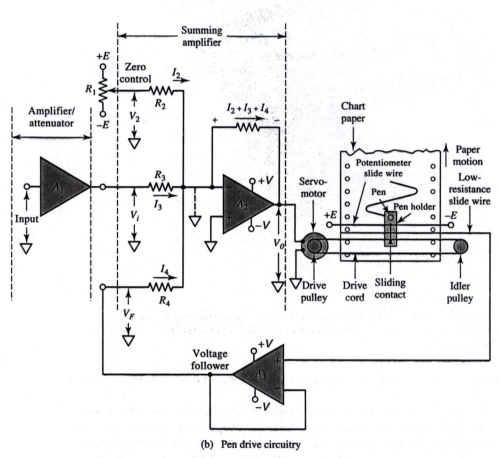

(b) Pen drive circuitry

Figure 13-4 A potentiometric strip-chart recorder moves the pen across the chart by means of a drive cord controlled by a servomotor. The summing amplifier receives the input voltage (V_i) and the feedback voltage (V_F), and sets the pen at the position where $V_F = V_i$.

Graphic Recording Instruments   Chap. 13

If $R_2 = R_3 = R_4 = R_5$,

$$V_o = -(V_2 + V_i + V_F) \qquad (13\text{-}1)$$

Thus, the output voltage of the summing amplifier is the sum of the input voltages. The negative sign indicates that the output is inverted, that is, positive inputs give a negative output, and vice versa.

Now consider what occurs when input voltage V_i is zero, and the zero control is set to the center point on R_1 to give $V_2 = 0$. If the pen holder is at the center point on the slide wire, $V_F = 0$. In this situation, the amplifier output is also zero, and there is no voltage applied to the servomotor. Consequently, the pen remains stationary at the center point on the chart. If the pen had been away from the center point on the slide wire, V_F would be some positive or negative voltage level. This would produce an output from the amplifier to drive the servomotor. The servomotor would move the pen holder in the direction necessary to make V_F approach zero. When V_F becomes zero, the motor is no longer energized and the pen remains stationary.

With the pen stationary at the center of the chart, suppose the zero control potentiometer (R_1) is now adjusted to give $V_2 = +E/2$. With V_F and V_i both zero, input current I_2 flows through R_5, giving output $V_o = -E/2$. This energizes the servomotor, causing it to drive the pen holder towards the $-E$ terminal of the slide wire. When $V_F = -E/2$, Equation 13-1 applies and the amplifier output becomes

$$V_o = -(V_2 + V_i + V_F)$$
$$= -\left(\frac{E}{2} + 0 - \frac{E}{2}\right)$$
$$= 0$$

Therefore, when $V_F = -V_2$, there is no amplifier output to drive the servomotor, and once again the pen remains stationary. If R_1 is adjusted to give $V_2 = -E/2$ or to $V_2 = -E$, the pen is driven to the points at which $V_F = +E/2$ or $+E$, respectively. It is seen that R_1 is actually a *pen zero-position control* that enables the pen to be set to any point across the chart.

Finally, consider the input voltage V_i, which is applied via an *amplifier/attenuator* input stage. As in the case of oscilloscopes, this input stage allows small voltages to be amplified and large voltage levels to be attenuated, to obtain convenient input levels for operating the circuitry. When V_i is a positive quantity, the amplifier output causes the servomotor to drive the pen holder in the direction of the $-E$ slide wire terminal. When V_F becomes equal to $-V_i$, the amplifier output is zero, and the pen is stationary. When V_i is a slowly changing quantity, the pen moves continuously, keeping $V_F = -V_i$ and tracing the waveform of V_i on the moving chart paper. By adjusting the zero control, the waveform of V_i can be traced symmetrically about the centerline on the chart paper, or at any other point across the chart.

The accuracy of the potentiometric strip-chart recorder depends on supply voltages $\pm E$. The instrument should be calibrated periodically to check the accuracy. This normally involves application of precisely measured positive and negative dc input levels. The deflection produced by each input level is checked, and the supply voltages adjusted as necessary to produce the appropriate deflection.

A major disadvantage of the potentiometric strip-chart recorder is that its frequency response is very low. Typically, 10 Hz is the maximum input frequency that may be traced. An important advantage of this instrument is that it can be much more accurate than the galvanometric type. Typical specified accuracies are ±0.2% for potentiometric type, and ±2% for galvanometric instruments.

13-2 *X-Y* RECORDER

An *X-Y recorder* uses a single stationary sheet of chart paper, and records by moving the pen simultaneously in both the *X* and *Y* directions. The recorder mechanism is the same as that of the potentiometric strip-chart recorder illustrated in Figure 13-4 except that two complete self-balancing potentiometers are involved. Figure 13-5 shows that the pen holder slides along a movable *pen carriage*. A servomotor, idler pulley, and drive cord control the pen position. The entire pen carriage and drive system can be made to slide across the chart paper under the control of another servomotor. Note that the servomotor controlling the position of the pen carriage uses a crossed drive cord which is mounted on four idler pulleys. Two potentiometer slide wires, two low resistance slide wires, and the associated control circuitry are required. These are not illustrated in Figure 13-5.

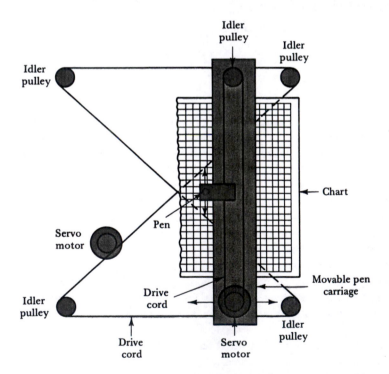

Figure 13-5   An *X-Y* recorder has two potentiometric recorder movements operating at right angles to each other to set the *X* and *Y* positions of the pen.

The control circuits for the X input and Y input are each essentially the same as shown in Figure 13-4(b). Each has a zero control, a summing amplifier, and a voltage follower for feedback from the slide wire. A typical input amplifier stage is illustrated in Figure 13-6. The circuit is a *noninverting amplifier* (see Section 4-2). With switches S_1 and S_2 in the positions shown, the amplifier gain is 10. Thus, an input of 0.1 V produces a 1 V output, which causes the recorder pen to deflect by 1 cm. So, with S_2 at 0.1 V/cm, a 0.1 V input produces a 1 cm deflection. With S_2 in the 0.01 V/cm or 1 V/cm positions, the amplifier gain is altered to give the required pen deflection for each input range. When S_1 is switched from *Range* to *Vernier*, the amplifier gain can be adjusted from a minimum of 1 to a maximum of 11, giving continuous (noncalibrated) V/cm adjustment.

The representative X-Y recorder illustrated in Figure 13-7 has three input terminals (+, –, and ground) for each (X and Y) section. The *Range* and *Vernier* controls for the X and Y inputs are as discussed for the amplifier, and the *Zero* controls provide a means of setting the pen to any desired starting point on the chart (as explained for Figure 13-4).

The row of switches above the X and Y controls in Figure 13-7 are identified as *LINE, SERVO, CHART,* and *PEN.* The *LINE* switch is simply an *on/off* control for the ac supply, and the *SERVO* switch disconnects the X and Y inputs when in the *off* position. When switched to *on,* pen deflection is produced proportional to the X and Y input volt-

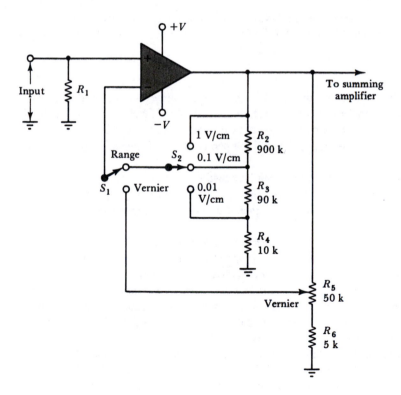

Figure 13-6 Input amplifier for an X-Y recorder. Switch S_2 selects the V/cm deflection sensitivity, and S_1 selects range or vernier.

Figure 13-7 X-Y recorder. Note the *RANGE, ZERO,* and *VERNIER* controls for the X and Y inputs. Also, note the *LINE, SERVO, CHART,* and *PEN* switches. (Courtesy of Hewlett-Packard.)

age levels. The *CHART* switch controls a high-voltage source which electrostatically holds the chart in position for the pen to trace an input signal. The *PEN down/up* control activates a solenoid that pulls the pen down onto the chart when switched to *down.* In the *up* position, the solenoid is not energized, and the pen is lifted off the paper by the action of a spring.

In some *X-Y* recorders a time base can be substituted in place of the *X* input. This is similar to the time base in an oscilloscope, except that it is normally not repetitive. The time base is usually triggered manually to trace a *Y* input waveform over a given time period.

The input resistance of an *X-Y* recorder typically varies from 11 kΩ to 1 MΩ, depending on the selected (X,Y) input voltage range. Accuracy is usually around 0.2% of full scale, and the *common-mode rejection ratio* is likely to be 90 dB or greater. *Common-mode* inputs are voltages that appear at both *X* input (+ and −) terminals at the same time, or both *Y* input terminals. It is important that such inputs do not produce pen deflection. A 90 dB common-mode rejection ratio implies that common-mode inputs are attenuated by 90 dB, which is approximately the same as dividing by 32 000.

Slewing speed is another item listed on an *X-Y* recorder specification. This defines the fastest rate at which the pen can be deflected, and might typically be stated as 50 cm/s. The slewing speed limits the frequency response of the instrument. For most *X-Y* recorders, the highest signal frequency that can be traced is around 10 Hz.

13-3 PLOTTING DEVICE CHARACTERISTICS ON AN *XY* RECORDER

Diode Characteristics

Large-scale characteristics for many types of electronic devices can be traced on an *X-Y* recorder. The method is much faster and more convenient than point-by-point plotting, and gives more satisfactory results than a photograph of a CRT display. Figure 13-8(a) shows the circuit for obtaining the forward characteristic of a semiconductor diode. The typical resultant characteristic is also shown. R_1 is a 1 kΩ resistor connected to the verti-

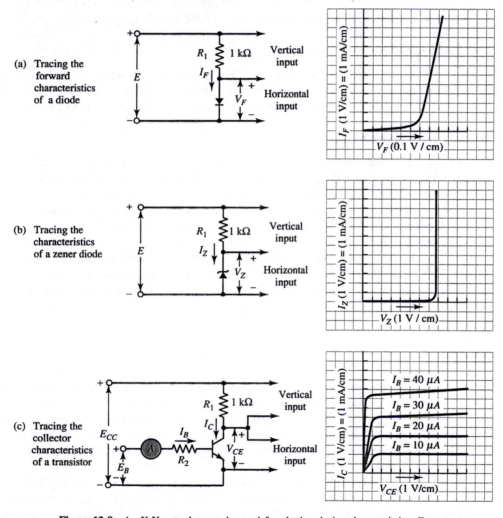

Figure 13-8 An X-Y recorder may be used for plotting device characteristics. For a diode, the voltage drop across resistor (R_l) is proportional to the forward current (I_F), so I_F is plotted versus forward voltage (V_F).

cal input of the XY recorder. When the diode forward current is 1 mA, a 1 V drop is produced across R_1. With the vertical deflection sensitivity at 1 V/cm, the 1 V input produces a 1 cm vertical deflection on the chart. So the vertical scale on the characteristic can be identified as 1 mA/cm. The diode terminals are connected to the horizontal input, and the horizontal input range is set to 0.1 V/cm. For a silicon diode V_F is approximately 0.7 V; thus the maximum horizontal deflection is around 7 cm. The vertical and horizontal ordinates of the graph can be drawn by adjusting the X and Y zero controls to move the pen. To trace the characteristic, supply voltage E is first set to zero, and the pen position is adjusted to the desired zero point on the graph. The supply voltage is then manually adjusted (using a power supply) from zero to the level necessary to trace the characteristic to the desired forward current level. At this point the pen RECORD/LIFT switch is set to LIFT before E is reduced to zero. If this is not done, a double trace may result.

Zener Diode Characteristics

The characteristics of a zener diode [Figure 13-8(b)] are traced in essentially the same way as for an ordinary diode. In this case, the horizontal scale should be set to 1 V/cm in order to give a convenient deflection for breakdown voltages ranging from 3 V to 12 V.

Transistor Characteristics

To trace the collector characteristics of a transistor, two power supplies are required, as illustrated in Figure 13-8(c). With the pen lifted and E_{CC} around 5 V, E_B is adjusted to give a convenient level of bias current (e.g., $I_B = 10$ μA). E_{CC} is next reduced to zero, the pen is lowered onto the chart, and then E_{CC} is gradually increased to produce the transistor collector characteristic for $I_B = 10$ μA. The pen is now lifted off the chart again and E_B is adjusted to increase I_B to 20 μA. E_{CC} is reduced to zero once again, the pen is lowered, and E_{CC} is again gradually increased. This causes the pen to trace the characteristic for I_B = 20 μA. The process described is repeated for other levels of I_B to give any number of characteristics. The characteristics of field effect transistors and other devices may also be traced on an X-Y recorder.

13-4 PLOTTERS

A *plotter* can be essentially the same as a potentiometric strip chart recorder, with the exception that it normally plots on a single sheet of paper (like an X-Y recorder) instead of a strip of paper. The sheet is roller-fed back and forward as required, and the pen is moved horizontally to produce the drawing. Figure 13-9 shows such an instrument. A plotter can be used for the same kinds of applications as an X-Y recorder. Typically, plotters are used for drafting technical diagrams.

The instrument illustrated in Figure 13-10 (the HP 7090A) is described by the manufacturer as a *measurement plotting system* capable of digital plotting and analog recording. It can simultaneously sample on each of three channels, converting from analog inputs to digital with a 12-bit resolution, and store the samples in 1000 word per channel memories. The maximum sampling rate is 33.3 kS/s, giving a bandwidth of dc to 3 kHz. Data can be transferred from the memories for plotting, or it can be plotted in real time

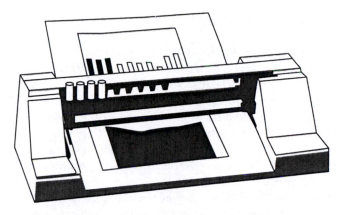

Figure 13-9 A plotter uses a pen that moves horizontally across a roller-fed sheet of paper.

from the ADCs at up to 500 points/s. Real-time X-Y and Y-T analog inputs from other instruments can be recorded. Plain paper can be used, instead of graph paper, because the instrument draws user-defined axes and grids on paper. The print outs can also be annotated with setup conditions, date, time, and so on. The sensitivity is specified as 5 mV to 100 V full scale, and accuracy is ±0.15% to ±0.26% depending on the selected range.

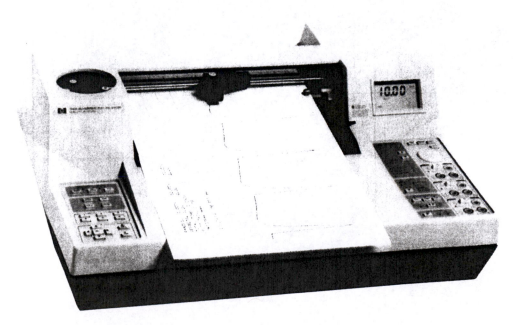

Figure 13-10 Three-channel plotter with digital storage. (Courtesy of Hewlett-Packard.)

13-5 DIGITAL WAVEFORM RECORDER/ANALYZER

Most digital storage oscilloscopes (see Section 10-4) have an output connector for a strip chart recorder, or a computer printer, to give a *hard copy* of stored waveforms. The *waveform analyzer/recorder* shown in Figure 13-11 (the Hioki 8850) is essentially a digital storage oscilloscope combined with a strip chart recorder. Input waveforms on three channels can be sampled at 20 MS/s and stored in 8-bit 64 k word/channel memories. The waveforms can be recalled from memory for display on the screen and/or printing out on paper as desired. As well as functioning as a memory recorder, this instrument can be used directly for real-time recording of input waveforms. It can also perform as an X-Y recorder and as an FFT analyzer (see Section 14-3).

The printing method uses a *line head,* or an array of electrodes which print thermally on the paper. The printing density is approximately 6 dots/mm, which produces a continuous line, and the maximum paper speed is 2 cm/s. The printer can be set to *automatically* print stored waveforms as they are recalled from memory, to print *manually* at the touch of a button, to *copy* the display on the CRT screen, or to *partial* print portions of waveforms identified by cursors on the screen.

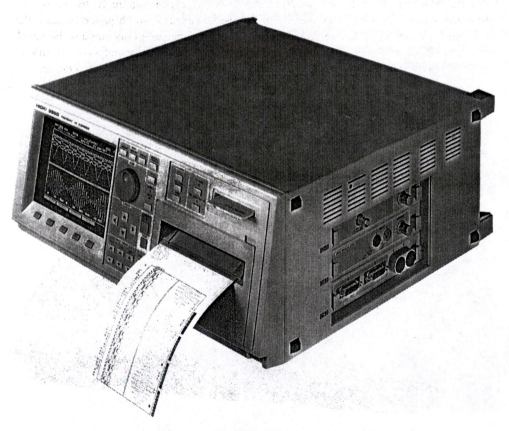

Figure 13-11 Waveform analyzer/recorder combining a digital storage oscilloscope with a strip chart recorder. (Courtesy of Hioki-RCC, Inc.)

13-1 Sketch the basic construction of a pen-type galvanometric strip chart recorder. Briefly explain the instrument operation.

13-2 Sketch the basic construction of a light-beam type galvanometric strip chart recorder. Explain briefly.

13-3 Compare the performance of light-beam and pen-type galvanometric strip-chart recorders.

13-4 Draw a sketch to show the basic construction of a potentiometric strip-chart recorder. Also sketch a summing amplifier control circuit for the moving system. Explain the operation of the complete system. State the typical upper frequency limit and accuracy for this type of pen recorder.

13-5 Sketch the mechanical system of a potentiometric *X-Y* recorder. Explain its operation.

13-6 Sketch the circuit of an input amplifier for an *X-Y* recorder. Explain the operation of the circuit.

13-7 List the controls normally found on an *X-Y* recorder. Explain the function of each control.

13-8 Sketch the circuits used when plotting diode and zener diode characteristics on an *X-Y* recorder. Explain the operation of each circuit and the recording procedure. Also explain why the traced characteristics may be marked in mA/cm and V/cm.

13-9 Sketch the circuit used for plotting the collector characteristics of a transistor on an *X-Y* recorder. Explain the circuit operation and the recording procedure.

13-10 Briefly describe a plotter, and discuss its applications.

13-11 Describe a digital waveform analyzer/recorder, and discuss its applications.

Waveform Analyzing Instruments

14

Objectives

You will be able to:

1. Sketch the basic circuitry and block diagram for a distortion meter. Explain the operation of the instrument, and discuss its performance.
2. Sketch basic block diagrams for swept TRF and swept superhetrodyne spectrum analyzers, and explain the operation of each instrument. Draw spectrum analyzer displays for various input waveforms. Discuss spectrum analyzer performance.
3. Explain the operating principle of a digital spectrum analyzer. Sketch its basic block diagram, and discuss its performance.
4. Briefly explain audio analyzers, wave analyzers, and modulation analyzers.

Introduction

A distortion meter measures the total harmonic distortion content in an input waveform. The rms level of the distortion is usually measured as a percentage of the rms level of the complete waveform. Alternatively, a decibel measurement may be made.

A spectrum analyzer uses a CRT display to show signal amplitude plotted to a base of frequency. The input waveform is first separated into its individual frequency components, then each component is displayed as a vertical line on the screen. The height of each line represents amplitude, and its horizontal position represents frequency. A digital spectrum analyzer samples a waveform, determines its equation, and then mathematically analyzes the equation to calculate the component waves for an amplitude versus frequency display.

14-1 DISTORTION METER

Harmonic Distortion

A sine-wave input to an electronic circuit may produce an output wave that is distorted, instead of being purely sinusoidal. Regardless of how severe the distortion may be, it can be shown that all repetitive waveforms consist of a *fundamental* frequency component and a number of *harmonics*. The fundamental is a sine wave with the same frequency (f) as the distorted repetitive wave. The harmonics are sine waves with frequencies that are multiples of the fundamental frequency; the second harmonic has a frequency $2f$, the third harmonic frequency is $3f$, and so on.

Distortion in a waveform can be measured in terms of its harmonic content. One method of determining the harmonic content is to suppress the fundamental component and measure the rms value of the combined harmonics (see Figure 14-1). The *total harmonic distortion (THD)* may then be expressed as a percentage of the rms value of the complete waveform (i.e., fundamental and harmonics). Alternatively, a decibel scale (see Section 4-6) may be used to indicate the distortion content.

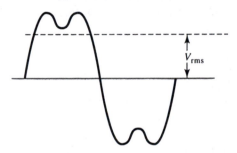

(a) Waveform with harmonic distortion

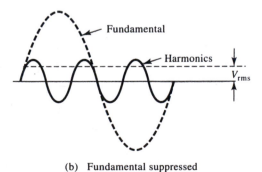

(b) Fundamental suppressed

Figure 14-1 To determine the total harmonic distortion (THD) in a waveform, the rms level of the complete waveform is first measured, then the fundamental is suppressed, and the rms level of the harmonics is measured. The THD is the harmonics voltage expressed as a percentage of the complete waveform voltage.

Rejection Amplifier

The basic component of a *fundamental-suppression distortion meter* is a *notch filter.* The filter must heavily attenuate the fundamental frequency component of the input waveform, while passing all harmonics with no alteration in amplitude or phase. Its suppression frequency must also be adjustable over a wide frequency range.

The *rejection amplifier* circuit in Figure 14-2 performs the function of a notch filter. It has two stages of amplification, identified as *preamplifier* and *bridge amplifier,* and a Wein bridge circuit between the two. Negative feedback is provided from the output of the bridge amplifier to the input stage of the preamplifier. The two amplification stages produce a high open-loop gain, and the NFB stabilizes the closed-loop gain at approximately 1 dB.

In Section 11-1 it is explained that the Wein bridge balances at only one frequency, giving a minimum null-detector voltage at that frequency. In the rejection amplifier, the bridge output is taken from the null-detector terminals, and the input (from the preamplifier) is applied to the bridge supply terminals. The frequency-dependent components of the bridge may be adjusted for any desired fundamental frequency. When the bridge is balanced, the fundamental frequency output to the bridge amplifier is attenuated by approximately 80 dB. All harmonics of the fundamental are passed without any added attenuation or distortion.

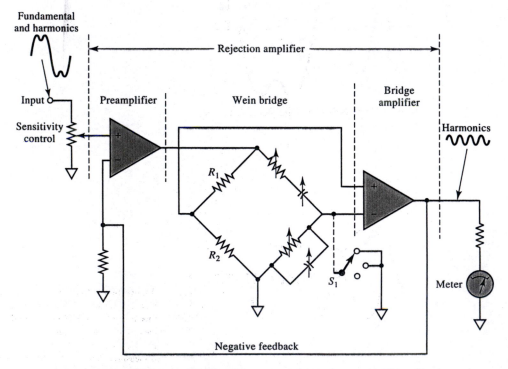

Figure 14-2   Rejection amplifier circuit for use in a distortion meter. The Wein bridge is tuned to balance at the fundamental frequency of the input waveform, and thus filter out (or reject) the fundamental frequency component.

Distortion Meter Block Diagram and Controls

The block diagram of a distortion meter is illustrated in Figure 14-3, and the front panel controls are shown in Figure 14-4. In Figure 14-3 it is seen that the input waveform passes via the *1 MΩ attenuator* and the *impedance converter* before arriving at the rejection amplifier. The attenuator reduces the signal amplitude to a suitable level for processing, and the impedance converter is simply a unity-gain amplifier for interfacing the attenuator and rejection amplifier. Attenuations of up to 50 dB can be selected in 10 dB steps by means of the *Sensitivity* control on the instrument front panel (Figure 14-4). The *Vernier* knob (*sensitivity vernier* in Figure 14-3) provides fine adjustment of signal attenuation. The *post attenuator* following the rejection amplifier is simply a meter range selector. It is controlled by the *Meter Range* switch on the front panel, which also controls the *1:1 and 1000:1 attenuator.*

S_1 is a five-wafer rotary switch: five switches ganged to be controlled by one knob. The knob is identified as *Function* on the front panel, and it has three positions: *Voltmeter, Set Level,* and *Distortion.* When S_1 is set at *Voltmeter* (position 1), the *1:1 and 1000:1 attenuator* is switched into the circuit, the rejection amplifier is bypassed, and the complete input waveform (fundamental plus harmonics) is applied to the meter. In this condition, the instrument functions as an ac voltmeter.

With S_1 in the *set level* position, the input wave passes to the rejection amplifier via the *1 MΩ attenuator* and the *sensitivity vernier.* For this condition of S_1, one of the (null-detector) output terminals of the bridge is grounded (see Figure 14-2), shorting-out frequency-dependent components of the bridge. Consequently, only the output from the junction of R_1 and R_2 is passed to the bridge amplifier and the meter circuits. This is the complete input waveform only slightly attenuated by R_1 and R_2. The input *sensitivity* con-

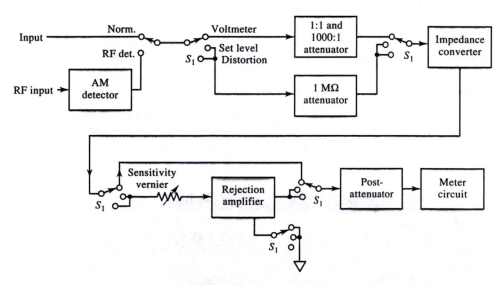

Figure 14-3 Block diagram for a *fundamental-suppression* distortion meter. With the switches as shown, the instrument functions as an ac voltmeter. At the S_1 *set level* position, the attenuators are adjusted to give a convenient meter reading. At the *distortion* position, the voltage level of the harmonics is measured.

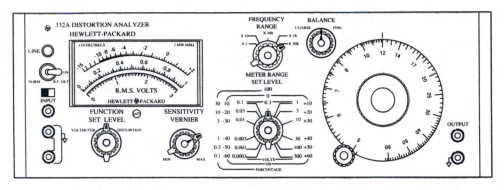

Figure 14-4 Front panel for the HP332A fundamental-suppression distortion meter. (Courtesy of Hewlett-Packard.)

trol can now be adjusted to give a convenient meter reading representing the rms level of the waveform.

Setting S_1 to its third position removes the ground from the bridge (see Figure 14-2), causing the fundamental frequency to be suppressed, so that only the harmonics get through to the meter. The rms level of the harmonics can now be measured in relation to the deflection produced by the complete waveform.

Figure 14-5 shows a HP339A *distortion measurement set,* which is capable of measuring total harmonic distortion over a fundamental frequency range of 10 Hz to 110 kHz. Measurement accuracy is ±1 dB at 20 Hz, and +1 dB − 4 dB at 110 kHz. The instrument includes a *tracking oscillator,* which is synchronized with the fundamental frequency of the distortion meter. The voltage measurement is true rms (see Section 15-2) over a 1 mV to 300 V range.

14-2 SPECTRUM ANALYZER

Swept TRF Spectrum Analyzer

A spectrum analyzer separates an ac signal into its various frequency components and displays each component as a vertical line on a CRT screen. The amplitude of each vertical line in the display represents the amplitude of each frequency component, and the horizontal position of each line defines the frequency.

Figure 14-5 HP339A distortion measurement set. (Courtesy of Hewlett-Packard.)

Consider Figure 14-6(a), which shows the block diagram of a *swept TRF* (tuned radio frequency) *spectrum analyzer.* A sweep generator produces a linear ramp, which provides horizontal deflection voltage for the CRT. The ramp is also applied to a *voltage-tunable bandpass filter.* This is a filter with a very narrow passband. The center frequency of the passband is swept from a minimum frequency (f_1) to a maximum (f_2) as the ramp voltage sweeps from minimum to maximum amplitude. An input signal with a frequency (f_s) would pass through the bandpass filter only during the brief time that the filter passband is tuned to f_s. When the signal (with frequency f_s) emerges from the filter, it is converted to a dc voltage level by the *detector* and applied as an input to the vertical deflection amplifier of the CRT. Thus, during the time that the filter passband sweeps through f_s, a vertical line representing the signal amplitude is traced on the CRT screen.

The horizontal position of the vertical line is determined by the amplitude of the sweep generator ramp voltage at the instant that f_s is passing through the filter. As explained earlier, the ramp voltage also dictates the instantaneous filter frequency. The horizontal sweep of the electron beam across the CRT screen represents the changing filter frequency, from f_1 to f_2. Consequently, the horizontal position of each vertical line in the display can be identified as a particular signal frequency. If f_s is exactly halfway between f_1 and f_2, the vertical line representing f_s should be exactly in the center of the screen [see Figure 14-6(b)].

Suppose that two different signals are applied simultaneously to the input of the filter. Assume that the signal frequencies are $f_a = 200$ kHz and $f_b = 300$ kHz and that the amplitudes are $V_a = 2$ V and $V_b = 1$ V. If the center frequency of the filter is swept from 100 kHz to 500 kHz, the display shown in Figure 14-7(a) on p. 395 would result. Signal *a* is displayed at a point one-fourth of the way across the screen because its 200 kHz frequency is one-fourth of the way between $f_1 = 100$ kHz and $f_2 = 500$ kHz. Similarly, signal *b* is displayed at the center of the screen because 300 kHz is halfway between 100 kHz and 500 kHz. The vertical line representing signal *a* is twice as large as the line representing *b* because $V_a = 2$ V and $V_b = 1$ V.

Figure 14-7(b) illustrates the result of applying an amplitude modulated signal to a spectrum analyzer. The display consists of a single, large, vertical line, representing the carrier frequency amplitude, together with two smaller lines representing the two side-band frequency components. The display allows the amplitudes and frequencies of the sidebands to be investigated in relation to the carrier.

The input signal shown in Figure 14-7(c) appears to be a pure sine wave. When displayed on an ordinary oscilloscope, the signal would not reveal any obvious distortion. However, when applied to a spectrum analyzer, harmonic distortion might well be revealed, as illustrated. In the display shown, small second, fourth, and sixth harmonic components are present.

An ordinary oscilloscope displays the amplitude of the input signal plotted to a base of time [see input in Figure 14-7(c)]. Thus, it is said to operate in the *time domain.* A spectrum analyzer displays the signal amplitude plotted to a base of frequency. So a spectrum analyzer operates in the *frequency domain.*

Swept Superheterodyne Spectrum Analyzer

The *swept superheterodyne spectrum analyzer* block diagram shown in Figure 14-8 on p. 396 is basically similar to the swept TRF system in Figure 14-6. The difference is that the voltage-tunable bandpass filter is now replaced with a *voltage-tunable oscillator* (VTO)

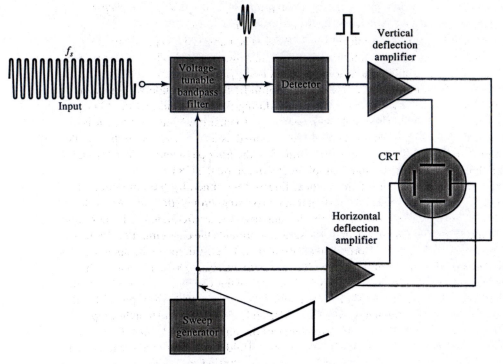

(a)   Block diagram of swept *TWF* spectrum analyzer

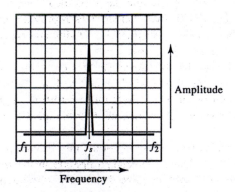

(b)   Display produced by a single input frequency

Figure 14-6   Block diagram, waveforms, and display for a *swept trf* spectrum analyzer. The bandpass filter is swept over the desired frequency range at the same time as the CRT trace is moved from left to right. Individual frequency components are detected and fed to the CRT to produce vertical deflection on the screen.

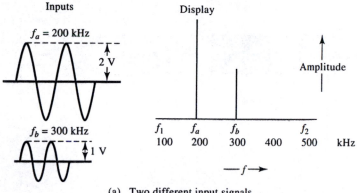

(a) Two different input signals
    applied simultaneously

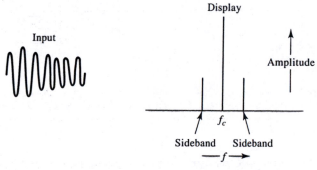

(b) Amplitude-modulated waveform is separated into its
    carrier frequency component and two sidebands.

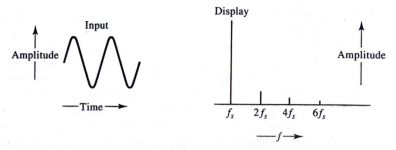

(c) An apparently perfect sine wave input might
    be shown to have harmonic distortion.

Figure 14-7 Spectrum analyzer displays produced by various input waveforms.

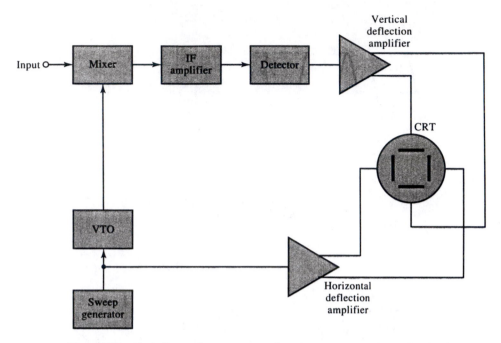

Figure 14-8 Block diagram for a *swept superhetrodyne* spectrum analyzer. The signal (f_s) is mixed with the output (f_o) of the voltage-tunable oscillator (VTO) to produce ($f_o \pm f_s$). The IF amplifier passes one of these two for detection and vertical deflection at the CRT screen.

(see Section 11-5), *a frequency mixer,* and an *intermediate-frequency* (IF) *amplifier.* The ramp voltage from the sweep generator is applied, once again, as a horizontal deflecting voltage to the CRT. The ramp is also applied to the VTO to produce a VTO output frequency that sweeps from a minimum (f_1) to a maximum (f_2). The VTO output is applied to one input of the mixer, and the other mixer input terminal receives the signal to be analyzed. If the signal frequency is f_s and the VTO frequency is f_o, the mixer output is the sum and difference of the two frequencies: $f_m = (f_o \pm f_s)$. These are applied to the IF amplifier, which passes and amplifies only one intermediate frequency.

Assume that the IF amplifier passes only a 100 kHz output frequency component from the mixer and that the VTO frequency (f_o) sweeps from 100 kHz to 200 kHz. If the signal frequency is $f_s = 50$ kHz, the mixer output when $f_o = 100$ kHz would be $f_m = 100$ kHz ± 50 kHz = (150 kHz or 50 kHz). Neither of these two frequencies is passed by the 100 kHz IF amplifier. Similarly, at the maximum VTO frequency of 200 kHz, $f_m = 200$ kHz ± 50 kHz = (250 kHz or 150 kHz). Again, neither frequency can pass through the IF amplifier. When the VTO output has a frequency of 150 kHz, and $f_s = 50$ kHz, then $f_m = 150$ kHz ± 50 kHz = (100 kHz or 200 kHz). The 100 kHz component is passed and amplified by the IF amplifier, and it produces a vertical line on the CRT. The VTO output at 150 kHz is exactly halfway between its extremes of 100 kHz and 200 kHz. Therefore, the vertical line representing $f_s = 50$ kHz occurs at the center of the CRT screen (see Figure 14-9).

When $f_s = 25$ kHz, a passable IF input of 100 kHz occurs at $f_o = 125$ kHz. This produces a vertical line one-fourth of the way across the screen horizontally. Similarly, when

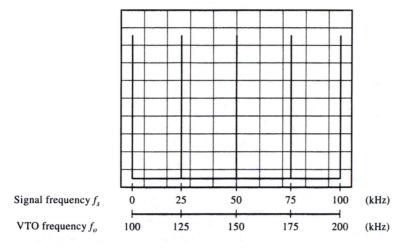

| Signal frequency f_s | 0 | 25 | 50 | 75 | 100 | (kHz) |
| VTO frequency f_o | 100 | 125 | 150 | 175 | 200 | (kHz) |

Figure 14-9 Signal frequency, VTO frequency, and display for a swept superhetrodyne spectrum analyzer that has a 100 kHz IF amplifier frequency.

$f_s = 75$ kHz, the 100 kHz IF is produced with $f_o = 175$ kHz to display a vertical line three-fourths of the way across the screen, as shown in Figure 14-9. When $f_s = 0$, and when $f_s = 100$ kHz, lines are produced at the left- and right-hand side, respectively, of the CRT screen. Thus, as displayed in Figure 14-9, signal frequencies from 0 to 100 kHz can be investigated. A change of IF amplifier or VTO will change the range of signal frequencies that may be displayed.

The major advantage of the superheterodyne spectrum analyzer over the TRF type is that the IF amplifier improves the instrument sensitivity. Also, the detector can give a better performance, since it has to operate at only one (IF) frequency.

Spectrum Analyzer Controls and Specifications

The spectrum analyzer *tuning* control selects the display frequency, either as a starting frequency at the left side of the screen or as a center frequency on the screen. The tuning control might also be used to set a negative marker spike anywhere on the screen to identify a particular signal to be investigated (see Figure 14-10).

Displayed frequency range is set by the *frequency span/div* control. At a 1 MHz selection, each horizontal division on the screen represents an increase in frequency (from left to right) of 1 MHz. The *reference level* control is used to adjust the amplitude of a main signal (or reference signal) to fill the screen vertically. The setting of the reference level control identifies the absolute power of the reference signal in dBm (see Section 4-6). All other signal levels may be measured by their amplitude relation to the (full-scale) reference level.

The *frequency range* of a spectrum analyzer defines the extremes of signal frequency that may be investigated. Typical frequency ranges are 20 Hz to 40 MHz, and 100 Hz to 22 GHz. The *resolution bandwidth (RBW)* specifies the smallest signal frequency separation that can be identified. Frequencies that are closer than this would tend to be displayed as one signal. For an analyzer with a frequency range of 20 Hz to 40 MHz, the RBW might be 3 Hz at the low end of the band and 30 kHz at the high end.

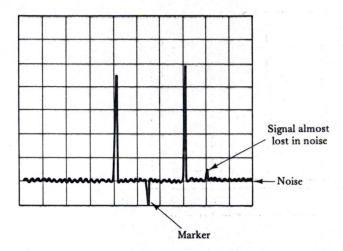

Signal almost
lost in noise

Noise

Marker

Figure 14-10 Spectrum analyzer display
showing noise and inverted marker spike.

The *sensitivity* of the instrument defines the minimum signal amplitude that may be observed. This is limited by the thermal (or circuit) noise generated within the spectrum analyzer (see Figure 14-10). Since the thermal noise is directly proportional to the bandwidth, the narrowest RBW gives the greatest sensitivity. The average noise level might typically be −120 dBm at the narrowest RBW, and the minimum detectable signal amplitude is 3 dB above this level. Typical input signal amplitudes range from 3 mV peak to 30 V peak.

Spurious responses are harmonic and other types of noise generated by large amplitude signals. Normally, the largest amplitude signal that can be applied to a spectrum analyzer is defined as that which produces a spurious response with an amplitude 3 dB greater than the thermal noise. The *dynamic range,* typically around 80 dB, is the difference between the minimum and maximum signal levels (as defined above).

14-3 DIGITAL SPECTRUM ANALYZER

If the equation for a waveform is known, it can be mathematically processed by Fourier analysis to evaluate its component waves. In a digital spectrum analyzer, the waveform to be analyzed is sampled, and the samples are digitized and fed to a computer. The computer is programmed to determine the waveform equation from the samples, and to analyze the equation to calculate the component waves. The component waveforms are stored in the computer memory, then, when retrieved from memory, they are converted back to analog form for display on a CRT screen as amplitude versus frequency. The computer can also readily assess the peak values, rms values, and phase of the component waves.

The sampling and conversion techniques used in a digital spectrum analyzer are essentially the same as those used in a digital oscilloscope (see Section 10-4). Figure 14-11 shows the basic block diagram of a digital spectrum analyzer. In the simplest approach to sampling the waveform to be analyzed, the attenuator (in Figure 14-11) operates as a switch that applies the input waveform to the ADC for a fixed time period. This can be thought of as opening a *rectangular window* on the waveform for the sampling to take place. The waveform applied to the computer during this time is then assumed to be repetitive. The rectangular window is satisfactory for investigation of transients, but can add

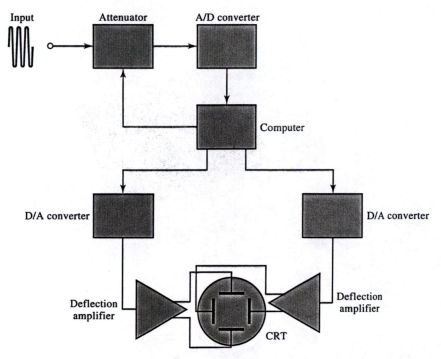

Figure 14-11 Basic block diagram for a digital (or Fourier) spectrum analyzer. The input waveform is sampled, and its equation is determined by the computer, which analyzes the equation to calculate the component waves.

considerable distortion to sinusoidal and other wave shapes. To minimize distortion, methods of slowly varying the attenuation of the waveform have been developed, as an alternative to switching it *on* and *off*. These can be thought of as slowly opening and closing the window on the waveform. The *Hann* window opens slowly to a peak (or wide open) level for a brief time, then closes in the same way. Other windows are available with different open/close characteristics, which are each suitable for particular waveform types.

Aliasing, or generation of waveforms that are not part of the original input can occur in a digital spectrum analyzer, as in a digital storage oscilloscope. This is avoided by the use of filters, and by ensuring that the input is sampled at a rate greater than two samples per cycle.

The algorithm used by the computer in a digital spectrum analyzer is known as a *fast Fourier transform (FFT)*. Therefore, this type of instrument is also known as an *FFT analyzer,* or as a *Fourier analyzer.* Recall from Section 14-2 that analog spectrum analyzers measure each component wave one at a time by sweeping through the selected frequency range. Digital spectrum analyzers measure all frequencies simultaneously, so they are able to investigate changing, or dynamic, waveforms. Consequently, they are also termed *dynamic signal analyzers.* Analog spectrum analyzers are sometimes referred to as *real-time analyzers,* as compared to the digital type, which display stored signals.

Fourier analyzers are essentially low-frequency instruments. A typical input frequency range is 250 µHz to 100 kHz, and a typical input voltage amplitude might be 4 mV to 30 V peak. These instruments can be employed for investigating the characteristics

Figure 14-12 HP3560A portable dynamic signal analyzer. (Courtesy of Hewlett-Packard.)

of audio amplifier, filters, and loudspeakers. Machinery vibration analysis is a major application of Fourier analyzers. Motion transducers are used to convert mechanical vibrations into electrical signals, which can then be studied as waveforms. For example, the relationship between vibration and the frequency of rotation of components within the machinery can be investigated.

A portable dynamic signal analyzer is illustrated in Figure 14-12. This is a two-channel instrument with a signal frequency range of 31.25 mHz to 40 kHz.

14-4 ADDITIONAL WAVEFORM ANALYZING INSTRUMENTS

Audio Analyzer

An *audio analyzer* is basically a distortion meter, as described in Section 14-1, with several additional measurement facilities for complete performance analysis of audio amplifiers, receivers, and other audio components. These include a frequency meter, a dc voltmeter, a true rms ac voltmeter, and an audio oscillator. For testing radio receiver sensitivity a *SINAD* function may also be included. SINAD refers to *Signal, Noise, And Distortion,* and it is a measure of the input signal level that gives an audio output at a specified (dB) level above the total noise plus distortion.

Wave Analyzer

The function of a *wave analyzer* can be described as the mirror image of that of a distortion meter. The distortion meter uses a tunable notch filter to remove a particular frequency, and then measures the combined level of the remaining component waves. The wave analyzer uses a tunable narrow bandpass filter to select a particular frequency component for measurement. Other names for this instrument are *selective level meter (SLM)* and *frequency-selective voltmeter.*

Modulation Analyzer

This instrument is used for investigating amplitude-, frequency-, and phase-modulated RF signals. Also termed a *precision receiver,* it can measure carrier frequency and power, as well as modulating frequency, modulation depth, frequency deviation, and so on. It is applied to testing transmitters, RF signal generators, attenuators, and other communication circuits.

REVIEW QUESTIONS

14-1 Describe the basic function of a distortion meter, and discuss its frequency range.

14-2 Sketch the circuit of a rejection amplifier for use with a fundamental-suppression distortion meter. Carefully explain the circuit operation, and show how the filtering process can be disabled to pass the complete waveform.

14-3 Draw the complete block diagram for a fundamental-suppression distortion meter, and explain its operation.

14-4 Referring to the block diagram drawn for Question 14-3, discuss the distortion meter controls.

14-5 Sketch the block diagram for a swept TRF spectrum analyzer. Show the waveforms at various points in the system, and explain its operation.

14-6 Sketch and explain the spectrum analyzer displays that are likely to be produced by the following inputs:
   (a) A pure sine wave with a frequency halfway between the extremes of the swept frequency range.
   (b) Two pure sine-wave inputs with different frequencies and amplitudes.
   (c) An amplitude-modulated sine waveform.
   (d) A sine wave with a small amount of harmonic distortion.

14-7 Sketch the block diagram for a swept superhetrodyne spectrum analyzer. Explain the system operation.

14-8 For a swept superhetrodyne spectrum analyzer, discuss the relationship between display, signal, and VTO frequency.

14-9 Define and briefly discuss each of the following quantities in reference to a spectrum analyzer: frequency span/div, reference level, frequency range, resolution bandwidth, sensitivity, spurious response, minimum signal level, maximum signal level, dynamic range.

14-10 Explain the operating principle of a digital spectrum analyzer. Draw the basic block diagram for a digital spectrum analyzer, and describe its operation.

14-11 Discuss the frequency range and applications of a Fourier analyzer.

14-12 Briefly explain audio analyzers, wave analyzers, and modulation analyzers.

Miscellaneous Meters

15

Objectives

You will be able to:

1. Briefly explain thermocouples, and draw sketches to show their typical construction. Also, sketch circuits of various thermocouple instruments, and explain their operation.

2. Sketch the circuit and waveforms for a peak detector, and explain its operation. Discuss the applications of a peak detector voltmeter.

3. Explain the need for a true rms voltmeter. Sketch a true rms voltmeter circuit using a nonlinear circuit, and a circuit of a thermocouple-type true-rms voltmeter. Explain the operation of each instrument, and compare their performance.

4. Discuss the problems that occur in measuring very low voltage levels. Draw the block diagram and waveforms for a chopper-stabilized amplifier as used in a low-level instrument, and explain its operation. Show how a guard terminal is used for low-level measurements.

Introduction

The usual measuring circuits employed in electronic voltmeters (analog and digital) are not suitable for many specialized applications. Consequently, special instruments have been developed for these applications. Thermocouples, which are junctions of dissimilar metals that produce a voltage when heated, can be employed in ammeters, voltmeters, and wattmeters. Because thermocouple voltages are proportional to the rms level of the heating current, they can be used in true-rms electronic voltmeters. These can measure rms levels for all voltages regardless of how distorted the waveform might be.

Voltage measurements on high-frequency waveforms are usually made with a peak detector voltmeter. As the name implies, this instrument measures the waveform peak level and then converts it to an rms value for display. Measurement of very low voltage levels is affected by very high source resistances, noise voltages, and bias voltage drifts in the measuring circuits. These problems are dealt with by the use of very high input resistance devices, ac-coupled amplifiers, and instrument guard terminals.

15-1 THERMOCOUPLE INSTRUMENTS

Thermocouples

A junction of two dissimilar metals develops an electromotive force (emf) when heated. By using a current to heat the junction, an emf is produced which is proportional to the heating effect of the current. Since the heating effect of a current is directly proportional to the rms value of the current (regardless of its waveform), the generated emf can be used as a measure of the rms level of the current.

Figure 15-1 illustrates the principle of the *thermocouple instrument.* The thermocouple consists of the junction of two dissimilar metal wires welded to a heating wire. The current to be measured passes through the heater and thus heats the junction. A millivoltmeter measures the voltage developed across the junction. The scale of the meter is calibrated to indicate the actual rms current in the heater.

Two types of thermocouples are illustrated in Figure 15-2. Figure 15-2(a) shows a thermocouple enclosed in a vacuum tube to protect the junction from loss of heat. The junction may be directly welded to the heater, or it may be thermally (but not electrically) connected to the heater by a bead of ceramic material.

In Figure 15-2(b) a thermocouple with a flat heating conductor is shown. The ends of the thermocouple wires are connected to copper pads which are electrically insulated from the heater. Although electrically insulated, the copper pads are in thermal contact with the large copper terminal blocks of the heater. This has the effect of keeping the ends of the thermocouple wires at the same ambient temperature as the terminal blocks. Thus, the thermocouple junction is heated, but the ends of the wires are at the normal (relatively cold) temperature of the terminals. When no current flows, both ends of each thermocouple wire are maintained at ambient temperature, and no voltage is generated. When current flows, the junction of the dissimilar metal wires is heated, while the opposite ends remain at the ambient temperature. These opposite ends are electrically connected together through the milli-

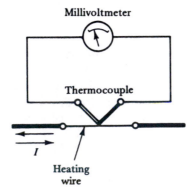

Figure 15-1 Basic thermocouple instrument. The current to be measured is passed through a wire that heats the junction of two dissimilar metals, thus producing a measurable emf.

Miscellaneous Meters   Chap. 15

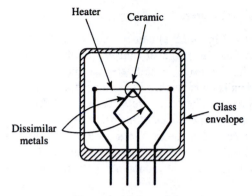

(a) Thermocouple in a vacuum tube

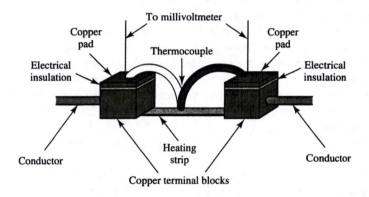

(b) Compensated thermocouple

Figure 15-2 Two types of thermocouple. The vacuum tube in (a) protects the junction from heat loss. In (b), the unjoined ends of the dissimilar metals are held at ambient temperature by the copper pads. This minimizes emfs generated by changes in ambient temperature.

voltmeter, so they can be termed a *cold junction*. This condition, of a hot junction and a cold junction in the thermocouple circuit, is the requirement for maximum emf generation. Since the thermocouple wires are maintained at the same ambient temperature when no current is flowing, the emf generated at the heated junction results only from the heating effect of the current. The device just described [Figure 15-2(b)] is termed a *compensated thermocouple,* meaning that it is compensated against the effects of any change in ambient temperature.

Some of the most common materials used as thermocouple pairs are iron–constantan, copper–constantan, chromel–alumel, and platinum–platinum/rhodium. Thermocouple junctions can survive very high temperatures and are used as temperature transducers. However, when used in a measuring instrument, the typical maximum heater temperature is about 300°C. The typical maximum thermocouple output at this temperature is around 12 mV. Heating element currents range from 2 mA to 50 mA.

Thermocouple Ammeters and Voltmeters

A thermocouple instrument (as in Figure 15-1) can be used directly as an ammeter, and shunts can be employed to expand its range of current measurement. By connecting multiplier resistors in series with the heater, a voltmeter can be constructed. The sensitivity of thermocouple voltmeters is considerably lower than that of a PMMC voltmeter. However, thermocouple instruments indicate the true rms level of the measured voltage or current. They can also be used as transfer instruments, calibrated on dc and then employed to measure either ac or dc. Furthermore, thermocouple instruments can be used from dc to very high frequencies, 50 MHz and higher. The frequency limit is not due to the thermocouple, but to the capacitance and inductance of connecting leads and series resistors.

Thermocouple Bridge

In Figure 15-3 eight thermocouple pairs are arranged in a bridge configuration. Two series-connected pairs are situated between A and B. Another two are connected between each of B and C, C and D, D and A. Junctions e, f, g, and h are secured to terminals which project from an insulated base. These are *cold* junctions, maintained at ambient temperature. Junctions i through p are *hot* junctions, heated by current flowing from A to C (or vice versa) through each of the two chains of series-connected thermocouples. When current flows, the hot junctions generate voltages with the polarity indicated on the diagram. Thus, if each junction produces an emf of 6 mV, the total emf developed across the millivoltmeter connected to B and D is 4×6 mV = 24 mV. Note that no portion of the main circuit current I flows through the millivoltmeter because of the bridge configuration.

It is seen that a thermocouple bridge can generate more emf than a single junction acting alone. Also, the weakest part of the *separately heated* thermocouple is the heater itself, which tends to be rapidly destroyed when only 50% overloaded. The bridge instrument (with directly heated thermocouples) is quite rugged because the current to be measured flows directly through the thermocouples.

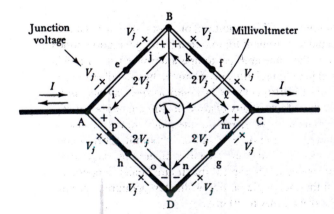

Figure 15-3 Thermocouple bridge instrument. The meter voltage is zero when no current flows. Current through the thermocouples generates junction voltages which add up to produce a meter reading.

Miscellaneous Meters   Chap. 15

Thermocouple Wattmeter

Since thermocouple instruments can be operated to very high frequencies, a thermocouple wattmeter is useful for power measurements at frequencies far beyond the range of electrodynamic wattmeters. The basic circuit of a thermocouple wattmeter is shown in Figure 15-4. A *current transformer* is employed to produce a secondary current i_i, which is directly proportional to the load current I. The *voltage transformer* secondary voltage is directly proportional to the supply voltage E. Suppose, at a given instant, that the instantaneous current i_i is flowing from left to right through the thermocouple heaters, as illustrated. At the same instant, the current directly proportional to the supply voltage E flows into the center tap of the secondary winding of the current transformer, then splits into equal current levels i_v. These currents (i_v) flow from the center tap through each half of the current transformer secondary winding, as illustrated, then through R_1 and R_2 in opposite directions. The currents again combine at the junction of R_1 and R_2 where they flow back to the transformer secondary winding.

The current $i_i + i_v$ in R_1 heats its adjacent thermocouple and produces an output e_1 with polarity: $+$ on the left, $-$ on the right. Current $i_i - i_v$ flowing in R_2 heats its thermocouple and produces an output e_2 with polarity opposite to that of e_1: $-$ on the left, $+$ on the right. The millivoltmeter indicates $e_1 - e_2$. Since e_1 and e_2 are proportional to the square of the current in each heater:

$$e_1 \propto (i_i + i_v)^2$$

$$i_i \propto I \qquad \text{and} \qquad i_v \propto E$$

so

$$e_1 \propto (I + E)^2$$

or

$$e_1 \propto I^2 + 2\,EI + E^2$$

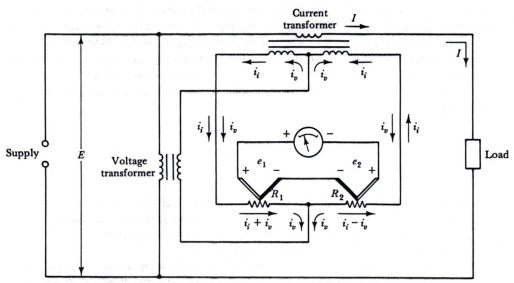

Figure 15-4 Basic circuit of a thermocouple wattmeter. The junction heater currents are proportional to the rms levels of I and E, giving a meter reading directly proportional to $EI \cos \phi$.

Also,
$$e_2 \propto (i_i - i_v)^2 \propto (I - E)^2$$

or
$$e_2 \propto I^2 - 2EI + E^2$$

millivoltmeter reading $= e_1 - e_2$
$$\propto (I^2 + 2EI + E^2) - (I^2 - 2EI + E^2)$$
$$\propto 4EI$$

Therefore, the millivoltmeter indicates a voltage proportional to EI, which is the power delivered to the load. When there is a phase difference between E and I, the power dissipated in each of the thermocouple heaters is proportional to $(I + E \cos \phi)^2$ and $(I - E \cos \phi)^2$. This gives a millivoltmeter indicates proportional to $EI \cos \phi$ (i.e., proportional to true power).

In practice, the thermocouple wattmeter circuit is usually a little more complicated than that shown in Figure 15-4. Instead of two thermocouples, a thermocouple bridge may be employed with the heating currents flowing directly through the thermocouple junctions.

15-2 PEAK RESPONSE VOLTMETER

The rectifier voltmeter circuits discussed in Sections 3-5 and 4-4 respond to the average value of the rectified waveform. In every case the instruments are calibrated to indicate rms voltages. Instead of passing the rectified waveform directly to the deflection meter, it is possible to detect the peak level of the input and apply this as a constant voltage to the meter. Figure 15-5 shows a *peak detector circuit,* as used with a *peak response voltmeter.*

Capacitor C_1 in series with the input terminal (in Figure 15-5) blocks dc input voltages and passes the ac voltage. When the input goes to its positive peak level, D_1 is forward biased and C_1 charges up to $V_p - V_F$ with the polarity shown. (The voltage developed across D_1 obviously cannot go above the level of $+V_F$ with respect to ground.) C_1 retains its charge, and when the input voltage goes negative diode D_1 is reverse biased. At the negative peak of the input, the voltage at the anode of D_1 is the sum of the input and the capacitor voltages:

$$e = -V_p - V_{C1}$$
$$= -V_p - (V_p - V_F)$$
$$\approx -2V_p + V_F$$

The alternating voltage (e) developed across the diode has exactly the same waveform and amplitude as the input. The important difference is that the positive peak of the waveform is now *clamped* to $+V_F$ above ground level (see the waveforms in Figure 15-5). Capacitor C_1 and diode D_1 constitute a *clamping circuit.*

The alternating voltage passed to R_1 and C_2 is the same as the input except that any dc input voltage is blocked and the ac wave is virtually all below ground level. Neglecting the diode voltage drop, this wave now has a (negative) average value of ap-

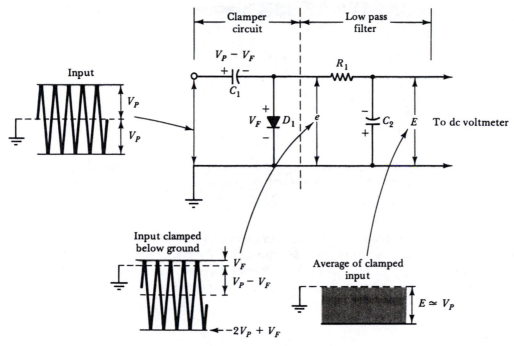

Figure 15-5 Peak detector circuit consisting of a clamper circuit and a low-pass filter. The peak voltage of a high-frequency waveform is converted to a dc quantity that can be measured on a dc voltmeter.

proximately one-half its peak-to-peak voltage (see illustration). Resistor R_1 and capacitor C_2 function as a low-pass filter to block the alternating component of the voltage and to pass the dc (average) level. The dc level which is (approximately) the peak value of the input (one-half the peak-to-peak voltage) is now passed to a dc voltmeter that is calibrated to indicate rms quantities.

The peak detector circuitry obviously passes a constant voltage level to the voltmeter, rather than a rectified waveform. This is not significant at low or medium frequencies within the upper- and lower-frequency limits of the moving-coil instrument. However, at high frequencies beyond the frequency range of the meter, the peak detector still functions correctly and produces a measurable output voltage. Therefore, the *peak response voltmeter* is a high-frequency instrument. Its maximum frequency of measurement generally exceeds 500 MHz.

Electronic ac voltmeters that use peak detection always have the detector circuitry in a separate *probe* on one end of a coaxial cable which plugs in to the voltmeter terminals (see Figure 15-6). This puts the detector right at the point of measurement, instead of at the opposite end of connecting cables which might have considerable capacitance and inductance. Such cable capacitance and inductance (prior to the detector) could have a serious effect on the measured voltage. With the high-frequency voltage converted to a direct quantity, the connecting cable impedance is no longer of any consequence.

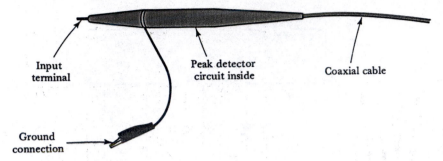

Input
terminal

Peak detector
circuit inside

Coaxial cable

Ground
connection

Figure 15-6 A peak detector is usually contained in a probe, to eliminate the capacitance of connecting cables that could affect the measured voltage.

Any dc electronic voltmeter can be made to operate as a high-frequency peak response instrument by use of a suitable peak detector probe. The output of the peak detector must be potentially divided to 0.707 of the peak level to give an rms indication on the dc meter. As in the case of rectifier voltmeters, the peak response instrument is accurate only for pure sine-wave inputs.

15-3 TRUE RMS METERS

Disadvantage of Average Responding Instruments

Full-wave and half-wave rectifier instruments operate satisfactorily only when measuring inputs which have purely sinusoidal waveforms. The meter current is proportional to the average level of the rectifier wave, and the scale is calibrated to indicate the rms level on the assumption that the input is a pure sine wave. When the waveform of the measured quantity is not sinusoidal, average-responding instruments do *not* indicate the *true rms* *(TRMS)* value. In fact, such instruments are virtually useless for measuring quantities that have nonsinusoidal waveforms.

Waveform Crest Factor

The *crest factor* of a waveform is the ratio of its peak value to its rms value:

$$\text{crest factor} = \frac{\text{peak value}}{\text{rms value}} \tag{15-1}$$

The crest factor for a pure sine wave is 1.414, but nonsinusoidal waveforms can have much larger crest factors. The rms level of waveforms with a crest factor of 2 or 3 can be determined by most rms measuring instruments. Waveforms with higher crest factors are more difficult to measure. The maximum waveform crest factor is usually specified for all rms measuring instruments.

410

Miscellaneous Meters   Chap. 15

TRMS Meter Using Nonlinear Circuit

A true-rms meter using a nonlinear diode-resistor circuit is illustrated in Figure 15-7. The input voltage is first full-wave rectified, and then applied to the circuit input, as shown. The voltage (V_{C1}) developed across capacitor C_1 is applied to the meter circuit, and this is also the voltage at the cathodes of the diodes. The diode anode voltages (V_{A1}, V_{A2}, and V_{A3}) are individually derived from the input by means of potential dividers.

When the input amplitude is low, V_{C1} remains lower than the diode anode voltages. Consequently, all three diodes become forward biased, so that current flows via D_1, D_2, D_3, and R_7, to charge the capacitor. At some higher input amplitude, V_{C1} becomes larger than V_{A3}, causing D_3 to be reverse biased. The charging current to C_1 is now reduced, because it flows only via D_1, D_2, and R_7. Similarly, D_2 and D_1 become reverse biased in turn as V_{C1} increases. The result of this action is that V_{C1} (the meter voltage) grows in a nonlinear fashion. Appropriate selection of resistors can produce a meter voltage that is proportional to the rms level of many common distorted waveforms. The specification for one such instrument states that the probable error is only ±2% for most waveforms with crest factors not exceeding 3.

Waveforms with a dc Component

Many waveforms have a dc component, which usually shows as the wave not being symmetrical above and below ground level. Determination of the rms level of such a waveform sometimes requires separate dc and ac measurements. First the dc level is measured on a dc voltmeter (with the ac quantity filtered out). Then the ac rms level is measured on a capacitor-coupled ac voltmeter. The rms value of the original waveform is then determined as the square root of the sum of the squares of the two readings,

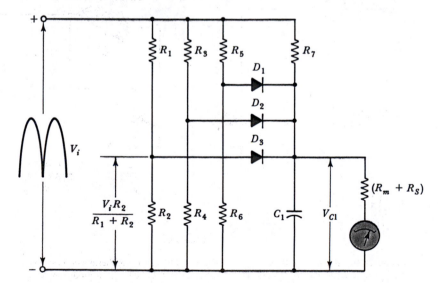

Figure 15-7 True rms voltmeter using a diode-resistor nonlinear circuit. This gives a reasonably accurate measurement of the rms level of many common distorted waveforms.

$$V_{rms} = \sqrt{(V_{dc})^2 + (V_{ac})^2}$$

(15-2)

The procedure above is not necessary with the true-rms instrument described next.

Thermocouple-Type True-rms Meter

It is explained in Section 15-1 that the output voltage from a thermocouple is directly proportional to the rms level of the current through its heater regardless of the current waveform. In Figure 15-8, an input voltage (E_1) with a nonsinusoidal waveform is amplified to $A_v E_1$ by means of *video amplifier A_1*. A video amplifier operates over a wide frequency range, from audio to very high frequencies, to faithfully amplify the input waveform. The current (I_1) passed from A_1 to resistor R_1 has the same waveform as E_1. Consequently, the heat generated in R_1 is directly proportional to the true rms level of the input voltage.

Now look at amplifier A_2 in Figure 15-8. A_2 functions as a noninverting dc amplifier which passes direct current I_2 through resistor R_2. R_2 is the heater for thermocouple T_2, and the thermocouple output voltage e_2 is directly proportional to the heating effect of I_2.

Recall from the noninverting amplifier description in Section 4-2 that the feedback from the output always produces a voltage at the op-amp inverting input equal to that at the noninverting input. Voltage e_1 is applied to the noninverting input of A_2, and e_2 is the

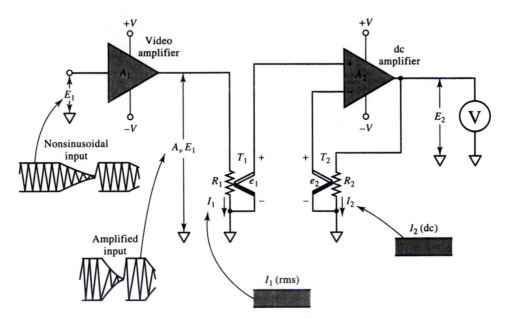

Figure 15-8 Basic circuit for a true rms electronic voltmeter using thermocouples. Voltage e_1, generated by the heating effect of $A_v E_1$, is balanced by e_2, which is produced by direct current I_2. DC output voltage E_2 is directly proportional to the rms level of input E_1.

Miscellaneous Meters   Chap. 15

Figure 15-9 Fluke 45 dual display multimeter which can make two measurements on one input signal. (© 1988, 1990, John Fluke Mfg. Co., Inc. All rights reserved. Reproduced with permission.)

feedback to the inverting input terminal. Output voltage E_2 from A_2 settles at the level that makes e_2 exactly equal to e_1. If e_2 becomes smaller than e_1, the output of A_2 rises, and I_2 increases until its heating effect raises e_2 to the same level as e_1. If e_2 becomes greater than e_1, A_2 output falls, I_2 is reduced, and consequently, e_2 falls toward e_1. Thus, if the thermocouples are similar devices, the heating effect of direct current I_2 must equal that of alternating current I_1. This means that I_2 equals the rms value of I_1. Therefore, dc voltage output E_2 is directly proportional to the true rms level of ac input E_1 for all input waveforms.

Representative True rms Meter

Figure 15-9 shows a true rms voltmeter (a Fluke 45) which is also a dual display instrument capable of simultaneously making two measurements on the same input. An example of this might be the frequency and voltage of an ac signal, or the dc voltage and ac ripple from a power supply. The voltage measurement range is 300 mV to 1000 V dc, and 300 mV to 750 V ac with a maximum frequency of 100 kHz. For true rms measurements, the maximum waveform crest factor is 3. The input impedance is 1 MΩ in parallel with a capacitance less than 100 pF.

15-4 LOW-LEVEL VOLTMETER/AMMETER

Low-Level Voltage Measurements

Measurement of voltage levels in the millivolt range (or greater) does not usually present any problem. But special instruments are required for measurements in the microvolt and nanovolt ranges. Low-level voltage sources normally have very high output resistances, so the measured voltage could be seriously affected by meter input currents. Consequently, very high input resistances must be offered by the input devices of low-level electronic voltmeters. This dictates the use of a junction field-effect transistor (JFET), or a metal-oxide semiconductor field-effect transistor (MOSFET). Small drifts in bias conditions of the voltmeter amplifier can have the same effect as changes in input voltage; they are amplified, and produce serious errors in measured quantities. Other sources of error in low-level measurements are conductor junction voltages, thermal noise voltages, emfs induced by magnetic fields, and ground loops.

Chopper-Stabilized Amplifier

One method of minimizing the effect of bias voltage drift is to amplify the input voltages by means of a capacitor-coupled amplifier. Coupling capacitors at input and output, and between the stages of an amplifier, pass alternating voltage signals but block direct voltages. They also block dc bias drift voltages. A *chopper-stabilized*, or *modulated*, dc amplifier actually amplifies dc voltages by converting them into ac quantities.

The diagram in Figure 15-10 shows the operating system and voltage waveforms for a chopper-stabilized dc voltmeter. The input stage may be a voltage-follower circuit for high input resistance. The low-level dc input voltage is passed from the voltage follower to a switching circuit controlled by an oscillator operating at a constant frequency. As the switching circuit opens and closes, it alternately blocks and passes the voltage to be measured. This results in a pulse waveform or *chopped dc voltage*, as illustrated.

Because the chopped dc voltage is an alternating quantity, it is passed by the coupling capacitor, and amplified by the ac amplifier. Once amplified, the waveform is precision rectified (see Section 4-4) and filtered. The rectified output is made up of a dc volt-

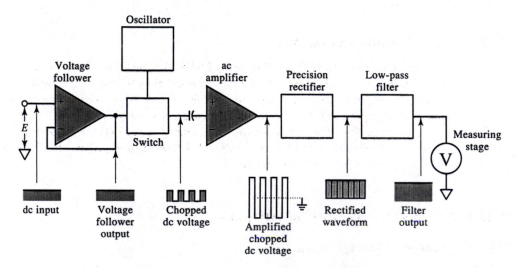

Figure 15-10 Chopper-stabilized dc amplifier used in many low-level voltmeters. The dc input voltage is chopped to convert it to an ac quantity. The chopped waveform is amplified by an ac amplifier, then precision rectified and filtered for measurement.

Miscellaneous Meters   Chap. 15

age (the amplified input) and a chopping-frequency ac component. The dc voltage is passed by the low-pass filter, and the ac quantity is severely attenuated. After filtering, the amplified dc voltage is passed to the voltage-measuring stage.

Electrometers

Electrometer is another name applied to a voltmeter that measures low-level voltages which have very high source resistances. The front panels of two such instruments are shown in Figure 15-11. The digital instrument in Figure 15-11(a) has a sensitivity of 10 nV and an input resistance greater than 1 GΩ on its lowest (2 mV) range. The center-zero scale on the analog instrument in Figure 15-11(b) allows it to be used as a null detector, as well as a voltmeter. Its lowest voltage range is ±1 μV (full scale), and its input resis-

(a) Digital nanovoltmeter

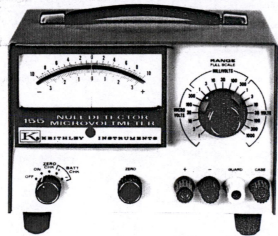

(b) Analog null detector microvoltmeter

Figure 15-11   Digital nanovoltmeter (a) and analog null-detector microvoltmeter (b). The digital instrument has a sensitivity of 10 nV on its 2 mV range. The analog meter has a minimum range of ± 1 μV full scale (Courtesy of Kiethley Instruments Inc.)

tance is 1 MΩ (for input voltages under 100 mV). Note that as well as + and − input terminals, this instrument has *case* and *guard* terminals. The case terminal is for grounding the instrument case, and the guard terminal eliminates insulation resistance problems, as explained below.

Guard Terminal

A low-level voltmeter with a coaxial cable connecting it to a voltage source is shown in Figure 15-12(a) on p. 417. High instrument input resistances and high signal source resis-

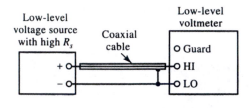

(a) Low-level voltmeter connected for measuring low-level high-resistance source

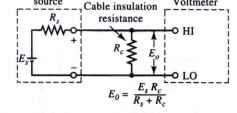

$$E_0 = \frac{E_s R_c}{R_s + R_c}$$

(b) Equivalent circuit, showing that R_c affects the measured voltage

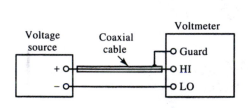

(c) Cable screen guarded

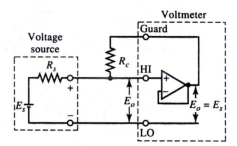

(d) Equivalent circuit when screen is guarded

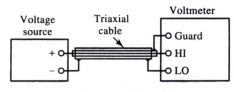

(e) Use of a triaxial cable

Figure 15-12 The insulation resistance of a connecting cable can cause errors when measuring a voltage that has a very high source resistance. A guard terminal can be used to eliminate the effect of the insulation resistance.

Miscellaneous Meters   Chap. 15

tances are subject to pickup of electromagnetic noise voltages, so coaxial cables are normally used to *screen* the circuit. A problem with this arrangement is that the cable insulation resistance (R_c) might be not very much greater than the source resistance (R_s). So, as illustrated by the equivalent circuit in Figure 15-12(b), the source voltage (E_s) is potentially divided by R_s and R_c before being measured.

In Figure 15-12(c), the outer screen of the coaxial cable is connected to the guard terminal instead of the *LO* input terminal. The equivalent circuit in Figure 15-12(d) shows that the guard terminal is at the same potential as the meter input (E_o), and is supplied from a low-impedance (voltage-follower) source. Now both ends of the cable insulation resistance (R_c) are at the same potential, and R_c has no effect on the measured voltage. The cable screen still blocks noise voltages from the central conductor.

A further advantage of the use of a guard terminal is that it minimizes the capacitive load on the voltage source, and thus reduces the time that must be allowed for the instrument to settle at its final reading. Without the feedback effect of the guard terminal, the source voltage has to charge the cable capacitance and the instrument input capacitance, and this could take considerable time.

Triaxial cable might well be used in the situation described above [see Figure 15-12(e)]. This is cable with a central conductor core, and two concentric, insulated screens. The core is used for connecting the voltage source to the instrument *HI* input terminal, and the inner screen is connected to the guard terminal. The outer screen is grounded, and used for connecting the source to the instrument *LO* input.

REVIEW QUESTIONS

15-1 Briefly describe a thermocouple. Draw a sketch to show the construction of a basic thermocouple instrument. Explain its operation.

15-2 Draw sketches of two types of thermocouples. Discuss the characteristics of each. State typical thermocouple heater current levels and maximum output voltage.

15-3 Sketch the circuit of a thermocouple bridge. Explain its operation and its advantages over a single thermocouple.

15-4 Sketch the basic circuit of a thermocouple wattmeter. Explain its operation, and show that its output is proportional to *EI*.

15-5 Discuss the application of a peak response voltmeter. Sketch a peak detector circuit as used with a peak response voltmeter. Explain its operation.

15-6 Discuss the disadvantage of average responding instruments when measuring nonsinusoidal waveforms. Define crest factor.

15-7 Sketch a diode-resistor nonlinear circuit for use with a TRMS meter. Explain its operation, and discuss its disadvantages.

15-8 Describe the procedure that must be used with many TRMS meters, when determining the true rms level of a nonsinusoidal waveform with a dc component.

15-9 Sketch the basic circuit and waveforms for a thermocouple-type TRMS meter. Carefully explain the instrumentation operation. Discuss a typical crest factor for a thermocouple-type TRMS instrument.

15-10 Discuss the difficulties encountered when measuring very low-level voltages which have very high source resistances. Using illustrations, explain how a guard terminal can be employed to eliminate error caused by coaxial connecting cables.

15-11 Sketch the basic block diagram of a chopper-stabilized dc amplifier as used in many low-level voltmeters. Sketch the waveforms at various points in the system, and explain its operation.

Laboratory Power Supplies

16

Objectives

You will be able to:

1. Sketch unregulated dc power supply circuits. Explain the operation of the circuits, and discuss their performance.
2. Define the following terms: line effect, load effect, line regulation, load regulation.
3. Sketch the characteristic of a zener diode, and discuss its application as a voltage reference source.
4. Sketch the circuits of a simple zener diode voltage regulator, an op-amp voltage follower regulator, and an adjustable output op-amp voltage regulator. Explain the operation of each circuit, and discuss its performance.
5. Show how the load current may be limited in a dc voltage regulator. Discuss short-circuit protection and foldback current limiting.
6. Discuss typical laboratory-type dc power supplies for use with electronic circuits.
7. Show how dc power supplies may be operated in series and parallel combinations for various applications.
8. Show how power supplies may be tested to determine line effect and load effect.

Introduction

A basic dc power supply uses a transformer to reduce the ac input to an appropriate voltage level, which is then rectified and smoothed by a capacitor filter or an *LC* filter circuit. A voltage regulator circuit must be added where a more stable dc output voltage is required. A regulator also allows the output voltage to be adjustable. All dc voltage regulators use a zener diode as a voltage reference source, and an operational amplifier is usually employed as a feedback amplifier to stabilize the output voltage. Current-limiting circuits are also normally included to protect the regulator against output short circuits.

16-1 UNREGULATED DC POWER SUPPLIES

Transformer–Rectifier Circuit

Unregulated dc power supplies normally consist of a transformer, a full-wave bridge rectifier, and a filter circuit, as shown in Figure 16-1(a). The bridge rectifier circuit is shown again in Figure 16-1(b) in a form often used in circuit diagrams. Apart from the difference in appearance, the circuits are identical.

The transformer changes the 115 V (primary) ac supply to an rms (secondary) voltage which is related to the required dc output voltage. The bridge rectifier converts the transformer secondary voltage into a train of sinusoidal repetitive positive half-cycles

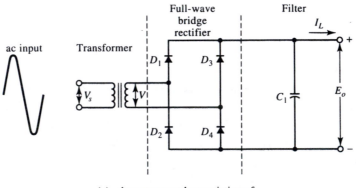

(a) dc power supply consisting of transformer, bridge rectifier, and reservoir capacitor

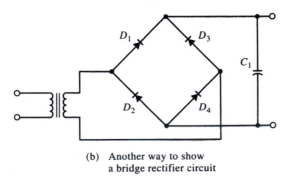

(b) Another way to show a bridge rectifier circuit

Figure 16-1 Unregulated dc power supply. The ac input is transformed to the appropriate level, and then full-wave rectified. The rectified output is smoothed, or converted to a dc voltage, by the reservoir capacitor.

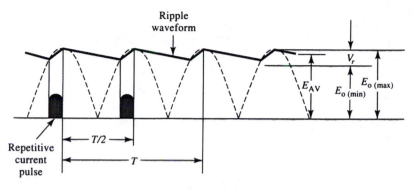

Figure 16-2 A full-wave-rectified waveform is smoothed by the effect of the reservoir capacitor retaining most of its charge during the time between the rectified peaks.

(Figure 16-2). The filter circuit, which is simply a *reservoir capacitor*, smoothes the rectified waveform to approximate a dc voltage.

Rectified and Smoothed Waveform

Consider the power supply waveforms illustrated in Figure 16-2. The rectified positive half-cycles are shown in dashed-line form, and the solid line represents the output (capacitor) voltage. C_1 is charged up to the peak of the rectifier output voltage [$E_{o(max)}$], and if there is no load current, the capacitor voltage remains constant. When a load current (I_L) is drawn from the supply, C_1 partially discharges between voltage peaks [to $E_{o(min)}$]. This gives an output *ripple voltage* with an amplitude V_r.

The capacitor acts as a *reservoir* to supply load current as required, and the transformer and bridge rectifier circuit supply recharging current to C_1 as a series of repetitive pulses (see Figure 16-2). The output voltage from the power supply has an average dc level E_{av}, with the ripple voltage superimposed. The amplitude of the ripple voltage depends on the size of the reservoir capacitor and the level of load current. With a given capacitor size, increasing the load current increases the ripple voltage amplitude.

Choke-Capacitor Filter

The use of an inductor, or *choke* (L in Figure 16-3), together with the reservoir capacitor further reduces the amplitude of the output ripple voltage. A *bleeder resistor (R)* is sometimes included in the circuit to maintain a minimum current flow in the choke when there is no output load current. This keeps the choke operating and thus helps to minimize the change in voltage drop across the choke when load current is demanded. When no *voltage regulator* (see Section 16-2) is employed, choke-capacitor filtering gives the most constant dc output voltage with the lowest ripple content.

Figure 16-4 shows the circuit of a high-current unregulated dc power supply. The transformer has two secondary windings which can be switched (by S_1) to operate either in series or in parallel. When the windings are connected in series, the dc output voltage

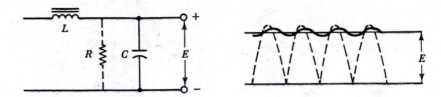

Figure 16-3 Choke-capacitor filter for reducing the output ripple amplitude from an un-regulated dc power supply.

is 12 V. Parallel operation gives an output of only 6 V. However, the 12 V output might be capable of supplying 10 A, while with parallel secondary windings the 6 V output may supply 20 A.

Source Effect and Load Effect

The ac supply voltage (E_s), (also termed the *line voltage* or *source voltage*) does not always remain absolutely constant; a ±10% variation is not unusual. When a power supply input voltage varies, there is always some variation in the dc output. One way of stating this uses the term *source effect*. If the output voltage change (ΔE_o) is 100 mV when the source voltage changes by a specified amount (usually 10%), the source effect is 100 mV. Alternatively, ΔE_o might be expressed as a percentage of E_o. In this case, the term *line regulation* is sometimes used instead of source effect.

$$\boxed{\text{Source effect} = \Delta E_o \text{ for } 10\% \text{ change in } E_s} \qquad (16\text{-}1)$$

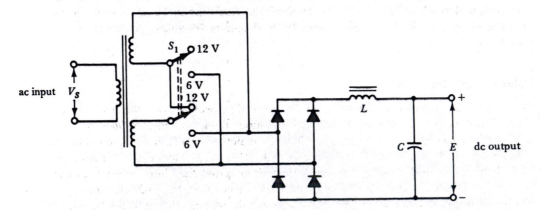

Figure 16-4 Circuit of a 12 V/6 V unregulated dc power supply. A 12 V output is produced when the transformer secondary windings are connected in series. Parallel connection of the secondary windings gives a 6 V output.

$$\text{Line regulation} = \frac{(\Delta E_o \text{ for 10\% change in } E_s) \times 100\%}{E_o} \qquad (16\text{-}2)$$

Power supply output voltage is also affected by changes in load current (I_L). The output voltage falls when I_L increases, and rises when I_L decreases. The *load effect* defines how the power supply output voltage changes when the load current is increased from zero to its specified maximum. If ΔE_o is 100 mV when I_L changes from zero to $I_{L(max)}$, the load effect is 100 mV. As in the case of source effect, ΔE_o can also be stated as a percentage of E_o, and in this case the term *load regulation* might be used.

$$\text{Load effect} = \Delta E_o \text{ for } \Delta I_{L(max)} \qquad (16\text{-}3)$$

$$\text{Load regulation} = \frac{[\Delta E_o \text{ for } \Delta I_{L(max)}] \times 100\%}{E_o} \qquad (16\text{-}4)$$

Example 16-1

The output of a dc power supply falls from 12 V to 11.95 V when the ac input drops by 10%. The output also falls from 12 V to 11.9 V when the load current goes from zero to its maximum level. Determine the source and load effects, and the line and load regulation.

Solution

Equation 16-1, $\qquad$ source effect $= \Delta E_o$ for 10% change in E_s

$$= 12 \text{ V} - 11.95 \text{ V}$$

$$= 50 \text{ mV}$$

Equation 16-2, $\qquad$ line regulation $= \dfrac{(\Delta E_o \text{ for 10\% change in } E_s) \times 100\%}{E_o}$

$$= \frac{50 \text{ mV} \times 100\%}{12 \text{ V}}$$

$$\simeq 0.42\%$$

Equation 16-3, $\qquad$ load effect $= \Delta E_o$ for $\Delta I_{L(max)}$

$$= 12 \text{ V} - 11.9 \text{ V}$$

$$= 100 \text{ mV}$$

Equation 16-4,

$$\text{load regulation} = \frac{[\Delta E_o \text{ for } \Delta I_{L(max)}] \times 100\%}{E_o}$$

$$= \frac{100 \text{ mV} \times 100\%}{12 \text{ V}}$$

$$= 0.83\%$$

16-2 DC VOLTAGE REGULATORS

Zener Diode

DC voltage regulators are based on the use of the *zener diode* (or *breakdown diode*) as a constant-voltage reference source. The characteristics and circuit symbol for a zener diode are shown in Figure 16-5. The device is seen to have a reverse breakdown voltage that remains substantially constant when the current is held constant. It should be noted

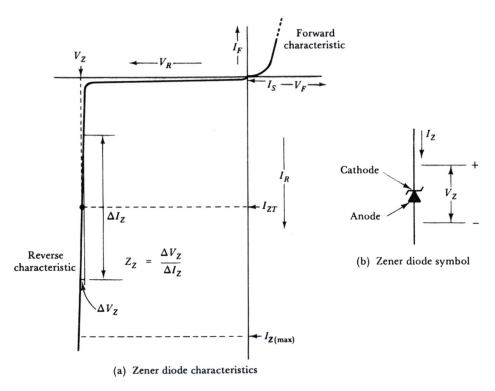

(a) Zener diode characteristics

(b) Zener diode symbol

Figure 16-5 Characteristics and circuit symbol for a zener diode, or breakdown diode. When operated in reverse breakdown mode, the diode voltage is a very stable quantity that can be used as a reference voltage for a dc voltage regulator.

424   Laboratory Power Supplies   Chap. 16

that the reverse current must be limited to the desired level by means of a series-connected resistor.

The most important parameters of a zener diode are *breakdown voltage* V_z, *test current* I_{zT} (current at the specified V_z), and *dynamic impedance* Z_z. The dynamic impedance is the ratio $\Delta V_z/\Delta I_z$ (see the characteristic), and it may be used to calculate the change in V_z that occurs when there is a change in diode reverse current. For the types of zener diode used in voltage regulators, typical voltages range from 2.7 V to 12 V. The usual diode reverse current is 10 mA to 20 mA.

Basic Voltage Regulator

The circuit in Figure 16-6(a) shows how a zener diode may be employed directly for voltage regulation. The output voltage remains substantially constant at V_z as the input voltage and load current vary. However, the load current is limited to a level less than the maximum diode reverse current, and the diode must pass this current when the load current is zero. Figure 16-6(b) shows how the zener diode circuit may be combined with a transistor to deliver almost any desired level of output current.

Zener diode D_1 in Figure 16-6(b) is supplied with reverse current via resistor R_1. D_1 provides a bias voltage V_z at the base of transistor Q_1, which functions as an emitter follower. The transistor emitter voltage is the circuit output, which is $E_o = V_z - V_{BE}$. Resistor R_E maintains a minimum transistor emitter current when the regulator load current is zero. The maximum load current that may be supplied is limited to the maximum transistor emitter current. For greater load currents, two transistors may be *Darlington connected,* as illustrated in Figure 16-6(c). In this case, Q_1 is usually a high-current device, and Q_2 is a low-current transistor to supply the base current for Q_1.

Operational Amplifier Voltage Regulators

The voltage regulator circuit in Figure 16-7 uses an operational amplifier connected as a voltage follower (see Section 4-2). This eliminates the V_{BE} voltage drop that occurs with the emitter follower, and gives an output voltage equal to V_z. The load current taken from the zener diode circuit is also limited to the op-amp input bias current (typically around 50 nA). Capacitor C_1 in Figure 16-7 shorts out the high-frequency oscillations that can occur in a circuit that uses a high-gain amplifier. It also acts as an output reservoir capacitor to supply very fast load current demands. C_1 is usually on the order of 30 μF to 100 μF.

The op-amp regulator circuit in Figure 16-8(a) is similar to the one in Figure 16-7, except that with the presence of resistors R_2 and R_3, the op-amp circuit becomes a noninverting amplifier (see Section 4-2). The output voltage is now greater than the zener diode voltage. Because of the negative feedback effect, the op-amp inverting input terminal voltage remains equal to the noninverting input, which is V_z.

Consequently,

$$E_o = V_z \times \frac{R_2 + R_3}{R_3} \qquad \text{(16-5)}$$

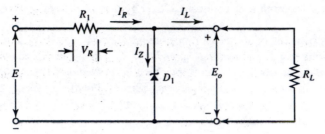

(a) Simple Zener diode voltage regulator

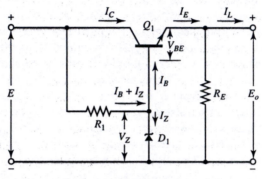

(b) Emitter-follower voltage
    regulator or series
    regulator

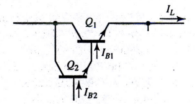

(c) Darlington–connected
    transistors

Figure 16-6   A zener diode may be used directly as a dc voltage regulator. Alternatively,
it may be used as a reference voltage at the base of an emitter-follower transistor circuit.
This gives a load current much larger than the zener diode current. "Darlington-connect-
ed" transistors further increase the load current capability of the regulator.

The regulator in Figure 16-8(b) is a further development of the circuit in Figure 16-8(a). Resistor R_1 is now connected to the output, so that the zener diode current is not affected by changes in the supply voltage. This further ensures the stability of voltage V_z. Potentiometer R_4 is included between R_2 and R_3 to make E_o
adjustable.

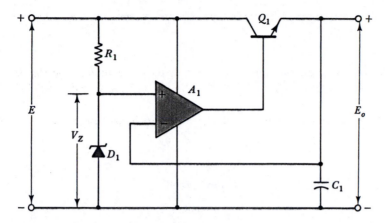

Figure 16-7 Op-amp voltage follower regulator. The action of the operation amplifier maintains E_o equal to V_z, and the series transistor supplies the required load current.

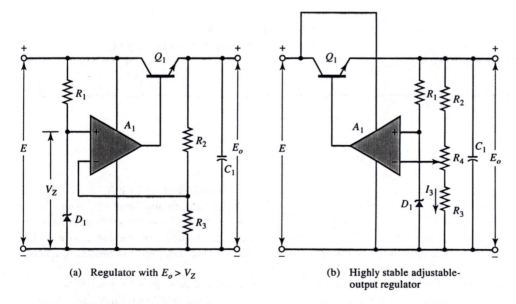

(a) Regulator with $E_o > V_Z$

(b) Highly stable adjustable-
output regulator

Figure 16-8 When the operational amplifier in a dc voltage regulator is connected as a noninverting amplifier, the output voltage is greater than the zener diode reference voltage. Inclusion of a potentiometer makes E_o adjustable.

Example 16-2

Determine the maximum and minimum output voltages available from the regulator circuit in Figure 16-8(b) when $V_z = 6$ V, $R_2 = 5.6$ kΩ, $R_3 = 5.6$ kΩ, and $R_4 = 3$ kΩ.

Solution   *When the moving contact is at the bottom of R_4,*

$$V_{R3} = V_z = 6 \text{ V}$$

$$I_3 = \frac{V_z}{R_3} = \frac{6 \text{ V}}{5.6 \text{ k}\Omega}$$

$$= 1.07 \text{ mA}$$

$$E_o = I_3 (R_2 + R_3 + R_4)$$
$$= 1.07 \text{ mA}(5.6 \text{ k}\Omega + 5.6 \text{ k}\Omega + 3 \text{ k}\Omega)$$
$$= 15.2 \text{ V}$$

When the moving contact is at the top of R_4,

$$V_{R3} + V_{R4} = V_z = 6 \text{ V}$$

$$I_3 = \frac{V_z}{R_3 + R_4} = \frac{6 \text{ V}}{5.6 \text{ k}\Omega + 3 \text{ k}\Omega}$$

$$= 0.7 \text{ mA}$$

$$E_o = I_3 (R_2 + R_3 + R_4)$$
$$= 0.7 \text{ mA}(5.6 \text{ k}\Omega + 5.6 \text{ k}\Omega + 3 \text{ k}\Omega)$$
$$= 9.9 \text{ V}$$

16-3 OUTPUT CURRENT LIMITING

Power supplies used in laboratories are subject to overloads and short-circuits that can occur in experimental circuitry. Current-limiting circuits are absolutely necessary in this circumstance. Transistor Q_2 and resistor R_5 in Figure 16-9 constitute a current-limiting circuit. Resistor R_6 must also be included because, as will be explained, the base voltage of Q_1 must be pulled down, and this cannot be done when Q_1 base is directly connected to the low impedance output terminal of the op-amp.

Load current I_L flows through resistor R_5, causing a voltage drop with the polarity shown. For normal current levels the voltage drop across R_5 is not large enough to forward bias the base-emitter junction of Q_2. However, when I_L reaches its maximum (designed for) level, V_{R5} biases Q_2 on. This causes current I_6 to flow through resistor R_6, producing a voltage drop across R_6 as shown. Q_2 is actually driven into saturation. So, the

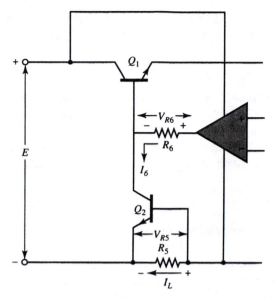

Figure 16-9 Current-limiting circuit for a dc voltage regulator. When I_L reaches its (designed) limit, V_{R5} biases Q_2 *on*, producing V_{R6}. This pulls the base of Q_1 down, and thus pulls the regulator output voltage down to zero.

base of transistor Q_1 is *pulled down* to the level of the negative supply terminal, causing the output voltage to go to zero.

The output terminals of the regulator can actually be short-circuited, and the short-circuit current is limited to the maximum level for which the current-limiting circuit is designed. Figure 16-10(a) is a graph of output voltage plotted against load current for a regulator with the type of current limiting illustrated in Figure 16-9. It is seen that E_o re-

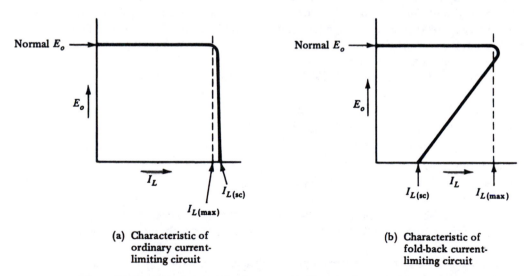

(a) Characteristic of ordinary current-limiting circuit

(b) Characteristic of fold-back current-limiting circuit

Figure 16-10 E_o/I_L characteristics for current-limiting circuits used with dc voltage regulators. With the ordinary current limiter, $I_{L(SC)}$ is just a little larger than the normal $I_{L(max)}$ for the regulator. Fold-back current limiting reduces $I_{L(SC)}$ to a level much lower than $I_{L(max)}$.

mains constant until $I_{L(\max)}$ is approached. Then, the output voltage falls to zero, while the output current remains at the short-circuit level ($I_{L(SC)}$). In this condition, the power dissipation in the series transistor (Q_1) is

$$P_D = EI_{L(SC)} \qquad \text{or} \qquad P_D \approx EI_{L(\max)}$$

To minimize the series transistor power dissipation during short circuit, another, slightly more complicated, current-limiting circuit is employed. This gives the regulator the *fold-back* short-circuit characteristic illustrated in Figure 16-10(b). In this case $I_{L(SC)}$ is substantially less than $I_{L(\max)}$, and the power dissipation in Q_1 during short circuit is substantially reduced. A regulator with this short-circuit characteristic is said to have *fold-back current limiting*.

16-4 POWER SUPPLY PERFORMANCE AND SPECIFICATIONS

Two typical laboratory-type dc power supplies for use with electronic circuitry are shown in Figure 16-11. The HPE3610A in Figure 16-11(a) produces a single-voltage output, and either output terminal can be grounded to give a *plus* or *minus* voltage. Dual digital meters are provided for monitoring output voltage and current simultaneously. The supply can be set to function as a constant-voltage source or as a constant-current source. The output can be (0 to 8 V) and (0 to 3 A) or (0 to 15 V) and (0 to 2 A), depending on the model selected.

The HP6236B *triple-output* power supply in Figure 16-11(b) produces (0 to +6 V) with 2.5 A maximum load current, and (0 to ±20 V) with 0.5 A maximum current. The +20 V and −20 V outputs track within 1%, to provide balanced output voltages. The tracking can also be made variable by moving the *Tracking Ratio* control off the *Fixed* position. This adjusts the negative supply voltage to a lower level than the positive output. When the positive/negative ratio has been established, the two output voltages will maintain the same ratio when the output is adjusted by the *Voltage ±20 V* control. Dual meters and a *Meter* output selection switch facilitate simultaneous monitoring of the output voltage and current at any output terminal.

A partial specification for the HP6236B dc power supply is as follows:

- *Input:* 115 V ac ±10%, 47 Hz to 63 Hz
- *Outputs:* 0 to 6 V, 2.5 A; 0 to ±20 V, tracking
- *Source effect:* 0.01% + 2 mV
- *Load effect:* 0.01% + 2 mV
- *Ripple and noise:* 350 μV rms, 1.5 mV peak-to-peak

Ripple and noise is sometimes listed under the term *PARD* on a power supply specification. This comes from *periodic and random deviation,* but it simply means ripple and noise voltage.

Other items that occur on a power supply specification are:

- *Voltage resolution:* smallest change in output voltage that can be produced when adjusting the fine voltage control knob

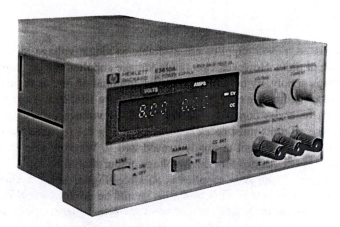

(a) Single-voltage-output dc power supply

(b) Triple-output dc power supply

Figure 16-11 Typical laboratory power supplies for use with electronics circuitry. The single-output supply (a) can have either terminal grounded. The triple-output unit (b) has a +6 V output and ± 20 V tracking outputs. (Courtesy of Hewlett-Packard)

- *Current resolution:* applies to a constant current supply, similar to voltage resolution
- *Transient recovery time:* time for output voltage to return to within 10 mV of its normal level after a specified load current change has suddenly occurred
- *Internal impedance:* output impedance offered to sinusoidal changes in load current over a specified frequency range
- *Stability or drift:* maximum output voltage change (or current change) during an 8-hour period after a 30-minute warm-up time
- *Temperature coefficient:* output change per degree Celsius change in ambient temperature

16-5 DC POWER SUPPLY USE

As already discussed, power supplies are available as single-voltage, dual-voltage, and multivoltage output units. However, two single-voltage supplies can be combined to produce a plus–minus supply, by connecting the plus terminal of one to the minus terminal of the other, and grounding the two, as illustrated in Figure 16-12(a). Also, several single-voltage units can be series connected, to give a higher voltage supply [see Figure 16-12(b)]. In this case all units must be capable of supplying the required load current, and the combined output should not normally exceed 240 V. Where higher voltages are required, a power supply designed for a high-voltage output should be used.

Parallel operation of power supplies is possible [Figure 16-12(c)] where a single unit cannot supply a required load current. This should be done only where current controls are available to ensure that each unit supplies its correct share of the current.

Two or more power supplies may be connected for *tracking operation* [Figure 16-12(d)]. In this case one supply becomes a *master* and the others are *slaves*. When the output voltage of the master increases or decreases, the other outputs vary in proportion. This kind of operation is achieved by using the output of the master supply (or a portion of it) in place of the internal (zener diode) reference sources in each of the other units.

The normal procedure for using a dc power supply with a single output voltage is listed below. With multioutput units, the procedure is repeated for each output:

1. With the output terminals open circuited, switch *on* the ac input and adjust the output voltage to the desired level.
2. Where the unit has a front panel current control: (a) Set the control for maximum current if there is no concern about passing too much current through the circuit to be supplied. (b) When the current is to be limited to a particular level, short-circuit the output terminals and adjust the current to the desired level.
3. Open-circuit the output terminals (if previously shorted), and set the meter to indicate output voltage.
4. Switch *off* the ac input, and connect the circuit to be supplied.
5. Switch *on* the ac input and observe (on the meter) that the output voltage remains at the selected level. If the voltage has fallen below the previously set level, the circuit is drawing excessive current.

16-6 POWER SUPPLY TESTING

Determination of the line effect for a given power supply requires careful monitoring of the output voltage while the ac input is adjusted by ±10%. Similarly, to measure the load effect, the power supply output voltage must be monitored while a load is switched *on* and *off*. A problem with both of these measurements is that the output voltage changes are very small, on the order of millivolts. A digital voltmeter can be used to measure the output voltage changes if it has sufficient digits. For example, to monitor millivolt changes in a 20 V output requires a meter that indicates 20.000 V; a five-digit instrument. Where such an instrument is not available, a null method must be employed.

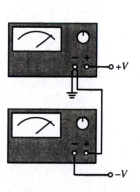

(a) Two power supplies connected to
     produce a $\pm V$ output

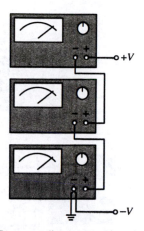

(b) Power supplies connected in series
     to produce a higher output voltage

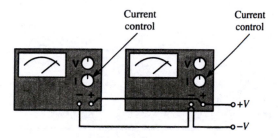

Current
control

Current
control

(c) Power supplies connected in parallel
     for high output current; each unit
     must have a current control

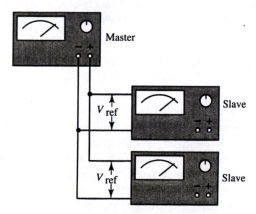

Master

V_{ref} Slave

V_{ref} Slave

(d) Power supplies connected for tracking
     (master-slave) operation

Figure 16-12 Single-output voltage power supplies may be connected to produce plus/minus voltages, operated in series for higher-voltage output, parallel connected for higher load current, or operated as tracking supplies.

Sec. 16-6   Power Supply Testing

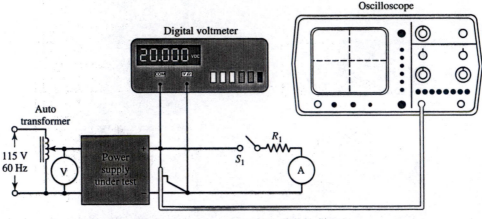

(a) Power supply testing using a digital voltmeter

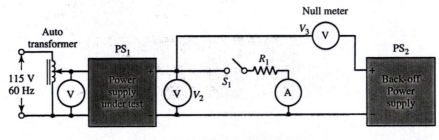

(b) Null method for testing a power supply

Figure 16-13 A digital voltmeter (with a sufficient number of digits) may be used to measure ΔV_o when the supply voltage and load current change. Where a suitable DVM is not available, a nulling method must be used for measuring the output voltage changes.

The line effect and load effect testing process using a digital voltmeter is illustrated in Figure 16-13(a). For line effect measurement, the ac input voltage is adjusted by ±10%, using the variable-voltage transformer. The output voltage is noted for input levels of 115 V, 115 V + 10%, and 115 V − 10%. The output voltage change ($\pm\Delta V_o$) is the line effect. For load effect determination, resistor R_1 is adjusted to its maximum resistance, and the input is set to 115 V. Switch S_1 is now closed, and R_1 is adjusted for maximum load current. The output voltage is measured with S_1 closed, and then with S_1 open. The load effect is the output voltage change from S_1 closed to S_1 open; that is, from $I_{L(\text{max})}$ to I_L equal zero. Note that an oscilloscope is connected to the output in Figure 16-13(a). This can be used to measure the ripple voltage amplitude at maximum load current.

Figure 16-13(b) shows the null meter method of determining line and load effects. PS_1 is the power supply under test, and the output voltage of PS_2 is adjusted to equal the PS_1 output. In this situation, voltmeter V_3 (the null meter) should read zero. V_3 can be a multirange analog voltmeter with a lowest range of about 50 mV. Initially, V_3 should be set to its highest range; then, with the PS_2 output adjusted to equal that of PS_1, V_3 should be switched to progressively lower ranges until it indicates zero on its lowest range. Now, when the supply voltage is varied, or the load is switched *on* and *off*, ΔV_o for PS_1 will be measured by voltmeter V_3.

REVIEW QUESTIONS

16-1 Draw the circuit diagram of an unregulated dc power supply using a bridge rectifier and smoothing capacitor. Sketch the rectified and output waveforms, and explain the circuit operation.

16-2 Discuss the relationships between output ripple waveform and load current for the dc power supply in Question 16-1.

16-3 Show how a choke should be used to improve the performance of an unregulated dc power supply. Explain.

16-4 Sketch the circuit of an unregulated dc power supply which is capable of producing either a 6 V output or a 12 V output voltage.

16-5 Sketch the characteristics and circuit symbol for a zener diode. Identify the most important parameters of the device.

16-6 Draw the circuit diagram of a simple zener diode voltage regulator, and explain its operation.

16-7 Sketch the circuit for an emitter-follower voltage regulator, or series regulator. Explain the circuit operation, and discuss the effect of the transistor on the regulator performance.

16-8 Show how two transistors may be connected as a 'Darlington pair' to increase the load current capability of a series regulator. Explain.

16-9 Sketch the circuit of an IC op-amp voltage-follower regulator. Explain the circuit operation, and discuss the effect of the operational amplifier on the regulator performance.

16-10 Show how a voltage-follower regulator can be modified to produce an adjustable output voltage which is greater than the zener diode reference voltage. Explain the circuit operation.

16-11 Show how a dc voltage regulator circuit should be modified to limit the maximum output current. Explain the operation of the current-limiting circuit.

16-12 Sketch the E_o/I_L characteristics for normal, and fold-back, current limiting in a dc voltage regulator. Discuss the effect of both types on the regulator series transistor.

16-13 List some of the most important items in a dc power supply specification. Explain each item briefly.

16-14 Show how dc power supplies should be connected for a plus/minus output voltage, series operation, parallel operation, and tracking voltages. Discuss each of the arrangements.

16-15 Draw a diagram to show how a dc power supply may be tested for line effect and load effect using a suitable digital voltmeter. Explain the test procedure.

16-16 Draw a diagram to show the null meter method for testing a dc power supply. Explain the test procedure.

PROBLEMS

16-1 An unregulated dc power supply, as in Figure 16-1, produces an average output of 15 V when the load current is zero. The output falls to a 14 V average level when the load is increased to 150 mA. The output also drops from 15 V to 14.5 V when the ac supply voltage is reduced by 10%. Determine the load effect and the source effect for the regulator.

16-2 Calculate the line regulation and load regulation for the regulator in Problem 16-1.

16-3 The simple dc voltage regulator circuit in Figure 16-6(a) has $E = 18$ V, $R_1 = 180\ \Omega$, $R_L = 3\ k\Omega$, and D_1 has $V_z = 7.5$ V and $Z_z = 6\ \Omega$. Calculate the source effect and load effect for the regulator.

16-4 The voltage-follower regulator circuit in Figure 16-7 has $R_1 = 470\ \Omega$, $V_z = 9$ V, and $Z_z = 5\ \Omega$. The supply voltage is 19 V when $I_L = 0$, and 18 V when $I_L = 32$ mA. Calculate the source effect and load effect for the regulator. Also determine the line regulation and load regulation.

16-5 The dc voltage regulator in Figure 16-8(a) has the following components: $R_1 = 820$ Ω, $R_2 = 1\ k\Omega$, $R_3 = 1.2\ k\Omega$, and D_1 has $V_z = 8$ V and $Z_z = 4\ \Omega$. If $E = 20$ V, determine E_o and the zener diode current.

16-6 Calculate the source effect for the regulator in Problem 16-5. Also, determine the output ripple voltage if the input ripple amplitude is 2 V peak-to-peak.

16-7 The adjustable output voltage regulator in Figure 16-8(b) has $E = 25$ V, $R_1 = 270$ Ω, $R_2 = 2.2\ k\Omega$, $R_3 = 1\ k\Omega$, and $R_4 = 500\ \Omega$. If D_1 has $V_z = 5$ V, determine $E_{o(max)}$ and $E_{o(min)}$.

16-8 Calculate the new levels of $E_{o(max)}$ and $E_{o(min)}$ for the regulator in Problem 16-7 if R_4 is changed to $300\ \Omega$.

16-9 If the current-limiter circuit in Figure 16-9 has $R_5 = 3.3\ \Omega$, determine the approximate short-circuit current for the regulator.

16-10 An op-amp dc voltage regulator has an input of $E = 18$ V and an output of $E_o = 10$ V when the maximum load current is 200 mA. The current-limiter circuit gives a short-circuit current of 240 mA. Calculate the maximum power dissipation in the series transistor under normal and short-circuit conditions.

16-11 If fold-back current limiting is used in the circuit of Problem 16-10 to limit the short-circuit current to $I_{L(max)}/4$, determine the new power dissipation in the series transistor at short circuit.

16-12 A power supply test performed as shown in Figure 16-13(a) gave the following results:

Ac input (V)	E_o (V)	I_L (mA)
115	21.333	0
126.5	21.345	0
115	21.326	200

Calculate the source effect, load effect, line regulation, and load regulation.

Unit Conversion Factors

The following factors may be used for conversion between non-SI units and SI units.

TO CONVERT	TO	MULTIPLY BY
	Area Units	
acres	square meters (m^2)	4047
acres	hectares (ha)	0.4047
circular mils	square meters (m^2)	5.067×10^{-10}
square feet	square meters (m^2)	0.0929
square inches	square centimeters (cm^2)	6.452
square miles	hectares (ha)	259
square miles	square kilometers (km^2)	2.59
square yards	square meters (m^2)	0.8361
	Electric and Magnetic Units	
amperes/inch	amperes/meter (A/m)	39.37
gauses	teslas (T)	10^{-4}
gilberts	ampere (turns) (A)	0.7958
lines/sq. inch	teslas (T)	1.55×10^{-5}
Maxwells	webers (Wb)	10^{-8}
mhos	Siemens (S)	1
Oersteds	amperes/meter	79.577
	Energy and Work Units	
Btu	joules (J)	1054.8
Btu	kilowatt hours (kW.h)	2.928×10^{-4}
ergs	joules (J)	10^{-7}
ergs	kilowatt hours (kW.h)	0.2778×10^{-13}
foot-pounds	joules (J)	1.356
foot-pounds	kilogram meters (kgm)	0.1383

TO CONVERT	TO	MULTIPLY BY
Force Units		
dynes	grams (g)	1.02×10^{-3}
dynes	newtons (N)	10^{-5}
pounds	newtons (N)	4.448
poundals	newtons (N)	0.1383
grams	newtons (N)	9.807×10^{-3}
Illumination Units		
foot-candles	lux (lx)	10.764
Linear Units		
angstroms	meters (m)	1×10^{-10}
feet	meters (m)	0.3048
fathoms	meters (m)	1.8288
inches	centimeters (cm)	2.54
microns	meters (m)	10^{-6}
miles (nautical)	kilometers (km)	1.853
miles (statute)	kilometers (km)	1.609
mils	centimeters (cm)	2.54×10^{-3}
yards	meters (m)	0.9144
Power Units		
horsepower	watts (W)	745.7
Pressure Units		
atmospheres	kilopascals (kPa)	101.325
bars	kilopascals (kPa)	100
inches of mercury	kilopascals (kPa)	3.386
pounds/sq. inch	kilopascals (kPa)	6.895
Temperature Units		
degrees Fahrenheit (°F)	degrees celsius (°C)	(°F − 32)/1.8
degrees Fahrenheit (°F)	kelvin (K)	273.15 + (°F − 32)/1.8
Velocity Units		
miles/hour (mph)	kilometers/hour (km/h)	1.609
knots	kilometers/hour (km/h)	1.853
Volume Units		
bushels	cubic meters (m^3)	0.035 24
cubic feet	cubic meters (m^3)	0.028 32
cubic feet	liters (l)	28.32
cubic inches	cubic centimeters (cm^3)	16.387
cubic inches	liters (l)	0.016 39
cubic yards	cubic meters (m^3)	0.7646
gallons (U.S.)	cubic meters (m^3)	3.7853×10^{-3}
gallons (imperial)	cubic meters (m^3)	4.546×10^{-3}
gallons (U.S.)	liters (l)	3.7853
gallons (imperial)	liters (l)	4.546

TO CONVERT	TO	MULTIPLY BY
Volume Units		
gills	liters (*l*)	0.1183
pints (U.S.)	liters (*l*)	0.4732
pints (imperial)	liters (*l*)	0.5683
quarts (U.S.)	liters (*l*)	0.9463
quarts (imperial)	liters (*l*)	1.137
Weight Units		
ounces	grams (g)	28.35
pounds	kilograms (kg)	0.453 59
tons (long)	kilograms (kg)	1016
tons (short)	kilograms (kg)	907.18

Answers for Odd-Numbered Problems

CHAPTER 1

1–1   9999.9 km, 80.45 km/h, 11 149 cm^2

1–3   1.07 miles

1–5   20 mT

1–7   7.46 kWh

1–9   [MLT^{-2}], [ML^{2}T^{-2}], [ML^{2}T^{-3}]

1–11 [M^{-1}L^{-2}T^{4}I^2], [ML^{2}T^{-2}I^{-2}]

CHAPTER 2

2–1   1.9 mA, 3.8 V, 76 mA

2–3   354.8 Ω, 290.3 Ω

2–5   ±10 mV

2–7   907.5 Ω, 1072.5 Ω

2–9   50.47 mA, 45.53 mA

2–11 396.8 Ω ±7.6%

2–13 1.69 W, 1.46 W

2–15 12 V, 2.2 mV, 2.32 mV, 1.57 mV

CHAPTER 3

3–1   5.7 μN.m

3–3   1.5 mV/mm, 2 MΩ

3–5   400 nA/mm, 2.5 MΩ

3–7   300 μA, 150 μA, 100 μA

3–9   20 A, 12 A, 24.9 A

3–11 1.33 MΩ, 399 kΩ, 65.8 kΩ, 13.3 kΩ/V

3–13 20 V, 40 V, 60 V

3–15 6.25 V, 6.6 V

3–17 60 V

3–19 9.7% FSD, 3% FSD

3–21 12 kΩ/V

3–23 97.6 kΩ

3–25 2.8 kΩ/V, 4 kΩ/V

3–27 ±6%, ±30%

3–29 0 Ω, 90 kΩ, 30 kΩ, 10 kΩ

3–31 75 Ω, 30 kΩ, 10 kΩ

3–33 10 kΩ, 28.57 Ω

3–35 5 kΩ, 5 kΩ

3–37 6.25 %

CHAPTER 4

4–1   1.15 kΩ, 200 kΩ

4–3   60 μA, 75 μA, 100 μA

4–5   3 V, 1.9 mA, 2.6 mA, 2.6 mA, 3.77 mA, (2.46 V to 4.35 V)

4–7   0 V, −0.7 V, 5.33 V

4–9   19.25 kΩ

4–11 3 V, 1.5 V

| 4–13 | R_1, 3 R_1 | 8–7 | 35.8 µH, 13.28 Ω, 16.9 |

4–13 R_1, 3 R_1
4–15   6.67%, 23%
4–17   1 V
4–19   500 mV, 250 mV

CHAPTER 5

5–1   7.6 V, 0.2 V
5–3   8.8 V, 0 V
5–5   7.7 V, −3.1 V
5–7   15 mA, 10.5 mA
5–9   375 kHz, 512 µs, 42.67 µs
5–11   1.11 V
5–13   0.0015%
5–15   1100, 0010

CHAPTER 6

6–1   1.999 ms, 400 Hz
6–3   2200, 5 ms, 11.58 ms, 5.26 ms, 1.000 V
6–5   1.257 V, 1.253 V
6–7   ±3.3%, ±0.001%, ±0.001%
6–9   1.56 ms, 2248
6–11   ±0.0012%, ±0.00013%

CHAPTER 7

7–1   990 Ω
7–3   9.52 kΩ, 9.21 kΩ, 9.83 kΩ
7–5   60 kΩ, 100 Ω
7–7   50 kΩ, 10 Ω
7–9   ±0.1%
7–11   0.049 Ω
7–13 2.67×10^{11} Ω, 1.04×10^{10} Ω

CHAPTER 8

8–1   31.9 kΩ‖5 nF
8–3   500 pF, 20 µF
8–5   0.183 µF, 364.3 Ω, 2.39

8–7   35.8 µH, 13.28 Ω, 16.9
8–9   1 mH to 10 mH, 0.0628 to 6.28
8–11   210 mH, 47.6 Ω, 467 Ω, 8.8 kΩ
8–13   20, 250 mV

CHAPTER 9

9–1   −0.7 V, −9.87 V, −1057 V, 0 V
9–3   two cycles of triangular wave
9–5 $2C_1$, R_3 ±10%
9–7   80 mV, 80 mV, 63°
9–9   15 V, 6 V, 16.7 Hz
9–11   0.2 V, 240 mV, 6.25 kHz, 20 µs, 14 µs, 18 µs
9–13   5 ms, 360 ns
9–15   250 Ω, 1 V/div, 0.1 V/div

CHAPTER 10

10–1   450 µs
10–3   200 MHz, 2000
10–5   48 µV
10–7   4 V
10–9   100, 10, 25, 15

CHAPTER 11

11–1   212 Hz, 3.18 kHz, 2.7 µF
11–3   3 kΩ, 54 kΩ, 3 kΩ
11–5   ±2 V
11–7   975 Hz
11–9   900 µs, 100 µs
11–11 800 µs
11–13 1.13 MHz, 2.25 MHz
11–15 235 MHz, 2.35 V, 117.5 MHz

CHAPTER 12

12–1   3.33%, 3%, 3.33%, 2%, 3.33%, 1%
12–3   ±(0.65% + 1 d)
12–5   19.8 mA
12–7   10 V, ±0.023%

12–9   100 Ω, 20 mA, 250 kΩ, 0.943 V
12–11   10 Ω, 100 mW

CHAPTER 16

16–1   1 V, 0.5 V
16–3   58 mV, 450 mV

16–5   14.7 V, 14.6 mA
16–7   18.5 V, 12.3 V
16–9   212 mA
16–11   0.9 W

Index